A꜀ᴘᴇx CALCULUS II

Version 4.0

Gregory Hartman, Ph.D.

Department of Applied Mathematics

Virginia Military Institute

Contributing Authors

Troy Siemers, Ph.D.

Department of Applied Mathematics

Virginia Military Institute

Brian Heinold, Ph.D.

Department of Mathematics and Computer Science

Mount Saint Mary's University

Dimplekumar Chalishajar, Ph.D.

Department of Applied Mathematics

Virginia Military Institute

Editor

Jennifer Bowen, Ph.D.

Department of Mathematics and Computer Science

The College of Wooster

Contents

PREFACE
A Note on Using this Text

Thank you for reading this short preface. Allow us to share a few key points about the text so that you may better understand what you will find beyond this page.

This text is Part II of a three–text series on Calculus. The first part covers material taught in many "Calc 1" courses: limits, derivatives, and the basics of integration, found in Chapters 1 through 6.1. The second text covers material often taught in "Calc 2:" integration and its applications, along with an introduction to sequences, series and Taylor Polynomials, found in Chapters 5 through 8. The third text covers topics common in "Calc 3" or "multivariable calc:" parametric equations, polar coordinates, vector–valued functions, and functions of more than one variable, found in Chapters 9 through 14. All three are available separately for free at www.apexcalculus.com. These three texts are intended to work together and make one cohesive text, *APEX Calculus*, which can also be downloaded from the website.

Printing the entire text as one volume makes for a large, heavy, cumbersome book. One can certainly only print the pages they currently need, but some prefer to have a nice, bound copy of the text. Therefore this text has been split into these three manageable parts, each of which can be purchased for about $15 at Amazon.com.

A result of this splitting is that sometimes a concept is said to be explored in an "earlier/later section," though that section does not actually appear in this particular text. Also, the index makes reference to topics, and page numbers, that do not appear in this text. This is done intentionally to show the reader what topics are available for study. Downloading the .pdf of *APEX Calculus* will ensure that you have all the content.

For Students: How to Read this Text

Mathematics textbooks have a reputation for being hard to read. High–level mathematical writing often seeks to say much with few words, and this style often seeps into texts of lower–level topics. This book was written with the goal of being easier to read than many other calculus textbooks, without becoming too verbose.

Each chapter and section starts with an introduction of the coming material, hopefully setting the stage for "why you should care," and ends with a look ahead to see how the just–learned material helps address future problems.

Please read the text; it is written to explain the concepts of Calculus. There are numerous examples to demonstrate the meaning of definitions, the truth of theorems, and the application of mathematical techniques. When you encounter a sentence you don't understand, read it again. If it still doesn't make sense, read on anyway, as sometimes confusing sentences are explained by later sentences.

You don't have to read every equation. The examples generally show "all" the steps needed to solve a problem. Sometimes reading through each step is helpful; sometimes it is confusing. When the steps are illustrating a new technique, one probably should follow each step closely to learn the new technique. When the steps are showing the mathematics needed to find a number to be used later, one can usually skip ahead and see how that number is being used, instead of getting bogged down in reading how the number was found.

Most proofs have been omitted. In mathematics, *proving* something is always true is extremely important, and entails much more than testing to see if it works twice. However, students often are confused by the details of a proof, or become concerned that they should have been able to construct this proof on their own. To alleviate this potential problem, we do not include the proofs to most theorems in the text. The interested reader is highly encouraged to find proofs online or from their instructor. In most cases, one is very capable of understanding what a theorem *means* and *how to apply it* without knowing fully *why* it is true.

Interactive, 3D Graphics

New to Version 3.0 was the addition of interactive, 3D graphics in the .pdf version. Nearly all graphs of objects in space can be rotated, shifted, and zoomed in/out so the reader can better understand the object illustrated.

As of this writing, the only pdf viewers that support these 3D graphics are Adobe Reader & Acrobat (and only the versions for PC/Mac/Unix/Linux computers, not tablets or smartphones). To activate the interactive mode, click on the image. Once activated, one can click/drag to rotate the object and use the scroll wheel on a mouse to zoom in/out. (A great way to investigate an image is to first zoom in on the page of the pdf viewer so the graphic itself takes up much of the screen, then zoom inside the graphic itself.) A CTRL-click/drag pans the object left/right or up/down. By right-clicking on the graph one can access a menu of other options, such as changing the lighting scheme or perspective. One can also revert the graph back to its default view. If you wish to deactivate the interactivity, one can right-click and choose the "Disable Content" option.

Thanks

There are many people who deserve recognition for the important role they have played in the development of this text. First, I thank Michelle for her support and encouragement, even as this "project from work" occupied my time and attention at home. Many thanks to Troy Siemers, whose most important contributions extend far beyond the sections he wrote or the 227 figures he coded in Asymptote for 3D interaction. He provided incredible support, advice and encouragement for which I am very grateful. My thanks to Brian Heinold and Dimplekumar Chalishajar for their contributions and to Jennifer Bowen for reading through so much material and providing great feedback early on. Thanks to Troy, Lee Dewald, Dan Joseph, Meagan Herald, Bill Lowe, John David, Vonda Walsh, Geoff Cox, Jessica Libertini and other faculty of VMI who have given me numerous suggestions and corrections based on their experience with teaching from the text. (Special thanks to Troy, Lee & Dan for their patience in teaching Calc III while I was still writing the Calc III material.) Thanks to Randy Cone for encouraging his tutors of VMI's Open Math Lab to read through the text and check the solutions, and thanks to the tutors for spending their time doing so. A very special thanks to Kristi Brown and Paul Janiczek who took this opportunity far above & beyond what I expected, meticulously checking every solution and carefully reading every example. Their comments have been extraordinarily helpful. I am also thankful for the support provided by Wane Schneiter, who as my Dean provided me with extra time to work on this project. Finally, a huge heap of thanks is to be bestowed on the numerous people I do not know who took the time to email me corrections and suggestions. I am blessed to have so many people give of their time to make this book better.

A꒒EX – Affordable Print and Electronic teXts

A꒒EX is a consortium of authors who collaborate to produce high–quality, low–cost textbooks. The current textbook–writing paradigm is facing a potential revolution as desktop publishing and electronic formats increase in popularity. However, writing a good textbook is no easy task, as the time requirements alone are substantial. It takes countless hours of work to produce text, write examples and exercises, edit and publish. Through collaboration, however, the cost to any individual can be lessened, allowing us to create texts that we freely distribute electronically and sell in printed form for an incredibly low cost. Having said that, nothing is entirely free; someone always bears some cost. This text "cost" the authors of this book their time, and that was not enough. *APEX Calculus* would not exist had not the Virginia Military Institute, through a generous Jackson–Hope grant, given the lead author significant time away from teaching so he could focus on this text.

Each text is available as a free .pdf, protected by a Creative Commons Attribution - Noncommercial 4.0 copyright. That means you can give the .pdf to anyone you like, print it in any form you like, and even edit the original content and redistribute it. If you do the latter, you must clearly reference this work and you cannot sell your edited work for money.

We encourage others to adapt this work to fit their own needs. One might add sections that are "missing" or remove sections that your students won't need. The source files can be found at github.com/APEXCalculus.

You can learn more at www.vmi.edu/APEX.

Version 4.0

Key changes from Version 3.0 to 4.0:

- Numerous typographical and "small" mathematical corrections (again, thanks to all my close readers!).

- "Large" mathematical corrections and adjustments. There were a number of places in Version 3.0 where a definition/theorem was not correct as stated. See www.apexcalculus.com for more information.

- More useful numbering of Examples, Theorems, etc. "Definition 11.4.2" refers to the second definition of Chapter 11, Section 4.

- The addition of Section 13.7: Triple Integration with Cylindrical and Spherical Coordinates

- The addition of Chapter 14: Vector Analysis.

5: INTEGRATION

We have spent considerable time considering the derivatives of a function and their applications. In the following chapters, we are going to starting thinking in "the other direction." That is, given a function $f(x)$, we are going to consider functions $F(x)$ such that $F'(x) = f(x)$. There are numerous reasons this will prove to be useful: these functions will help us compute area, volume, mass, force, pressure, work, and much more.

5.1 Antiderivatives and Indefinite Integration

Given a function $y = f(x)$, a *differential equation* is one that incorporates y, x, and the derivatives of y. For instance, a simple differential equation is:

$$y' = 2x.$$

Solving a differential equation amounts to finding a function y that satisfies the given equation. Take a moment and consider that equation; can you find a function y such that $y' = 2x$?

Can you find another?

And yet another?

Hopefully one was able to come up with at least one solution: $y = x^2$. "Finding another" may have seemed impossible until one realizes that a function like $y = x^2 + 1$ also has a derivative of $2x$. Once that discovery is made, finding "yet another" is not difficult; the function $y = x^2 + 123,456,789$ also has a derivative of $2x$. The differential equation $y' = 2x$ has many solutions. This leads us to some definitions.

Definition 5.1.1   **Antiderivatives and Indefinite Integrals**

Let a function $f(x)$ be given. An **antiderivative** of $f(x)$ is a function $F(x)$ such that $F'(x) = f(x)$.

The set of all antiderivatives of $f(x)$ is the **indefinite integral of** f, denoted by

$$\int f(x)\, dx.$$

Make a note about our definition: we refer to *an* antiderivative of f, as opposed to *the* antiderivative of f, since there is *always* an infinite number of them.

We often use upper-case letters to denote antiderivatives.

Knowing one antiderivative of f allows us to find infinitely more, simply by adding a constant. Not only does this give us *more* antiderivatives, it gives us *all* of them.

Theorem 5.1.1 Antiderivative Forms

Let $F(x)$ and $G(x)$ be antiderivatives of $f(x)$ on an interval I. Then there exists a constant C such that, on I,

$$G(x) = F(x) + C.$$

Given a function f defined on an interval I and one of its antiderivatives F, we know *all* antiderivatives of f on I have the form $F(x) + C$ for some constant C. Using Definition 5.1.1, we can say that

$$\int f(x)\, dx = F(x) + C.$$

Let's analyze this indefinite integral notation.

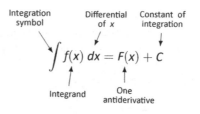

Figure 5.1.1: Understanding the indefinite integral notation.

Figure 5.1.1 shows the typical notation of the indefinite integral. The integration symbol, $\int$, is in reality an "elongated S," representing "take the sum." We will later see how *sums* and *antiderivatives* are related.

The function we want to find an antiderivative of is called the *integrand*. It contains the differential of the variable we are integrating with respect to. The $\int$ symbol and the differential dx are not "bookends" with a function sandwiched in between; rather, the symbol $\int$ means "find all antiderivatives of what follows," and the function $f(x)$ and dx are multiplied together; the dx does not "just sit there."

Let's practice using this notation.

Notes:

Example 5.1.1   Evaluating indefinite integrals

Evaluate $\int \sin x \, dx$.

SOLUTION   We are asked to find all functions $F(x)$ such that $F'(x) = \sin x$. Some thought will lead us to one solution: $F(x) = -\cos x$, because $\frac{d}{dx}(-\cos x) = \sin x$.

The indefinite integral of $\sin x$ is thus $-\cos x$, plus a constant of integration. So:

$$\int \sin x \, dx = -\cos x + C.$$

A commonly asked question is "What happened to the dx?" The unenlightened response is "Don't worry about it. It just goes away." A full understanding includes the following.

This process of *antidifferentiation* is really solving a *differential* question. The integral

$$\int \sin x \, dx$$

presents us with a differential, $dy = \sin x \, dx$. It is asking: "What is y?" We found lots of solutions, all of the form $y = -\cos x + C$.

Letting $dy = \sin x \, dx$, rewrite

$$\int \sin x \, dx \quad \text{as} \quad \int dy.$$

This is asking: "What functions have a differential of the form dy?" The answer is "Functions of the form $y + C$, where C is a constant." What is y? We have lots of choices, all differing by a constant; the simplest choice is $y = -\cos x$.

Understanding all of this is more important later as we try to find antiderivatives of more complicated functions. In this section, we will simply explore the rules of indefinite integration, and one can succeed for now with answering "What happened to the dx?" with "It went away."

Let's practice once more before stating integration rules.

Example 5.1.2   Evaluating indefinite integrals

Evaluate $\int (3x^2 + 4x + 5) \, dx$.

SOLUTION   We seek a function $F(x)$ whose derivative is $3x^2 + 4x + 5$. When taking derivatives, we can consider functions term–by–term, so we can likely do that here.

What functions have a derivative of $3x^2$? Some thought will lead us to a cubic, specifically $x^3 + C_1$, where C_1 is a constant.

Notes:

What functions have a derivative of $4x$? Here the x term is raised to the first power, so we likely seek a quadratic. Some thought should lead us to $2x^2 + C_2$, where C_2 is a constant.

Finally, what functions have a derivative of 5? Functions of the form $5x + C_3$, where C_3 is a constant.

Our answer appears to be

$$\int (3x^2 + 4x + 5)\, dx = x^3 + C_1 + 2x^2 + C_2 + 5x + C_3.$$

We do not need three separate constants of integration; combine them as one constant, giving the final answer of

$$\int (3x^2 + 4x + 5)\, dx = x^3 + 2x^2 + 5x + C.$$

It is easy to verify our answer; take the derivative of $x^3 + 2x^3 + 5x + C$ and see we indeed get $3x^2 + 4x + 5$.

This final step of "verifying our answer" is important both practically and theoretically. In general, taking derivatives is easier than finding antiderivatives so checking our work is easy and vital as we learn.

We also see that taking the derivative of our answer returns the function in the integrand. Thus we can say that:

$$\frac{d}{dx}\left(\int f(x)\, dx \right) = f(x).$$

Differentiation "undoes" the work done by antidifferentiation.

Theorem 2.7.3 gave a list of the derivatives of common functions we had learned at that point. We restate part of that list here to stress the relationship between derivatives and antiderivatives. This list will also be useful as a glossary of common antiderivatives as we learn.

_____

Notes:

Theorem 5.1.2   Derivatives and Antiderivatives

Common Differentiation Rules   Common Indefinite Integral Rules

1. $\frac{d}{dx}\left(cf(x)\right) = c \cdot f'(x)$

1. $\int c \cdot f(x)\, dx = c \cdot \int f(x)\, dx$

2. $\frac{d}{dx}\left(f(x) \pm g(x)\right) =$ $f'(x) \pm g'(x)$

2. $\int \left(f(x) \pm g(x)\right)\, dx =$ $\int f(x)\, dx \pm \int g(x)\, dx$

3. $\frac{d}{dx}\left(C\right) = 0$

3. $\int 0\, dx = C$

4. $\frac{d}{dx}\left(x\right) = 1$

4. $\int 1\, dx = \int dx = x + C$

5. $\frac{d}{dx}\left(x^n\right) = n \cdot x^{n-1}$

5. $\int x^n\, dx - \frac{1}{n+1}x^{n+1} + C$ $(n \neq -1)$

6. $\frac{d}{dx}\left(\sin x\right) = \cos x$

6. $\int \cos x\, dx = \sin x + C$

7. $\frac{d}{dx}\left(\cos x\right) = -\sin x$

7. $\int \sin x\, dx = -\cos x + C$

8. $\frac{d}{dx}\left(\tan x\right) = \sec^2 x$

8. $\int \sec^2 x\, dx = \tan x + C$

9. $\frac{d}{dx}\left(\csc x\right) = -\csc x \cot x$

9. $\int \csc x \cot x\, dx = -\csc x + C$

10. $\frac{d}{dx}\left(\sec x\right) = \sec x \tan x$

10. $\int \sec x \tan x\, dx = \sec x + C$

11. $\frac{d}{dx}\left(\cot x\right) = -\csc^2 x$

11. $\int \csc^2 x\, dx = -\cot x + C$

12. $\frac{d}{dx}\left(e^x\right) = e^x$

12. $\int e^x\, dx = e^x + C$

13. $\frac{d}{dx}\left(a^x\right) = \ln a \cdot a^x$

13. $\int a^x\, dx = \frac{1}{\ln a} \cdot a^x + C$

14. $\frac{d}{dx}\left(\ln x\right) = \frac{1}{x}$

14. $\int \frac{1}{x}\, dx = \ln |x| + C$

We highlight a few important points from Theorem 5.1.2:

- Rule #1 states $\int c \cdot f(x)\, dx = c \cdot \int f(x)\, dx$. This is the Constant Multiple Rule: we can temporarily ignore constants when finding antiderivatives, just as we did when computing derivatives (i.e., $\frac{d}{dx}\left(3x^2\right)$ is just as easy to compute as $\frac{d}{dx}\left(x^2\right)$). An example:

$$\int 5\cos x\, dx = 5 \cdot \int \cos x\, dx = 5 \cdot (\sin x + C) = 5\sin x + C.$$

In the last step we can consider the constant as also being multiplied by

Notes:

5, but "5 times a constant" is still a constant, so we just write "C".

- Rule #2 is the Sum/Difference Rule: we can split integrals apart when the integrand contains terms that are added/subtracted, as we did in Example 5.1.2. So:

$$\int (3x^2 + 4x + 5)\, dx = \int 3x^2\, dx + \int 4x\, dx + \int 5\, dx$$
$$= 3\int x^2\, dx + 4\int x\, dx + \int 5\, dx$$
$$= 3 \cdot \frac{1}{3}x^3 + 4 \cdot \frac{1}{2}x^2 + 5x + C$$
$$= x^3 + 2x^2 + 5x + C$$

In practice we generally do not write out all these steps, but we demonstrate them here for completeness.

- Rule #5 is the Power Rule of indefinite integration. There are two important things to keep in mind:

  1. Notice the restriction that $n \neq -1$. This is important: $\int \frac{1}{x}\, dx \neq$ "$\frac{1}{0}x^0 + C$"; rather, see Rule #14.

  2. We are presenting antidifferentiation as the "inverse operation" of differentiation. Here is a useful quote to remember:

     "Inverse operations do the opposite things in the opposite order."

     When taking a derivative using the Power Rule, we **first** *multiply* by the power, then **second** *subtract* 1 from the power. To find the antiderivative, do the opposite things in the opposite order: **first** *add* one to the power, then **second** *divide* by the power.

- Note that Rule #14 incorporates the absolute value of x. The exercises will work the reader through why this is the case; for now, know the absolute value is important and cannot be ignored.

Initial Value Problems

In Section 2.3 we saw that the derivative of a position function gave a velocity function, and the derivative of a velocity function describes acceleration. We can now go "the other way:" the antiderivative of an acceleration function gives a velocity function, etc. While there is just one derivative of a given function, there are infinitely many antiderivatives. Therefore we cannot ask "What is *the* velocity of an object whose acceleration is -32ft/s^2?", since there is more than one answer.

Notes:

We can find *the* answer if we provide more information with the question, as done in the following example. Often the additional information comes in the form of an *initial value*, a value of the function that one knows beforehand.

Example 5.1.3   Solving initial value problems
The acceleration due to gravity of a falling object is -32 ft/s². At time $t = 3$, a falling object had a velocity of -10 ft/s. Find the equation of the object's velocity.

SOLUTION   We want to know a velocity function, $v(t)$. We know two things:

- The acceleration, i.e., $v'(t) = -32$, and

- the velocity at a specific time, i.e., $v(3) = -10$.

Using the first piece of information, we know that $v(t)$ is an antiderivative of $v'(t) = -32$. So we begin by finding the indefinite integral of -32:

$$\int (-32)\, dt = -32t + C = v(t).$$

Now we use the fact that $v(3) = -10$ to find C:

$$v(t) = -32t + C$$
$$v(3) = -10$$
$$-32(3) + C = -10$$
$$C = 86$$

Thus $v(t) = -32t + 86$. We can use this equation to understand the motion of the object: when $t = 0$, the object had a velocity of $v(0) = 86$ ft/s. Since the velocity is positive, the object was moving upward.

When did the object begin moving down? Immediately after $v(t) = 0$:

$$-32t + 86 = 0 \quad \Rightarrow \quad t = \frac{43}{16} \approx 2.69\text{s}.$$

Recognize that we are able to determine quite a bit about the path of the object knowing just its acceleration and its velocity at a single point in time.

Example 5.1.4   Solving initial value problems
Find $f(t)$, given that $f''(t) = \cos t$, $f'(0) = 3$ and $f(0) = 5$.

SOLUTION   We start by finding $f'(t)$, which is an antiderivative of $f''(t)$:

$$\int f''(t)\, dt = \int \cos t\, dt = \sin t + C = f'(t).$$

Notes:

So $f'(t) = \sin t + C$ for the correct value of C. We are given that $f'(0) = 3$, so:

$$f'(0) = 3 \quad \Rightarrow \quad \sin 0 + C = 3 \quad \Rightarrow \quad C = 3.$$

Using the initial value, we have found $f'(t) = \sin t + 3$.

We now find $f(t)$ by integrating again.

$$f(t) = \int f'(t)\, dt = \int (\sin t + 3)\, dt = -\cos t + 3t + C.$$

We are given that $f(0) = 5$, so

$$-\cos 0 + 3(0) + C = 5$$
$$-1 + C = 5$$
$$C = 6$$

Thus $f(t) = -\cos t + 3t + 6$.

This section introduced antiderivatives and the indefinite integral. We found they are needed when finding a function given information about its derivative(s). For instance, we found a velocity function given an acceleration function.

In the next section, we will see how position and velocity are unexpectedly related by the areas of certain regions on a graph of the velocity function. Then, in Section 5.4, we will see how areas and antiderivatives are closely tied together. This connection is incredibly important, as indicated by the name of the theorem that describes it: The Fundamental Theorem of Calculus.

Notes:

Exercises 5.1

Terms and Concepts

1. Define the term "antiderivative" in your own words.

2. Is it more accurate to refer to "the" antiderivative of $f(x)$ or "an" antiderivative of $f(x)$?

3. Use your own words to define the indefinite integral of $f(x)$.

4. Fill in the blanks: "Inverse operations do the _____ things in the _____ order."

5. What is an "initial value problem"?

6. The derivative of a position function is a _____ function.

7. The antiderivative of an acceleration function is a _____ function.

8. If $F(x)$ is an antiderivative of $f(x)$, and $G(x)$ is an antiderivative of $g(x)$, give an antiderivative of $f(x) + g(x)$.

Problems

In Exercises 9 – 27, evaluate the given indefinite integral.

9. $\int 3x^3 \, dx$

10. $\int x^8 \, dx$

11. $\int (10x^2 - 2) \, dx$

12. $\int dt$

13. $\int 1 \, ds$

14. $\int \frac{1}{3t^2} \, dt$

15. $\int \frac{3}{t^2} \, dt$

16. $\int \frac{1}{\sqrt{x}} \, dx$

17. $\int \sec^2 \theta \, d\theta$

18. $\int \sin \theta \, d\theta$

19. $\int (\sec x \tan x + \csc x \cot x) \, dx$

20. $\int 5e^\theta \, d\theta$

21. $\int 3^t \, dt$

22. $\int \frac{5^t}{2} \, dt$

23. $\int (2l + 3)^2 \, dl$

24. $\int (t^2 + 3)(t^3 - 2t) \, dt$

25. $\int x^2 x^3 \, dx$

26. $\int e^\pi \, dx$

27. $\int a \, dx$

28. This problem investigates why Theorem 5.1.2 states that $\int \frac{1}{x} \, dx = \ln |x| + C$.

   (a) What is the domain of $y = \ln x$?
   (b) Find $\frac{d}{dx}\left(\ln x \right)$.
   (c) What is the domain of $y = \ln(-x)$?
   (d) Find $\frac{d}{dx}\left(\ln(-x) \right)$.
   (e) You should find that $1/x$ has two types of antiderivatives, depending on whether $x > 0$ or $x < 0$. In one expression, give a formula for $\int \frac{1}{x} \, dx$ that takes these different domains into account, and explain your answer.

In Exercises 29 – 39, find $f(x)$ described by the given initial value problem.

29. $f'(x) = \sin x$ and $f(0) = 2$

30. $f'(x) = 5e^x$ and $f(0) = 10$

31. $f'(x) = 4x^3 - 3x^2$ and $f(-1) = 9$

32. $f'(x) = \sec^2 x$ and $f(\pi/4) = 5$

33. $f'(x) = 7^x$ and $f(2) = 1$

34. $f''(x) = 5$ and $f'(0) = 7, f(0) = 3$

35. $f''(x) = 7x$ and $f'(1) = -1, f(1) = 10$

36. $f''(x) = 5e^x$ and $f'(0) = 3, f(0) = 5$

37. $f''(\theta) = \sin\theta$ and $f'(\pi) = 2, f(\pi) = 4$

38. $f''(x) = 24x^2 + 2^x - \cos x$ and $f'(0) = 5, f(0) = 0$

39. $f''(x) = 0$ and $f'(1) = 3, f(1) = 1$

Review

40. Use information gained from the first and second derivatives to sketch $f(x) = \dfrac{1}{e^x + 1}$.

41. Given $y = x^2 e^x \cos x$, find dy.

5.2   The Definite Integral

We start with an easy problem. An object travels in a straight line at a constant velocity of 5ft/s for 10 seconds. How far away from its starting point is the object?

We approach this problem with the familiar "Distance = Rate × Time" equation. In this case, Distance = 5ft/s × 10s = 50 feet.

It is interesting to note that this solution of 50 feet can be represented graphically. Consider Figure 5.2.1, where the constant velocity of 5ft/s is graphed on the axes. Shading the area under the line from $t = 0$ to $t = 10$ gives a rectangle with an area of 50 square units; when one considers the units of the axes, we can say this area represents 50 ft.

Now consider a slightly harder situation (and not particularly realistic): an object travels in a straight line with a constant velocity of 5ft/s for 10 seconds, then instantly reverses course at a rate of 2ft/s for 4 seconds. (Since the object is traveling in the opposite direction when reversing course, we say the velocity is a constant −2ft/s.) How far away from the starting point is the object – what is its *displacement*?

Here we use "Distance = Rate_1 × Time_1 + Rate_2 × Time_2," which is

$$\text{Distance} = 5 \cdot 10 + (-2) \cdot 4 = 42 \text{ ft.}$$

Hence the object is 42 feet from its starting location.

We can again depict this situation graphically. In Figure 5.2.2 we have the velocities graphed as straight lines on $[0, 10]$ and $[10, 14]$, respectively. The displacement of the object is

$$\text{"Area above the } t\text{–axis} \quad - \quad \text{Area below the } t\text{–axis,"}$$

which is easy to calculate as $50 - 8 = 42$ feet.

Now consider a more difficult problem.

Example 5.2.1   Finding position using velocity
The velocity of an object moving straight up/down under the acceleration of gravity is given as $v(t) = -32t + 48$, where time t is given in seconds and velocity is in ft/s. When $t = 0$, the object had a height of 0 ft.

1. What was the initial velocity of the object?

2. What was the maximum height of the object?

3. What was the height of the object at time $t = 2$?

SOLUTION   It is straightforward to find the initial velocity; at time $t = 0$, $v(0) = -32 \cdot 0 + 48 = 48$ ft/s.

Figure 5.2.1: The area under a constant velocity function corresponds to distance traveled.

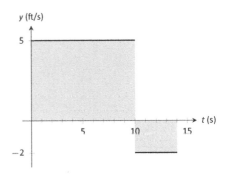

Figure 5.2.2: The total displacement is the area above the t–axis minus the area below the t–axis.

Notes:

To answer questions about the height of the object, we need to find the object's position function $s(t)$. This is an initial value problem, which we studied in the previous section. We are told the initial height is 0, i.e., $s(0) = 0$. We know $s'(t) = v(t) = -32t + 48$. To find s, we find the indefinite integral of $v(t)$:

$$\int v(t)\, dt = \int (-32t + 48)\, dt = -16t^2 + 48t + C = s(t).$$

Since $s(0) = 0$, we conclude that $C = 0$ and $s(t) = -16t^2 + 48t$.

To find the maximum height of the object, we need to find the maximum of s. Recalling our work finding extreme values, we find the critical points of s by setting its derivative equal to 0 and solving for t:

$$s'(t) = -32t + 48 = 0 \quad \Rightarrow \quad t = 48/32 = 1.5s.$$

(Notice how we ended up just finding when the velocity was 0ft/s!) The first derivative test shows this is a maximum, so the maximum height of the object is found at

$$s(1.5) = -16(1.5)^2 + 48(1.5) = 36\text{ft}.$$

The height at time $t = 2$ is now straightforward to compute: it is $s(2) = 32$ft.

While we have answered all three questions, let's look at them again graphically, using the concepts of area that we explored earlier.

Figure 5.2.3 shows a graph of $v(t)$ on axes from $t = 0$ to $t = 3$. It is again straightforward to find $v(0)$. How can we use the graph to find the maximum height of the object?

Recall how in our previous work that the displacement of the object (in this case, its height) was found as the area under the velocity curve, as shaded in the figure. Moreover, the area between the curve and the t–axis that is below the t–axis counted as "negative" area. That is, it represents the object coming back toward its starting position. So to find the maximum distance from the starting point – the maximum height – we find the area under the velocity line that is above the t–axis, i.e., from $t = 0$ to $t = 1.5$. This region is a triangle; its area is

$$\text{Area} = \frac{1}{2}\text{Base} \times \text{Height} = \frac{1}{2} \times 1.5s \times 48\text{ft/s} = 36\text{ft,}$$

which matches our previous calculation of the maximum height.

Finally, to find the height of the object at time $t = 2$ we calculate the total "signed area" (where some area is negative) under the velocity function from $t = 0$ to $t = 2$. This signed area is equal to $s(2)$, the displacement (i.e., signed distance) from the starting position at $t = 0$ to the position at time $t = 2$. That is,

$$\text{Displacement = Area above the } t\text{–axis} - \text{Area below } t\text{–axis.}$$

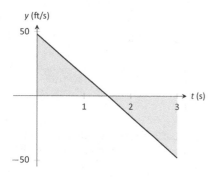

Figure 5.2.3: A graph of $v(t) = -32t + 48$; the shaded areas help determine displacement.

Notes:

The regions are triangles, and we find

$$\text{Displacement} = \frac{1}{2}(1.5\text{s})(48\text{ft/s}) - \frac{1}{2}(.5\text{s})(16\text{ft/s}) = 32\text{ft}.$$

This also matches our previous calculation of the height at $t = 2$.

Notice how we answered each question in this example in two ways. Our first method was to manipulate equations using our understanding of antiderivatives and derivatives. Our second method was geometric: we answered questions looking at a graph and finding the areas of certain regions of this graph.

The above example does not *prove* a relationship between area under a velocity function and displacement, but it does imply a relationship exists. Section 5.4 will fully establish fact that the area under a velocity function is displacement.

Given a graph of a function $y = f(x)$, we will find that there is great use in computing the area between the curve $y = f(x)$ and the x-axis. Because of this, we need to define some terms.

Definition 5.2.1   The Definite Integral, Total Signed Area

Let $y = f(x)$ be defined on a closed interval $[a, b]$. The **total signed area from $x = a$ to $x = b$ under f** is:
(area under f and above the x-axis on $[a, b]$) $-$ (area above f and under the x-axis on $[a, b]$).

The **definite integral of f on $[a, b]$** is the total signed area of f on $[a, b]$, denoted

$$\int_a^b f(x)\, dx,$$

where a and b are the **bounds of integration.**

By our definition, the definite integral gives the "signed area under f." We usually drop the word "signed" when talking about the definite integral, and simply say the definite integral gives "the area under f" or, more commonly, "the area under the curve."

The previous section introduced the indefinite integral, which related to antiderivatives. We have now defined the definite integral, which relates to areas under a function. The two are very much related, as we'll see when we learn the Fundamental Theorem of Calculus in Section 5.4. Recall that earlier we said that the "$\int$" symbol was an "elongated S" that represented finding a "sum." In the context of the definite integral, this notation makes a bit more sense, as we are adding up areas under the function f.

Notes:

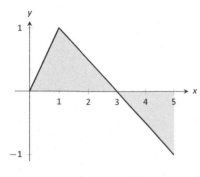

Figure 5.2.4: A graph of $f(x)$ in Example 5.2.2.

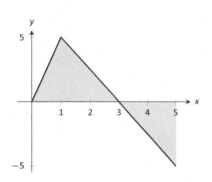

Figure 5.2.5: A graph of $5f$ in Example 5.2.2. (Yes, it looks just like the graph of f in Figure 5.2.4, just with a different y-scale.)

We practice using this notation.

Example 5.2.2   Evaluating definite integrals
Consider the function f given in Figure 5.2.4.

Find:

1. $\displaystyle\int_0^3 f(x)\ dx$

2. $\displaystyle\int_3^5 f(x)\ dx$

3. $\displaystyle\int_0^5 f(x)\ dx$

4. $\displaystyle\int_0^3 5f(x)\ dx$

5. $\displaystyle\int_1^1 f(x)\ dx$

SOLUTION

1. $\int_0^3 f(x)\ dx$ is the area under f on the interval $[0,3]$. This region is a triangle, so the area is $\int_0^3 f(x)\ dx = \frac{1}{2}(3)(1) = 1.5$.

2. $\int_3^5 f(x)\ dx$ represents the area of the triangle found under the x–axis on $[3,5]$. The area is $\frac{1}{2}(2)(1) = 1$; since it is found *under* the x–axis, this is "negative area." Therefore $\int_3^5 f(x)\ dx = -1$.

3. $\int_0^5 f(x)\ dx$ is the total signed area under f on $[0,5]$. This is $1.5 + (-1) = 0.5$.

4. $\int_0^3 5f(x)\ dx$ is the area under $5f$ on $[0,3]$. This is sketched in Figure 5.2.5. Again, the region is a triangle, with height 5 times that of the height of the original triangle. Thus the area is $\int_0^3 5f(x)\ dx = \frac{1}{2}(15)(1) = 7.5$.

5. $\int_1^1 f(x)\ dx$ is the area under f on the "interval" $[1,1]$. This describes a line segment, not a region; it has no width. Therefore the area is 0.

This example illustrates some of the properties of the definite integral, given here.

Notes:

Theorem 5.2.1 Properties of the Definite Integral

Let f and g be defined on a closed interval I that contains the values a, b and c, and let k be a constant. The following hold:

1. $\displaystyle\int_a^a f(x)\,dx = 0$

2. $\displaystyle\int_a^b f(x)\,dx + \int_b^c f(x)\,dx = \int_a^c f(x)\,dx$

3. $\displaystyle\int_a^b f(x)\,dx = -\int_b^a f(x)\,dx$

4. $\displaystyle\int_a^b \big(f(x) \pm g(x)\big)\,dx = \int_a^b f(x)\,dx \pm \int_a^b g(x)\,dx$

5. $\displaystyle\int_a^b k\cdot f(x)\,dx = k\cdot \int_a^b f(x)\,dx$

We give a brief justification of Theorem 5.2.1 here.

1. As demonstrated in Example 5.2.2, there is no "area under the curve" when the region has no width; hence this definite integral is 0.

2. This states that total area is the sum of the areas of subregions. It is easily considered when we let $a < b < c$. We can break the interval $[a, c]$ into two subintervals, $[a, b]$ and $[b, c]$. The total area over $[a, c]$ is the area over $[a, b]$ plus the area over $[b, c]$.

   It is important to note that this still holds true even if $a < b < c$ is not true. We discuss this in the next point.

3. This property can be viewed a merely a convention to make other properties work well. (Later we will see how this property has a justification all its own, not necessarily in support of other properties.) Suppose $b < a < c$. The discussion from the previous point clearly justifies

$$\int_b^a f(x)\,dx + \int_a^c f(x)\,dx = \int_b^c f(x)\,dx. \tag{5.1}$$

However, we still claim that, as originally stated,

$$\int_a^b f(x)\,dx + \int_b^c f(x)\,dx = \int_a^c f(x)\,dx. \tag{5.2}$$

Notes:

How do Equations (5.1) and (5.2) relate? Start with Equation (5.1):

$$\int_b^a f(x)\,dx + \int_a^c f(x)\,dx = \int_b^c f(x)\,dx$$

$$\int_a^c f(x)\,dx = -\int_b^a f(x)\,dx + \int_b^c f(x)\,dx$$

Property (3) justifies changing the sign and switching the bounds of integration on the $-\int_b^a f(x)\,dx$ term; when this is done, Equations (5.1) and (5.2) are equivalent.

The conclusion is this: by adopting the convention of Property (3), Property (2) holds no matter the order of a, b and c. Again, in the next section we will see another justification for this property.

4,5. Each of these may be non–intuitive. Property (5) states that when one scales a function by, for instance, 7, the area of the enclosed region also is scaled by a factor of 7. Both Properties (4) and (5) can be proved using geometry. The details are not complicated but are not discussed here.

Example 5.2.3   **Evaluating definite integrals using Theorem 5.2.1.**

Consider the graph of a function $f(x)$ shown in Figure 5.2.6. Answer the following:

1. Which value is greater: $\displaystyle\int_a^b f(x)\,dx$ or $\displaystyle\int_b^c f(x)\,dx$?

2. Is $\displaystyle\int_a^c f(x)\,dx$ greater or less than 0?

3. Which value is greater: $\displaystyle\int_a^b f(x)\,dx$ or $\displaystyle\int_c^b f(x)\,dx$?

Figure 5.2.6: A graph of a function in Example 5.2.3.

SOLUTION

1. $\int_a^b f(x)\,dx$ has a positive value (since the area is above the x–axis) whereas $\int_b^c f(x)\,dx$ has a negative value. Hence $\int_a^b f(x)\,dx$ is bigger.

2. $\int_a^c f(x)\,dx$ is the total signed area under f between $x = a$ and $x = c$. Since the region below the x–axis looks to be larger than the region above, we conclude that the definite integral has a value less than 0.

3. Note how the second integral has the bounds "reversed." Therefore $\int_c^b f(x)\,dx$ represents a positive number, greater than the area described by the first definite integral. Hence $\int_c^b f(x)\,dx$ is greater.

Notes:

The area definition of the definite integral allows us to use geometry to compute the definite integral of some simple functions.

Example 5.2.4   Evaluating definite integrals using geometry

Evaluate the following definite integrals:

$$1. \int_{-2}^{5} (2x - 4)\, dx \qquad 2. \int_{-3}^{3} \sqrt{9 - x^2}\, dx.$$

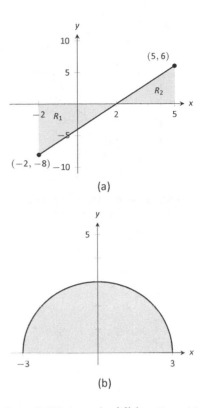

(a)

SOLUTION

1. It is useful to sketch the function in the integrand, as shown in Figure 5.2.7(a). We see we need to compute the areas of two regions, which we have labeled R_1 and R_2. Both are triangles, so the area computation is straightforward:

$$R_1 : \frac{1}{2}(4)(8) = 16 \qquad R_2 : \frac{1}{2}(3)6 = 9.$$

Region R_1 lies under the x–axis, hence it is counted as negative area (we can think of the triangle's height as being "-8"), so

$$\int_{-2}^{5} (2x - 4)\, dx = -16 + 9 = -7.$$

2. Recognize that the integrand of this definite integral describes a half circle, as sketched in Figure 5.2.7(b), with radius 3. Thus the area is:

$$\int_{-3}^{3} \sqrt{9 - x^2}\, dx = \frac{1}{2}\pi r^2 = \frac{9}{2}\pi.$$

(b)

Figure 5.2.7: A graph of $f(x) = 2x - 4$ in (a) and $f(x) = \sqrt{9 - x^2}$ in (b), from Example 5.2.4.

Example 5.2.5   Understanding motion given velocity

Consider the graph of a velocity function of an object moving in a straight line, given in Figure 5.2.8, where the numbers in the given regions gives the area of that region. Assume that the definite integral of a velocity function gives displacement. Find the maximum speed of the object and its maximum displacement from its starting position.

SOLUTION   Since the graph gives velocity, finding the maximum speed is simple: it looks to be 15ft/s.

At time $t = 0$, the displacement is 0; the object is at its starting position. At time $t = a$, the object has moved backward 11 feet. Between times $t = a$ and

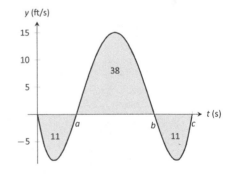

Figure 5.2.8: A graph of a velocity in Example 5.2.5.

Notes:

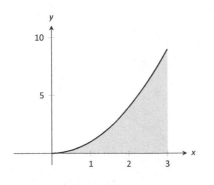

Figure 5.2.9: What is the area below $y = x^2$ on $[0, 3]$? The region is not a usual geometric shape.

$t = b$, the object moves forward 38 feet, bringing it into a position 27 feet forward of its starting position. From $t = b$ to $t = c$ the object is moving backwards again, hence its maximum displacement is 27 feet from its starting position.

In our examples, we have either found the areas of regions that have nice geometric shapes (such as rectangles, triangles and circles) or the areas were given to us. Consider Figure 5.2.9, where a region below $y = x^2$ is shaded. What is its area? The function $y = x^2$ is relatively simple, yet the shape it defines has an area that is not simple to find geometrically.

In the next section we will explore how to find the areas of such regions.

Notes:

Exercises 5.2

Terms and Concepts

1. What is "total signed area"?

2. What is "displacement"?

3. What is $\int_3^3 \sin x \, dx$?

4. Give a single definite integral that has the same value as

$$\int_0^1 (2x+3) \, dx + \int_1^2 (2x+3) \, dx.$$

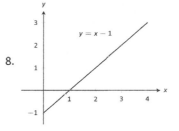

7.

(a) $\int_0^2 f(x) \, dx$

(d) $\int_0^1 4x \, dx$

(b) $\int_2^4 f(x) \, dx$

(e) $\int_2^3 (2x-4) \, dx$

(c) $\int_2^4 2f(x) \, dx$

(f) $\int_2^3 (4x-8) \, dx$

Problems

In Exercises 5 – 10, a graph of a function $f(x)$ is given. Using the geometry of the graph, evaluate the definite integrals.

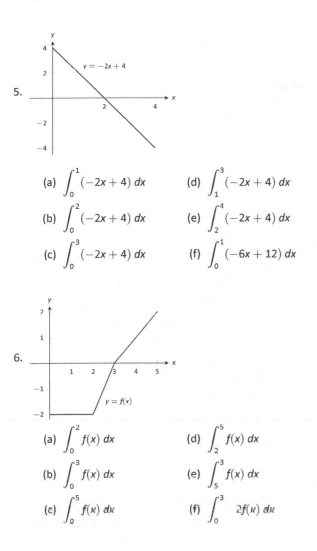

5.

(a) $\int_0^1 (-2x+4) \, dx$

(d) $\int_1^3 (-2x+4) \, dx$

(b) $\int_0^2 (-2x+4) \, dx$

(e) $\int_2^4 (-2x+4) \, dx$

(c) $\int_0^3 (-2x+4) \, dx$

(f) $\int_0^1 (-6x+12) \, dx$

8.

(a) $\int_0^1 (x-1) \, dx$

(d) $\int_2^3 (x-1) \, dx$

(b) $\int_0^2 (x-1) \, dx$

(e) $\int_1^4 (x-1) \, dx$

(c) $\int_0^3 (x-1) \, dx$

(f) $\int_1^4 ((x-1)+1) \, dx$

6.

(a) $\int_0^2 f(x) \, dx$

(d) $\int_2^5 f(x) \, dx$

(b) $\int_0^3 f(x) \, dx$

(e) $\int_5^3 f(x) \, dx$

(c) $\int_0^5 f(x) \, dx$

(f) $\int_0^3 2f(x) \, dx$

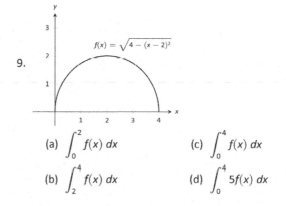

9.

(a) $\int_0^2 f(x) \, dx$

(c) $\int_0^4 f(x) \, dx$

(b) $\int_2^4 f(x) \, dx$

(d) $\int_0^4 5f(x) \, dx$

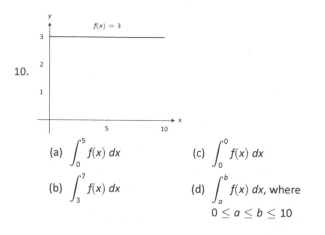

10.

(a) $\int_0^5 f(x)\,dx$

(c) $\int_0^0 f(x)\,dx$

(b) $\int_3^7 f(x)\,dx$

(d) $\int_a^b f(x)\,dx$, where

$0 \le a \le b \le 10$

In Exercises 11 – 14, a graph of a function $f(x)$ is given; the numbers inside the shaded regions give the area of that region. Evaluate the definite integrals using this area information.

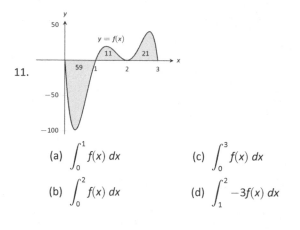

11.

(a) $\int_0^1 f(x)\,dx$

(c) $\int_0^3 f(x)\,dx$

(b) $\int_0^2 f(x)\,dx$

(d) $\int_1^2 -3f(x)\,dx$

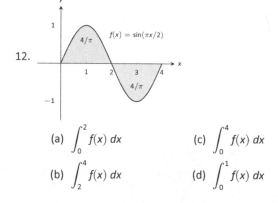

12.

(a) $\int_0^2 f(x)\,dx$

(c) $\int_0^4 f(x)\,dx$

(b) $\int_2^4 f(x)\,dx$

(d) $\int_0^1 f(x)\,dx$

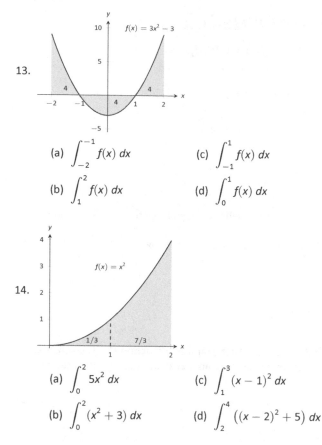

13.

(a) $\int_{-2}^{-1} f(x)\,dx$

(c) $\int_{-1}^1 f(x)\,dx$

(b) $\int_1^2 f(x)\,dx$

(d) $\int_0^1 f(x)\,dx$

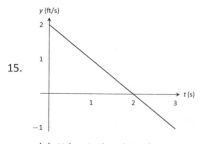

14.

(a) $\int_0^2 5x^2\,dx$

(c) $\int_1^3 (x-1)^2\,dx$

(b) $\int_0^2 (x^2+3)\,dx$

(d) $\int_2^4 \left((x-2)^2+5\right)\,dx$

In Exercises 15 – 16, a graph of the velocity function of an object moving in a straight line is given. Answer the questions based on that graph.

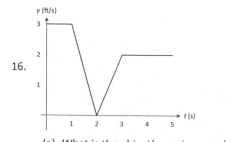

15.

(a) What is the object's maximum velocity?

(b) What is the object's maximum displacement?

(c) What is the object's total displacement on $[0,3]$?

16.

(a) What is the object's maximum velocity?

(b) What is the object's maximum displacement?

(c) What is the object's total displacement on $[0,5]$?

17. An object is thrown straight up with a velocity, in ft/s, given by $v(t) = -32t + 64$, where t is in seconds, from a height of 48 feet.

    (a) What is the object's maximum velocity?

    (b) What is the object's maximum displacement?

    (c) When does the maximum displacement occur?

    (d) When will the object reach a height of 0? (Hint: find when the displacement is -48ft.)

18. An object is thrown straight up with a velocity, in ft/s, given by $v(t) = -32t + 96$, where t is in seconds, from a height of 64 feet.

    (a) What is the object's initial velocity?

    (b) When is the object's displacement 0?

    (c) How long does it take for the object to return to its initial height?

    (d) When will the object reach a height of 210 feet?

In Exercises 19 – 22, let

- $\displaystyle\int_0^2 f(x)\,dx = 5,$

- $\displaystyle\int_0^3 f(x)\,dx = 7,$

- $\displaystyle\int_0^2 g(x)\,dx = -3,$ **and**

- $\displaystyle\int_2^3 g(x)\,dx = 5.$

Use these values to evaluate the given definite integrals.

19. $\displaystyle\int_0^2 \big(f(x) + g(x)\big)\,dx$

20. $\displaystyle\int_0^3 \big(f(x) - g(x)\big)\,dx$

21. $\displaystyle\int_2^3 \big(3f(x) + 2g(x)\big)\,dx$

22. Find nonzero values for a and b such that
$$\int_0^3 \big(af(x) + bg(x)\big)\,dx = 0$$

In Exercises 23 – 26, let

- $\displaystyle\int_0^3 s(t)\,dt = 10,$

- $\displaystyle\int_3^5 s(t)\,dt = 8,$

- $\displaystyle\int_3^5 r(t)\,dt = -1,$ **and**

- $\displaystyle\int_0^5 r(t)\,dt = 11.$

Use these values to evaluate the given definite integrals.

23. $\displaystyle\int_0^3 \big(s(t) + r(t)\big)\,dt$

24. $\displaystyle\int_5^0 \big(s(t) - r(t)\big)\,dt$

25. $\displaystyle\int_3^3 \big(\pi s(t) - 7r(t)\big)\,dt$

26. Find nonzero values for a and b such that
$$\int_0^5 \big(ar(t) + bs(t)\big)\,dt = 0$$

Review

In Exercises 27 – 30, evaluate the given indefinite integral.

27. $\displaystyle\int \big(x^3 - 2x^2 + 7x - 9\big)\,dx$

28. $\displaystyle\int \big(\sin x - \cos x + \sec^2 x\big)\,dx$

29. $\displaystyle\int \Big(\sqrt[3]{t} + \frac{1}{t^2} + 2^t\Big)\,dt$

30. $\displaystyle\int \Big(\frac{1}{x} - \csc x \cot x\Big)\,dx$

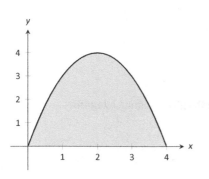

Figure 5.3.1: A graph of $f(x) = 4x - x^2$. What is the area of the shaded region?

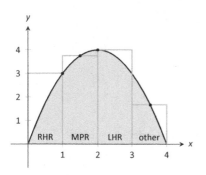

Figure 5.3.2: Approximating $\int_0^4 (4x - x^2)\, dx$ using rectangles. The heights of the rectangles are determined using different rules.

5.3   Riemann Sums

In the previous section we defined the definite integral of a function on $[a, b]$ to be the signed area between the curve and the x–axis. Some areas were simple to compute; we ended the section with a region whose area was not simple to compute. In this section we develop a technique to find such areas.

A fundamental calculus technique is to first answer a given problem with an approximation, then refine that approximation to make it better, then use limits in the refining process to find the exact answer. That is what we will do here.

Consider the region given in Figure 5.3.1, which is the area under $y = 4x - x^2$ on $[0, 4]$. What is the signed area of this region – i.e., what is $\int_0^4 (4x - x^2)\, dx$?

We start by approximating. We can surround the region with a rectangle with height and width of 4 and find the area is approximately 16 square units. This is obviously an *over–approximation*; we are including area in the rectangle that is not under the parabola.

We have an approximation of the area, using one rectangle. How can we refine our approximation to make it better? The key to this section is this answer: *use more rectangles.*

Let's use 4 rectangles with an equal width of 1. This *partitions* the interval $[0, 4]$ into 4 *subintervals*, $[0, 1]$, $[1, 2]$, $[2, 3]$ and $[3, 4]$. On each subinterval we will draw a rectangle.

There are three common ways to determine the height of these rectangles: the **Left Hand Rule**, the **Right Hand Rule**, and the **Midpoint Rule**. The **Left Hand Rule** says to evaluate the function at the left–hand endpoint of the subinterval and make the rectangle that height. In Figure 5.3.2, the rectangle drawn on the interval $[2, 3]$ has height determined by the Left Hand Rule; it has a height of $f(2)$. (The rectangle is labeled "LHR.")

The **Right Hand Rule** says the opposite: on each subinterval, evaluate the function at the right endpoint and make the rectangle that height. In the figure, the rectangle drawn on $[0, 1]$ is drawn using $f(1)$ as its height; this rectangle is labeled "RHR.".

The **Midpoint Rule** says that on each subinterval, evaluate the function at the midpoint and make the rectangle that height. The rectangle drawn on $[1, 2]$ was made using the Midpoint Rule, with a height of $f(1.5)$. That rectangle is labeled "MPR."

These are the three most common rules for determining the heights of approximating rectangles, but one is not forced to use one of these three methods. The rectangle on $[3, 4]$ has a height of approximately $f(3.53)$, very close to the Midpoint Rule. It was chosen so that the area of the rectangle is *exactly* the area of the region under f on $[3, 4]$. (Later you'll be able to figure how to do this, too.)

The following example will approximate the value of $\int_0^4 (4x - x^2)\, dx$ using these rules.

Notes:

Example 5.3.1   **Using the Left Hand, Right Hand and Midpoint Rules**

Approximate the value of $\int_0^4 (4x - x^2)\, dx$ using the Left Hand Rule, the Right Hand Rule, and the Midpoint Rule, using 4 equally spaced subintervals.

SOLUTION   We break the interval $[0, 4]$ into four subintervals as before. In Figure 5.3.3(a) we see 4 rectangles drawn on $f(x) = 4x - x^2$ using the Left Hand Rule. (The areas of the rectangles are given in each figure.)

Note how in the first subinterval, $[0, 1]$, the rectangle has height $f(0) = 0$. We add up the areas of each rectangle (height × width) for our Left Hand Rule approximation:

$$f(0) \cdot 1 + f(1) \cdot 1 + f(2) \cdot 1 + f(3) \cdot 1 =$$
$$0 + 3 + 4 + 3 = 10.$$

Figure 5.3.3(b) shows 4 rectangles drawn under f using the Right Hand Rule; note how the $[3, 4]$ subinterval has a rectangle of height 0.

In this example, these rectangle seem to be the mirror image of those found in part (a) of the Figure. This is because of the symmetry of our shaded region. Our approximation gives the same answer as before, though calculated a different way:

$$f(1) \cdot 1 + f(2) \cdot 1 + f(3) \cdot 1 + f(4) \cdot 1 =$$
$$3 + 4 + 3 + 0 = 10.$$

Figure 5.3.3(c) shows 4 rectangles drawn under f using the Midpoint Rule. This gives an approximation of $\int_0^4 (4x - x^2)\, dx$ as:

$$f(0.5) \cdot 1 + f(1.5) \cdot 1 + f(2.5) \cdot 1 + f(3.5) \cdot 1 =$$
$$1.75 + 3.75 + 3.75 + 1.75 = 11.$$

Our three methods provide two approximations of $\int_0^4 (4x - x^2)\, dx$: 10 and 11.

Summation Notation

It is hard to tell at this moment which is a better approximation: 10 or 11? We can continue to refine our approximation by using more rectangles. The notation can become unwieldy, though, as we add up longer and longer lists of numbers. We introduce **summation notation** to ameliorate this problem.

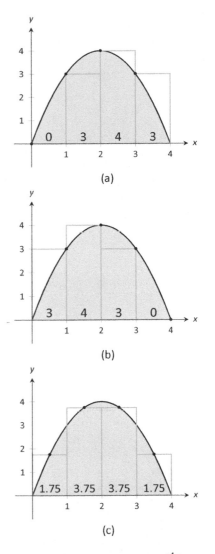

Figure 5.3.3: Approximating $\int_0^4 (4x - x^2)\, dx$ in Example 5.3.1. In (a), the Left Hand Rule is used; in (b), the Right Hand Rule is used; in (c), the Midpoint Rule is used.

Notes:

Suppose we wish to add up a list of numbers a_1, a_2, a_3, ..., a_9. Instead of writing

$$a_1 + a_2 + a_3 + a_4 + a_5 + a_6 + a_7 + a_8 + a_9,$$

we use summation notation and write

Figure 5.3.4: Understanding summation notation.

The upper case sigma represents the term "sum." The index of summation in this example is i; any symbol can be used. By convention, the index takes on only the integer values between (and including) the lower and upper bounds.

Let's practice using this notation.

Example 5.3.2   Using summation notation

Let the numbers $\{a_i\}$ be defined as $a_i = 2i - 1$ for integers i, where $i \geq 1$. So $a_1 = 1$, $a_2 = 3$, $a_3 = 5$, etc. (The output is the positive odd integers). Evaluate the following summations:

$$1.\ \sum_{i=1}^{6} a_i \qquad\qquad 2.\ \sum_{i=3}^{7} (3a_i - 4) \qquad\qquad 3.\ \sum_{i=1}^{4} (a_i)^2$$

SOLUTION

1.
$$\sum_{i=1}^{6} a_i = a_1 + a_2 + a_3 + a_4 + a_5 + a_6$$
$$= 1 + 3 + 5 + 7 + 9 + 11$$
$$= 36.$$

2. Note the starting value is different than 1:

$$\sum_{i=3}^{7} (3a_i - 4) = (3a_3 - 4) + (3a_4 - 4) + (3a_5 - 4) + (3a_6 - 4) + (3a_7 - 4)$$
$$= 11 + 17 + 23 + 29 + 35$$
$$= 115.$$

Notes:

3.

$$\sum_{i=1}^{4}(a_i)^2 = (a_1)^2 + (a_2)^2 + (a_3)^2 + (a_4)^2$$
$$= 1^2 + 3^2 + 5^2 + 7^2$$
$$= 84.$$

It might seem odd to stress a new, concise way of writing summations only to write each term out as we add them up. It is. The following theorem gives some of the properties of summations that allow us to work with them without writing individual terms. Examples will follow.

Theorem 5.3.1   **Properties of Summations**

1. $\sum_{i=1}^{n} c = c \cdot n$, where c is a constant.

2. $\sum_{i=m}^{n}(a_i \pm b_i) = \sum_{i=m}^{n} a_i \pm \sum_{i=m}^{n} b_i$

3. $\sum_{i=m}^{n} c \cdot a_i = c \cdot \sum_{i=m}^{n} a_i$

4. $\sum_{i=m}^{j} a_i + \sum_{i=j+1}^{n} a_i = \sum_{i=m}^{n} a_i$

5. $\sum_{i=1}^{n} i = \dfrac{n(n+1)}{2}$

6. $\sum_{i=1}^{n} i^2 = \dfrac{n(n+1)(2n+1)}{6}$

7. $\sum_{i=1}^{n} i^3 = \left(\dfrac{n(n+1)}{2}\right)^2$

Example 5.3.3   **Evaluating summations using Theorem 5.3.1**
Revisit Example 5.3.2 and, using Theorem 5.3.1, evaluate

$$\sum_{i=1}^{6} a_i = \sum_{i=1}^{6}(2i - 1).$$

Notes:

SOLUTION

$$\sum_{i=1}^{6}(2i-1) = \sum_{i=1}^{6}2i - \sum_{i=1}^{6}(1)$$

$$= \left(2\sum_{i=1}^{6}i\right) - 6$$

$$= 2\frac{6(6+1)}{2} - 6$$

$$= 42 - 6 = 36$$

We obtained the same answer without writing out all six terms. When dealing with small sizes of n, it may be faster to write the terms out by hand. However, Theorem 5.3.1 is incredibly important when dealing with large sums as we'll soon see.

Riemann Sums

Consider again $\int_0^4 (4x - x^2)\, dx$. We will approximate this definite integral using 16 equally spaced subintervals and the Right Hand Rule in Example 5.3.4. Before doing so, it will pay to do some careful preparation.

Figure 5.3.5 shows a number line of $[0,4]$ divided, or *partitioned*, into 16 equally spaced subintervals. We denote 0 as x_1; we have marked the values of x_5, x_9, x_{13} and x_{17}. We could mark them all, but the figure would get crowded. While it is easy to figure that $x_{10} = 2.25$, in general, we want a method of determining the value of x_i without consulting the figure. Consider:

$$x_i = x_1 + (i-1)\Delta x$$

So $x_{10} = x_1 + 9(4/16) = 2.25$.

If we had partitioned $[0,4]$ into 100 equally spaced subintervals, each subinterval would have length $\Delta x = 4/100 = 0.04$. We could compute x_{32} as

$$x_{32} = x_1 + 31(4/100) = 1.24.$$

(That was far faster than creating a sketch first.)

Notes:

![number line of [0,4] divided into 16 equally spaced subintervals with markings at 0, 1, 2, 3, 4 labeled x₁, x₅, x₉, x₁₃, x₁₇]

Figure 5.3.5: Dividing $[0,4]$ into 16 equally spaced subintervals.

Given any subdivision of $[0, 4]$, the first subinterval is $[x_1, x_2]$; the second is $[x_2, x_3]$; the ith subinterval is $[x_i, x_{i+1}]$.

When using the Left Hand Rule, the height of the ith rectangle will be $f(x_i)$.

When using the Right Hand Rule, the height of the ith rectangle will be $f(x_{i+1})$.

When using the Midpoint Rule, the height of the ith rectangle will be $f\left(\dfrac{x_i + x_{i+1}}{2}\right)$.

Thus approximating $\int_0^4 (4x - x^2)\, dx$ with 16 equally spaced subintervals can be expressed as follows, where $\Delta x = 4/16 = 1/4$:

Left Hand Rule: $\displaystyle\sum_{i=1}^{16} f(x_i)\Delta x$

Right Hand Rule: $\displaystyle\sum_{i=1}^{16} f(x_{i+1})\Delta x$

Midpoint Rule: $\displaystyle\sum_{i=1}^{16} f\left(\dfrac{x_i + x_{i+1}}{2}\right)\Delta x$

We use these formulas in the next two examples. The following example lets us practice using the Right Hand Rule and the summation formulas introduced in Theorem 5.3.1.

Example 5.3.4   Approximating definite integrals using sums
Approximate $\int_0^4 (4x - x^2)\, dx$ using the Right Hand Rule and summation formulas with 16 and 1000 equally spaced intervals.

SOLUTION   Using the formula derived before, using 16 equally spaced intervals and the Right Hand Rule, we can approximate the definite integral as

$$\sum_{i=1}^{16} f(x_{i+1})\Delta x.$$

We have $\Delta x = 4/16 = 0.25$. Since $x_i = 0 + (i-1)\Delta x$, we have

$$\begin{aligned} x_{i+1} &= 0 + \big((i+1) - 1\big)\Delta x \\ &= i\Delta x \end{aligned}$$

Notes:

Using the summation formulas, consider:

$$\int_0^4 (4x - x^2)\, dx \approx \sum_{i=1}^{16} f(x_{i+1})\Delta x$$

$$= \sum_{i=1}^{16} f(i\Delta x)\Delta x$$

$$= \sum_{i=1}^{16} \left(4i\Delta x - (i\Delta x)^2\right)\Delta x$$

$$= \sum_{i=1}^{16} (4i\Delta x^2 - i^2 \Delta x^3)$$

$$= (4\Delta x^2)\sum_{i=1}^{16} i - \Delta x^3 \sum_{i=1}^{16} i^2 \tag{5.3}$$

$$= (4\Delta x^2)\frac{16 \cdot 17}{2} - \Delta x^3 \frac{16(17)(33)}{6} \quad (\Delta x = 0.25)$$

$$= 10.625$$

We were able to sum up the areas of 16 rectangles with very little computation. In Figure 5.3.6 the function and the 16 rectangles are graphed. While some rectangles over–approximate the area, other under–approximate the area (by about the same amount). Thus our approximate area of 10.625 is likely a fairly good approximation.

Notice Equation (5.3); by changing the 16's to 1,000's (and appropriately changing the value of Δx), we can use that equation to sum up 1000 rectangles! We do so here, skipping from the original summand to the equivalent of Equation (5.3) to save space. Note that $\Delta x = 4/1000 = 0.004$.

$$\int_0^4 (4x - x^2)\, dx \approx \sum_{i=1}^{1000} f(x_{i+1})\Delta x$$

$$= (4\Delta x^2)\sum_{i=1}^{1000} i - \Delta x^3 \sum_{i=1}^{1000} i^2$$

$$= (4\Delta x^2)\frac{1000 \cdot 1001}{2} - \Delta x^3 \frac{1000(1001)(2001)}{6}$$

$$= 10.666656$$

Using many, many rectangles, we have a likely good approximation of $\int_0^4 (4x - x^2)\Delta x$. That is,

$$\int_0^4 (4x - x^2)\, dx \approx 10.666656.$$

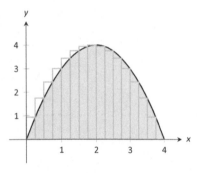

Figure 5.3.6: Approximating $\int_0^4 (4x - x^2)\, dx$ with the Right Hand Rule and 16 evenly spaced subintervals.

Notes:

Before the above example, we stated what the summations for the Left Hand, Right Hand and Midpoint Rules looked like. Each had the same basic structure, which was:

1. each rectangle has the same width, which we referred to as Δx, and

2. each rectangle's height is determined by evaluating f at a particular point in each subinterval. For instance, the Left Hand Rule states that each rectangle's height is determined by evaluating f at the left hand endpoint of the subinterval the rectangle lives on.

One could partition an interval $[a, b]$ with subintervals that do not have the same size. We refer to the length of the ith subinterval as Δx_i. Also, one could determine each rectangle's height by evaluating f at *any* point c_i in the ith subinterval. Thus the height of the ith subinterval would be $f(c_i)$, and the area of the ith rectangle would be $f(c_i)\Delta x_i$. These ideas are formally defined below.

Definition 5.3.1   Partition

A **partition** Δx of a closed interval $[a, b]$ is a set of numbers $x_1, x_2, \ldots x_{n+1}$ where
$$a = x_1 < x_2 < \ldots < x_n < x_{n+1} = b.$$

The length of the ith subinterval, $[x_i, x_{i+1}]$, is $\Delta x_i = x_{i+1} - x_i$. If $[a, b]$ is partitioned into subintervals of equal length, we let Δx represent the length of each subinterval.

The **size of the partition**, denoted $||\Delta x||$, is the length of the largest subinterval of the partition.

Summations of rectangles with area $f(c_i)\Delta x_i$ are named after mathematician Georg Friedrich Bernhard Riemann, as given in the following definition.

Definition 5.3.2   Riemann Sum

Let f be defined on a closed interval $[a, b]$, let Δx be a partition of $[a, b]$ and let c_i denote any value in the ith subinterval.
The sum
$$\sum_{i=1}^{n} f(c_i)\Delta x_i$$
is a **Riemann sum** of f on $[a, b]$.

Notes:

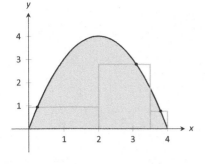

Figure 5.3.7: An example of a general Riemann sum to approximate $\int_0^4 (4x - x^2)\, dx$.

Figure 5.3.7 shows the approximating rectangles of a Riemann sum of $\int_0^4 (4x - x^2)\, dx$. While the rectangles in this example do not approximate well the shaded area, they demonstrate that the subinterval widths may vary and the heights of the rectangles can be determined without following a particular rule.

"Usually" Riemann sums are calculated using one of the three methods we have introduced. The uniformity of construction makes computations easier. Before working another example, let's summarize some of what we have learned in a convenient way.

Key Idea 5.3.1   Riemann Sum Concepts

Consider $\displaystyle\int_a^b f(x)\, dx \approx \sum_{i=1}^{n} f(c_i) \Delta x_i$.

1. When the n subintervals have equal length, $\Delta x_i = \Delta x = \dfrac{b - a}{n}$.

2. The ith term of an equally spaced partition is $x_i = a + (i - 1)\Delta x$. (Thus $x_1 = a$ and $x_{n+1} = b$.)

3. The Left Hand Rule summation is: $\displaystyle\sum_{i=1}^{n} f(x_i)\Delta x$.

4. The Right Hand Rule summation is: $\displaystyle\sum_{i=1}^{n} f(x_{i+1})\Delta x$.

5. The Midpoint Rule summation is: $\displaystyle\sum_{i=1}^{n} f\left(\dfrac{x_i + x_{i+1}}{2}\right)\Delta x$.

Let's do another example.

Example 5.3.5   Approximating definite integrals with sums
Approximate $\int_{-2}^{3}(5x + 2)\, dx$ using the Midpoint Rule and 10 equally spaced intervals.

SOLUTION   Following Key Idea 5.3.1, we have

$$\Delta x = \frac{3 - (-2)}{10} = 1/2 \quad \text{and} \quad x_i = (-2) + (1/2)(i - 1) = i/2 - 5/2.$$

Notes:

As we are using the Midpoint Rule, we will also need x_{i+1} and $\dfrac{x_i + x_{i+1}}{2}$. Since $x_i = i/2 - 5/2$, $\quad x_{i+1} = (i+1)/2 - 5/2 = i/2 - 2$. This gives

$$\frac{x_i + x_{i+1}}{2} = \frac{(i/2 - 5/2) + (i/2 - 2)}{2} = \frac{i - 9/2}{2} = i/2 - 9/4.$$

We now construct the Riemann sum and compute its value using summation formulas.

$$
\begin{aligned}
\int_{-2}^{3} (5x + 2)\, dx &\approx \sum_{i=1}^{10} f\left(\frac{x_i + x_{i+1}}{2}\right) \Delta x \\
&= \sum_{i=1}^{10} f(i/2 - 9/4)\, \Delta x \\
&= \sum_{i=1}^{10} \left(5(i/2 - 9/4) + 2\right) \Delta x \\
&= \Delta x \sum_{i=1}^{10} \left[\left(\frac{5}{2}\right)i - \frac{37}{4}\right] \\
&= \Delta x \left(\frac{5}{2}\sum_{i=1}^{10}(i) - \sum_{i=1}^{10}\left(\frac{37}{4}\right)\right) \\
&= \frac{1}{2}\left(\frac{5}{2} \cdot \frac{10(11)}{2} - 10 \cdot \frac{37}{4}\right) \\
&= \frac{45}{2} = 22.5
\end{aligned}
$$

Figure 5.3.8: Approximating $\int_{-2}^{3}(5x + 2)\, dx$ using the Midpoint Rule and 10 evenly spaced subintervals in Example 5.3.5.

Note the graph of $f(x) = 5x + 2$ in Figure 5.3.8. The regions whose area is computed by the definite integral are triangles, meaning we can find the exact answer without summation techniques. We find that the exact answer is indeed 22.5. One of the strengths of the Midpoint Rule is that often each rectangle includes area that should not be counted, but misses other area that should. When the partition size is small, these two amounts are about equal and these errors almost "cancel each other out." In this example, since our function is a line, these errors are exactly equal and they do cancel each other out, giving us the exact answer.

Note too that when the function is negative, the rectangles have a "negative" height. When we compute the area of the rectangle, we use $f(c_i)\Delta x$; when f is negative, the area is counted as negative.

Notice in the previous example that while we used 10 equally spaced intervals, the number "10" didn't play a big role in the calculations until the very end.

Notes:

Mathematicians love to abstract ideas; let's approximate the area of another region using n subintervals, where we do not specify a value of n until the very end.

Example 5.3.6   **Approximating definite integrals with a formula, using sums**
Revisit $\int_0^4 (4x-x^2)\,dx$ yet again. Approximate this definite integral using the Right Hand Rule with n equally spaced subintervals.

SOLUTION   Using Key Idea 5.3.1, we know $\Delta x = \frac{4-0}{n} = 4/n$. We also find $x_i = 0 + \Delta x(i-1) = 4(i-1)/n$. The Right Hand Rule uses x_{i+1}, which is $x_{i+1} = 4i/n$.

We construct the Right Hand Rule Riemann sum as follows. Be sure to follow each step carefully. If you get stuck, and do not understand how one line proceeds to the next, you may skip to the result and consider how this result is used. You should come back, though, and work through each step for full understanding.

$$
\begin{aligned}
\int_0^4 (4x - x^2)\,dx &\approx \sum_{i=1}^{n} f(x_{i+1})\Delta x \\
&= \sum_{i=1}^{n} f\left(\frac{4i}{n}\right)\Delta x \\
&= \sum_{i=1}^{n} \left[4\frac{4i}{n} - \left(\frac{4i}{n}\right)^2\right]\Delta x \\
&= \sum_{i=1}^{n} \left(\frac{16\Delta x}{n}\right)i - \sum_{i=1}^{n}\left(\frac{16\Delta x}{n^2}\right)i^2 \\
&= \left(\frac{16\Delta x}{n}\right)\sum_{i=1}^{n} i - \left(\frac{16\Delta x}{n^2}\right)\sum_{i=1}^{n} i^2 \\
&= \left(\frac{16\Delta x}{n}\right)\cdot\frac{n(n+1)}{2} - \left(\frac{16\Delta x}{n^2}\right)\frac{n(n+1)(2n+1)}{6} \quad \left(\begin{smallmatrix}\text{recall}\\ \Delta x = 4/n\end{smallmatrix}\right) \\
&= \frac{32(n+1)}{n} - \frac{32(n+1)(2n+1)}{3n^2} \quad \text{(now simplify)} \\
&= \frac{32}{3}\left(1 - \frac{1}{n^2}\right)
\end{aligned}
$$

The result is an amazing, easy to use formula. To approximate the definite integral with 10 equally spaced subintervals and the Right Hand Rule, set $n = 10$ and compute

$$
\int_0^4 (4x - x^2)\,dx \approx \frac{32}{3}\left(1 - \frac{1}{10^2}\right) = 10.56.
$$

Notes:

Recall how earlier we approximated the definite integral with 4 subintervals; with $n = 4$, the formula gives 10, our answer as before.

It is now easy to approximate the integral with 1,000,000 subintervals! Hand-held calculators will round off the answer a bit prematurely giving an answer of 10.66666667. (The actual answer is 10.666666666656.)

We now take an important leap. Up to this point, our mathematics has been limited to geometry and algebra (finding areas and manipulating expressions). Now we apply *calculus*. For any *finite* n, we know that

$$\int_0^4 (4x - x^2)\, dx \approx \frac{32}{3}\left(1 - \frac{1}{n^2}\right).$$

Both common sense and high–level mathematics tell us that as n gets large, the approximation gets better. In fact, if we take the *limit* as $n \to \infty$, we get the *exact area* described by $\int_0^4 (4x - x^2)\, dx$. That is,

$$\int_0^4 (4x - x^2)\, dx = \lim_{n\to\infty} \frac{32}{3}\left(1 - \frac{1}{n^2}\right)$$
$$= \frac{32}{3}(1 - 0)$$
$$= \frac{32}{3} = 10.\overline{6}$$

This is a fantastic result. By considering n equally–spaced subintervals, we obtained a formula for an approximation of the definite integral that involved our variable n. As n grows large – without bound – the error shrinks to zero and we obtain the exact area.

This section started with a fundamental calculus technique: make an approximation, refine the approximation to make it better, then use limits in the refining process to get an exact answer. That is precisely what we just did.

Let's practice this again.

Example 5.3.7   **Approximating definite integrals with a formula, using sums**
Find a formula that approximates $\int_{-1}^5 x^3\, dx$ using the Right Hand Rule and n equally spaced subintervals, then take the limit as $n \to \infty$ to find the exact area.

SOLUTION   Following Key Idea 5.3.1, we have $\Delta x = \frac{5-(-1)}{n} = 6/n$. We have $x_i = (-1) + (i-1)\Delta x$; as the Right Hand Rule uses x_{i+1}, we have $x_{i+1} = (-1) + i\Delta x$.

The Riemann sum corresponding to the Right Hand Rule is (followed by sim-

Notes:

plifications):

$$\int_{-1}^{5} x^3 \, dx \approx \sum_{i=1}^{n} f(x_{i+1}) \, \Delta x$$

$$= \sum_{i=1}^{n} f(-1 + i\Delta x) \, \Delta x$$

$$= \sum_{i=1}^{n} (-1 + i\Delta x)^3 \, \Delta x$$

$$= \sum_{i=1}^{n} \left((i\Delta x)^3 - 3(i\Delta x)^2 + 3i\Delta x - 1 \right) \Delta x \quad \text{(now distribute } \Delta x)$$

$$= \sum_{i=1}^{n} \left(i^3 \Delta x^4 - 3i^2 \Delta x^3 + 3i \Delta x^2 - \Delta x \right) \quad \text{(now split up summation)}$$

$$= \Delta x^4 \sum_{i=1}^{n} i^3 - 3\Delta x^3 \sum_{i=1}^{n} i^2 + 3\Delta x^2 \sum_{i=1}^{n} i - \sum_{i=1}^{n} \Delta x$$

$$= \Delta x^4 \left(\frac{n(n+1)}{2} \right)^2 - 3\Delta x^3 \frac{n(n+1)(2n+1)}{6} + 3\Delta x^2 \frac{n(n+1)}{2} - n\Delta x$$

(use $\Delta x = 6/n$)

$$= \frac{1296}{n^4} \cdot \frac{n^2(n+1)^2}{4} - 3\frac{216}{n^3} \cdot \frac{n(n+1)(2n+1)}{6} + 3\frac{36}{n^2} \frac{n(n+1)}{2} - 6$$

(now do a sizable amount of algebra to simplify)

$$= 156 + \frac{378}{n} + \frac{216}{n^2}$$

Once again, we have found a compact formula for approximating the definite integral with n equally spaced subintervals and the Right Hand Rule. Using 10 subintervals, we have an approximation of 195.96 (these rectangles are shown in Figure 5.3.9). Using $n = 100$ gives an approximation of 159.802.

Now find the exact answer using a limit:

$$\int_{-1}^{5} x^3 \, dx = \lim_{n \to \infty} \left(156 + \frac{378}{n} + \frac{216}{n^2} \right) = 156.$$

Limits of Riemann Sums

We have used limits to evaluate given definite integrals. Will this always work? We will show, given not–very–restrictive conditions, that yes, it will always work.

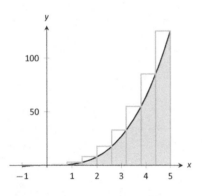

Figure 5.3.9: Approximating $\int_{-1}^{5} x^3 \, dx$ using the Right Hand Rule and 10 evenly spaced subintervals.

Notes:

The previous two examples demonstrated how an expression such as

$$\sum_{i=1}^{n} f(x_{i+1})\Delta x$$

can be rewritten as an expression explicitly involving n, such as $32/3(1 - 1/n^2)$.

Viewed in this manner, we can think of the summation as a function of n. An n value is given (where n is a positive integer), and the sum of areas of n equally spaced rectangles is returned, using the Left Hand, Right Hand, or Midpoint Rules.

Given a definite integral $\int_a^b f(x)\, dx$, let:

- $S_L(n) = \sum_{i=1}^{n} f(x_i)\Delta x$, the sum of equally spaced rectangles formed using the Left Hand Rule,

- $S_R(n) = \sum_{i=1}^{n} f(x_{i+1})\Delta x$, the sum of equally spaced rectangles formed using the Right Hand Rule, and

- $S_M(n) = \sum_{i=1}^{n} f\left(\dfrac{x_i + x_{i+1}}{2}\right)\Delta x$, the sum of equally spaced rectangles formed using the Midpoint Rule.

Recall the definition of a limit as $n \to \infty$: $\lim_{n\to\infty} S_L(n) = K$ if, given any $\varepsilon > 0$, there exists $N > 0$ such that

$$|S_L(n) - K| < \varepsilon \quad \text{when} \quad n \geq N.$$

The following theorem states that we can use any of our three rules to find the exact value of a definite integral $\int_a^b f(x)\, dx$. It also goes two steps further. The theorem states that the height of each rectangle doesn't have to be determined following a specific rule, but could be $f(c_i)$, where c_i is any point in the ith subinterval, as discussed before Riemann Sums were defined in Definition 5.3.2.

The theorem goes on to state that the rectangles do not need to be of the same width. Using the notation of Definition 5.3.1, let Δx_i denote the length of the ith subinterval in a partition of $[a, b]$ and let $||\Delta x||$ represent the length of the largest subinterval in the partition: that is, $||\Delta x||$ is the largest of all the Δx_i's. If $||\Delta x||$ is small, then $[a, b]$ must be partitioned into many subintervals, since all subintervals must have small lengths. "Taking the limit as $||\Delta x||$ goes to zero" implies that the number n of subintervals in the partition is growing to

Notes:

infinity, as the largest subinterval length is becoming arbitrarily small. We then interpret the expression

$$\lim_{||\Delta x||\to 0} \sum_{i=1}^{n} f(c_i)\Delta x_i$$

as "the limit of the sum of the areas of rectangles, where the width of each rectangle can be different but getting small, and the height of each rectangle is not necessarily determined by a particular rule." The theorem states that this Riemann Sum also gives the value of the definite integral of f over $[a,b]$.

Theorem 5.3.2   Definite Integrals and the Limit of Riemann Sums

Let f be continuous on the closed interval $[a,b]$ and let $S_L(n)$, $S_R(n)$, $S_M(n)$, Δx, Δx_i and c_i be defined as before. Then:

1. $\displaystyle\lim_{n\to\infty} S_L(n) = \lim_{n\to\infty} S_R(n) = \lim_{n\to\infty} S_M(n) = \lim_{n\to\infty} \sum_{i=1}^{n} f(c_i)\Delta x$,

2. $\displaystyle\lim_{n\to\infty} \sum_{i=1}^{n} f(c_i)\Delta x = \int_a^b f(x)\,dx$, and

3. $\displaystyle\lim_{||\Delta x||\to 0} \sum_{i=1}^{n} f(c_i)\Delta x_i = \int_a^b f(x)\,dx$.

We summarize what we have learned over the past few sections here.

- Knowing the "area under the curve" can be useful. One common example: the area under a velocity curve is displacement.

- We have defined the definite integral, $\int_a^b f(x)\,dx$, to be the signed area under f on the interval $[a,b]$.

- While we can approximate a definite integral many ways, we have focused on using rectangles whose heights can be determined using the Left Hand Rule, the Right Hand Rule and the Midpoint Rule.

- Sums of rectangles of this type are called Riemann sums.

- The exact value of the definite integral can be computed using the limit of a Riemann sum. We generally use one of the above methods as it makes the algebra simpler.

Notes:

We first learned of derivatives through limits then learned rules that made the process simpler. We know of a way to evaluate a definite integral using limits; in the next section we will see how the Fundamental Theorem of Calculus makes the process simpler. The key feature of this theorem is its connection between the indefinite integral and the definite integral.

Notes:

Exercises 5.3

Terms and Concepts

1. A fundamental calculus technique is to use _____ to refine approximations to get an exact answer.

2. What is the upper bound in the summation
$$\sum_{i=7}^{14}(48i - 201)?$$

3. This section approximates definite integrals using what geometric shape?

4. T/F: A sum using the Right Hand Rule is an example of a Riemann Sum.

Problems

In Exercises 5 – 12, write out each term of the summation and compute the sum.

5. $\displaystyle\sum_{i=2}^{4} i^2$

6. $\displaystyle\sum_{i=-1}^{3}(4i - 2)$

7. $\displaystyle\sum_{i=-2}^{2} \sin(\pi i/2)$

8. $\displaystyle\sum_{i=1}^{10} 5$

9. $\displaystyle\sum_{i=1}^{5} \frac{1}{i}$

10. $\displaystyle\sum_{i=1}^{6}(-1)^i i$

11. $\displaystyle\sum_{i=1}^{4}\left(\frac{1}{i} - \frac{1}{i+1}\right)$

12. $\displaystyle\sum_{i=0}^{5}(-1)^i \cos(\pi i)$

In Exercises 13 – 16, write each sum in summation notation.

13. $3 + 6 + 9 + 12 + 15$

14. $-1 + 0 + 3 + 8 + 15 + 24 + 35 + 48 + 63$

15. $\dfrac{1}{2} + \dfrac{2}{3} + \dfrac{3}{4} + \dfrac{4}{5}$

16. $1 - e + e^2 - e^3 + e^4$

In Exercises 17 – 24, evaluate the summation using Theorem 5.3.1.

17. $\displaystyle\sum_{i=1}^{10} 5$

18. $\displaystyle\sum_{i=1}^{25} i$

19. $\displaystyle\sum_{i=1}^{10}(3i^2 - 2i)$

20. $\displaystyle\sum_{i=1}^{15}(2i^3 - 10)$

21. $\displaystyle\sum_{i=1}^{10}(-4i^3 + 10i^2 - 7i + 11)$

22. $\displaystyle\sum_{i=1}^{10}(i^3 - 3i^2 + 2i + 7)$

23. $1 + 2 + 3 + \ldots + 99 + 100$

24. $1 + 4 + 9 + \ldots + 361 + 400$

Theorem 5.3.1 states

$$\sum_{i=1}^{n} a_i = \sum_{i=1}^{k} a_i + \sum_{i=k+1}^{n} a_i \text{ , so}$$

$$\sum_{i=k+1}^{n} a_i = \sum_{i=1}^{n} a_i - \sum_{i=1}^{k} a_i \,.$$

Use this fact, along with other parts of Theorem 5.3.1, to evaluate the summations given in Exercises 25 – 28.

25. $\displaystyle\sum_{i=11}^{20} i$

26. $\displaystyle\sum_{i=16}^{25} i^3$

27. $\displaystyle\sum_{i=7}^{12} 4$

28. $\displaystyle\sum_{i=5}^{10} 4i^3$

In Exercises 29 – 34, a definite integral $\int_a^b f(x)\, dx$ is given.

(a) Graph $f(x)$ on $[a, b]$.

(b) Add to the sketch rectangles using the provided rule.

(c) Approximate $\int_a^b f(x)\, dx$ by summing the areas of the rectangles.

29. $\int_{-3}^3 x^2\, dx$, with 6 rectangles using the Left Hand Rule.

30. $\int_0^2 (5 - x^2)\, dx$, with 4 rectangles using the Midpoint Rule.

31. $\int_0^\pi \sin x\, dx$, with 6 rectangles using the Right Hand Rule.

32. $\int_0^3 2^x\, dx$, with 5 rectangles using the Left Hand Rule.

33. $\int_1^2 \ln x\, dx$, with 3 rectangles using the Midpoint Rule.

34. $\int_1^9 \frac{1}{x}\, dx$, with 4 rectangles using the Right Hand Rule.

In Exercises 35 – 40, a definite integral $\int_a^b f(x)\, dx$ is given. As demonstrated in Examples 5.3.6 and 5.3.7, do the following.

(a) Find a formula to approximate $\int_a^b f(x)\, dx$ using n subintervals and the provided rule.

(b) Evaluate the formula using $n = 10, 100$ and $1,000$.

(c) Find the limit of the formula, as $n \to \infty$, to find the exact value of $\int_a^b f(x)\, dx$.

35. $\int_0^1 x^3\, dx$, using the Right Hand Rule.

36. $\int_{-1}^1 3x^2\, dx$, using the Left Hand Rule.

37. $\int_{-1}^3 (3x - 1)\, dx$, using the Midpoint Rule.

38. $\int_1^4 (2x^2 - 3)\, dx$, using the Left Hand Rule.

39. $\int_{-10}^{10} (5 - x)\, dx$, using the Right Hand Rule.

40. $\int_0^1 (x^3 - x^2)\, dx$, using the Right Hand Rule.

Review

In Exercises 41 – 46, find an antiderivative of the given function.

41. $f(x) = 5\sec^2 x$

42. $f(x) = \dfrac{7}{x}$

43. $g(t) = 4t^5 - 5t^3 + 8$

44. $g(t) = 5 \cdot 8^t$

45. $g(t) = \cos t + \sin t$

46. $f(x) = \dfrac{1}{\sqrt{x}}$

5.4   The Fundamental Theorem of Calculus

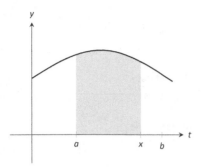

Let $f(t)$ be a continuous function defined on $[a, b]$. The definite integral $\int_a^b f(x)\,dx$ is the "area under f" on $[a, b]$. We can turn this concept into a function by letting the upper (or lower) bound vary.

Let $F(x) = \int_a^x f(t)\,dt$. It computes the area under f on $[a, x]$ as illustrated in Figure 5.4.1. We can study this function using our knowledge of the definite integral. For instance, $F(a) = 0$ since $\int_a^a f(t)\,dt = 0$.

We can also apply calculus ideas to $F(x)$; in particular, we can compute its derivative. While this may seem like an innocuous thing to do, it has far–reaching implications, as demonstrated by the fact that the result is given as an important theorem.

Figure 5.4.1: The area of the shaded region is $F(x) = \int_a^x f(t)\,dt$.

Theorem 5.4.1   The Fundamental Theorem of Calculus, Part 1

Let f be continuous on $[a, b]$ and let $F(x) = \int_a^x f(t)\,dt$. Then F is a differentiable function on (a, b), and

$$F'(x) = f(x).$$

Initially this seems simple, as demonstrated in the following example.

Example 5.4.1   Using the Fundamental Theorem of Calculus, Part 1

Let $F(x) = \displaystyle\int_{-5}^x (t^2 + \sin t)\,dt$. What is $F'(x)$?

SOLUTION   Using the Fundamental Theorem of Calculus, we have $F'(x) = x^2 + \sin x$.

This simple example reveals something incredible: $F(x)$ is an antiderivative of $x^2 + \sin x$! Therefore, $F(x) = \frac{1}{3}x^3 - \cos x + C$ for some value of C. (We can find C, but generally we do not care. We know that $F(-5) = 0$, which allows us to compute C. In this case, $C = \cos(-5) + \frac{125}{3}$.)

We have done more than found a complicated way of computing an antiderivative. Consider a function f defined on an open interval containing a, b and c. Suppose we want to compute $\int_a^b f(t)\,dt$. First, let $F(x) = \int_c^x f(t)\,dt$. Using

---

Notes:

the properties of the definite integral found in Theorem 5.2.1, we know

$$\int_a^b f(t)\, dt = \int_a^c f(t)\, dt + \int_c^b f(t)\, dt$$

$$= -\int_c^a f(t)\, dt + \int_c^b f(t)\, dt$$

$$= -F(a) + F(b)$$

$$= F(b) - F(a).$$

We now see how indefinite integrals and definite integrals are related: we can evaluate a definite integral using antiderivatives! This is the second part of the Fundamental Theorem of Calculus.

Theorem 5.4.2   The Fundamental Theorem of Calculus, Part 2

Let f be continuous on $[a, b]$ and let F be *any* antiderivative of f. Then

$$\int_a^b f(x)\, dx = F(b) - F(a).$$

Example 5.4.2   Using the Fundamental Theorem of Calculus, Part 2

We spent a great deal of time in the previous section studying $\int_0^4 (4x - x^2)\, dx$. Using the Fundamental Theorem of Calculus, evaluate this definite integral.

SOLUTION   We need an antiderivative of $f(x) = 4x - x^2$. All antiderivatives of f have the form $F(x) = 2x^2 - \frac{1}{3}x^3 + C$; for simplicity, choose $C = 0$.

The Fundamental Theorem of Calculus states

$$\int_0^4 (4x - x^2)\, dx = F(4) - F(0) = \left(2(4)^2 - \frac{1}{3}4^3\right) - (0 - 0) = 32 - \frac{64}{3} = 32/3.$$

This is the same answer we obtained using limits in the previous section, just with much less work.

Notation: A special notation is often used in the process of evaluating definite integrals using the Fundamental Theorem of Calculus. Instead of explicitly writing $F(b) - F(a)$, the notation $F(x)\Big|_a^b$ is used. Thus the solution to Example 5.4.2 would be written as:

$$\int_0^4 (4x - x^2)\, dx = \left(2x^2 - \frac{1}{3}x^3\right)\Big|_0^4 = \left(2(4)^2 - \frac{1}{3}4^3\right) - (0 - 0) = 32/3.$$

Notes:

The Constant C: *Any* antiderivative $F(x)$ can be chosen when using the Fundamental Theorem of Calculus to evaluate a definite integral, meaning any value of C can be picked. The constant *always* cancels out of the expression when evaluating $F(b) - F(a)$, so it does not matter what value is picked. This being the case, we might as well let $C = 0$.

Example 5.4.3   Using the Fundamental Theorem of Calculus, Part 2
Evaluate the following definite integrals.

1. $\displaystyle\int_{-2}^{2} x^3 \, dx$ 2. $\displaystyle\int_{0}^{\pi} \sin x \, dx$ 3. $\displaystyle\int_{0}^{5} e^t \, dt$ 4. $\displaystyle\int_{4}^{9} \sqrt{u} \, du$ 5. $\displaystyle\int_{1}^{5} 2 \, dx$

SOLUTION

1. $\displaystyle\int_{-2}^{2} x^3 \, dx = \frac{1}{4}x^4 \Big|_{-2}^{2} = \left(\frac{1}{4}2^4\right) - \left(\frac{1}{4}(-2)^4\right) = 0.$

2. $\displaystyle\int_{0}^{\pi} \sin x \, dx = -\cos x \Big|_{0}^{\pi} = -\cos \pi - (-\cos 0) = 1 + 1 = 2.$

   (This is interesting; it says that the area under one "hump" of a sine curve is 2.)

3. $\displaystyle\int_{0}^{5} e^t \, dt = e^t \Big|_{0}^{5} = e^5 - e^0 = e^5 - 1 \approx 147.41.$

4. $\displaystyle\int_{4}^{9} \sqrt{u} \, du = \int_{4}^{9} u^{\frac{1}{2}} \, du = \frac{2}{3}u^{\frac{3}{2}} \Big|_{4}^{9} = \frac{2}{3}\left(9^{\frac{3}{2}} - 4^{\frac{3}{2}}\right) = \frac{2}{3}(27 - 8) = \frac{38}{3}.$

5. $\displaystyle\int_{1}^{5} 2 \, dx = 2x \Big|_{1}^{5} = 2(5) - 2 = 2(5 - 1) = 8.$

   This integral is interesting; the integrand is a constant function, hence we are finding the area of a rectangle with width $(5 - 1) = 4$ and height 2. Notice how the evaluation of the definite integral led to $2(4) = 8$.

   In general, if c is a constant, then $\int_{a}^{b} c \, dx = c(b - a)$.

Understanding Motion with the Fundamental Theorem of Calculus

We established, starting with Key Idea 2.2.1, that the derivative of a position function is a velocity function, and the derivative of a velocity function is an acceleration function. Now consider definite integrals of velocity and acceleration functions. Specifically, if $v(t)$ is a velocity function, what does $\displaystyle\int_{a}^{b} v(t) \, dt$ mean?

Notes:

The Fundamental Theorem of Calculus states that

$$\int_a^b v(t)\, dt = V(b) - V(a),$$

where $V(t)$ is any antiderivative of $v(t)$. Since $v(t)$ is a velocity function, $V(t)$ must be a position function, and $V(b) - V(a)$ measures a change in position, or **displacement**.

Example 5.4.4   Finding displacement
A ball is thrown straight up with velocity given by $v(t) = -32t + 20$ft/s, where t is measured in seconds. Find, and interpret, $\int_0^1 v(t)\, dt$.

  SOLUTION   Using the Fundamental Theorem of Calculus, we have

$$\int_0^1 v(t)\, dt = \int_0^1 (-32t + 20)\, dt$$

$$= -16t^2 + 20t \Big|_0^1$$

$$= 4.$$

Thus if a ball is thrown straight up into the air with velocity $v(t) = -32t + 20$, the height of the ball, 1 second later, will be 4 feet above the initial height. (Note that the ball has *traveled* much farther. It has gone up to its peak and is falling down, but the difference between its height at $t = 0$ and $t = 1$ is 4ft.)

   Integrating a rate of change function gives total change. Velocity is the rate of position change; integrating velocity gives the total change of position, i.e., displacement.

   Integrating a speed function gives a similar, though different, result. Speed is also the rate of position change, but does not account for direction. So integrating a speed function gives total change of position, without the possibility of "negative position change." Hence the integral of a speed function gives *distance traveled.*

   As acceleration is the rate of velocity change, integrating an acceleration function gives total change in velocity. We do not have a simple term for this analogous to displacement. If $a(t) = 5$miles/h^2 and t is measured in hours, then

$$\int_0^3 a(t)\, dt = 15$$

means the velocity has increased by 15m/h from $t = 0$ to $t = 3$.

Notes:

The Fundamental Theorem of Calculus and the Chain Rule

Part 1 of the Fundamental Theorem of Calculus (FTC) states that given $F(x) = \int_a^x f(t)\, dt$, $F'(x) = f(x)$. Using other notation, $\dfrac{d}{dx}\big(F(x)\big) = f(x)$. While we have just practiced evaluating definite integrals, sometimes finding antiderivatives is impossible and we need to rely on other techniques to approximate the value of a definite integral. Functions written as $F(x) = \int_a^x f(t)\, dt$ are useful in such situations.

It may be of further use to compose such a function with another. As an example, we may compose $F(x)$ with $g(x)$ to get

$$F\big(g(x)\big) = \int_a^{g(x)} f(t)\, dt.$$

What is the derivative of such a function? The Chain Rule can be employed to state

$$\frac{d}{dx}\Big(F\big(g(x)\big)\Big) = F'\big(g(x)\big)g'(x) = f\big(g(x)\big)g'(x).$$

An example will help us understand this.

Example 5.4.5   The FTC, Part 1, and the Chain Rule

Find the derivative of $F(x) = \displaystyle\int_2^{x^2} \ln t\, dt$.

SOLUTION   We can view $F(x)$ as being the function $G(x) = \displaystyle\int_2^x \ln t\, dt$ composed with $g(x) = x^2$; that is, $F(x) = G\big(g(x)\big)$. The Fundamental Theorem of Calculus states that $G'(x) = \ln x$. The Chain Rule gives us

$$\begin{aligned}
F'(x) &= G'\big(g(x)\big)g'(x)\\
&= \ln(g(x))g'(x)\\
&= \ln(x^2)2x\\
&= 2x \ln x^2
\end{aligned}$$

Normally, the steps defining $G(x)$ and $g(x)$ are skipped.

Practice this once more.

Example 5.4.6   The FTC, Part 1, and the Chain Rule

Find the derivative of $F(x) = \displaystyle\int_{\cos x}^5 t^3\, dt$.

Notes:

SOLUTION   Note that $F(x) = -\displaystyle\int_5^{\cos x} t^3\, dt$. Viewed this way, the derivative of F is straightforward:

$$F'(x) = \sin x \cos^3 x.$$

Area Between Curves

Consider continuous functions $f(x)$ and $g(x)$ defined on $[a, b]$, where $f(x) \geq g(x)$ for all x in $[a, b]$, as demonstrated in Figure 5.4.2. What is the area of the shaded region bounded by the two curves over $[a, b]$?

The area can be found by recognizing that this area is "the area under f − the area under g." Using mathematical notation, the area is

$$\int_a^b f(x)\, dx - \int_a^b g(x)\, dx.$$

Properties of the definite integral allow us to simplify this expression to

$$\int_a^b \big(f(x) - g(x)\big)\, dx.$$

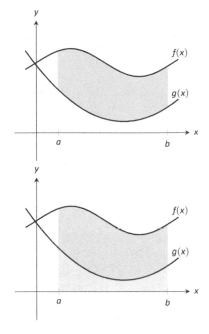

Figure 5.4.2: Finding the area bounded by two functions on an interval; it is found by subtracting the area under g from the area under f.

Theorem 5.4.3   Area Between Curves

Let $f(x)$ and $g(x)$ be continuous functions defined on $[a, b]$ where $f(x) \geq g(x)$ for all x in $[a, b]$. The area of the region bounded by the curves $y = f(x)$, $y = g(x)$ and the lines $x = a$ and $x = b$ is

$$\int_a^b \big(f(x) - g(x)\big)\, dx.$$

Example 5.4.7   Finding area between curves
Find the area of the region enclosed by $y = x^2 + x - 5$ and $y = 3x - 2$.

SOLUTION   It will help to sketch these two functions, as done in Figure 5.4.3. The region whose area we seek is completely bounded by these two functions; they seem to intersect at $x = -1$ and $x = 3$. To check, set $x^2 + x - 5 =$

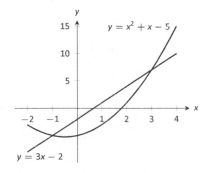

Figure 5.4.3: Sketching the region enclosed by $y = x^2 + x - 5$ and $y = 3x - 2$ in Example 5.4.7.

Notes:

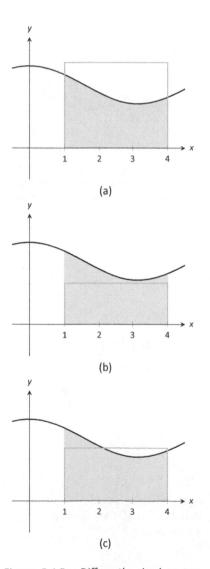

Figure 5.4.4: A graph of a function f to introduce the Mean Value Theorem.

(a)

(b)

(c)

Figure 5.4.5: Differently sized rectangles give upper and lower bounds on $\int_1^4 f(x)\,dx$; the last rectangle matches the area exactly.

$3x - 2$ and solve for x:

$$x^2 + x - 5 = 3x - 2$$
$$(x^2 + x - 5) - (3x - 2) = 0$$
$$x^2 - 2x - 3 = 0$$
$$(x - 3)(x + 1) = 0$$
$$x = -1,\ 3.$$

Following Theorem 5.4.3, the area is

$$\int_{-1}^{3} \left(3x - 2 - (x^2 + x - 5)\right) dx = \int_{-1}^{3} (-x^2 + 2x + 3)\, dx$$
$$= \left(-\frac{1}{3}x^3 + x^2 + 3x\right)\Bigg|_{-1}^{3}$$
$$= -\frac{1}{3}(27) + 9 + 9 - \left(\frac{1}{3} + 1 - 3\right)$$
$$= 10\frac{2}{3} = 10.\overline{6}$$

The Mean Value Theorem and Average Value

Consider the graph of a function f in Figure 5.4.4 and the area defined by $\int_1^4 f(x)\,dx$. Three rectangles are drawn in Figure 5.4.5; in (a), the height of the rectangle is greater than f on $[1, 4]$, hence the area of this rectangle is is greater than $\int_0^4 f(x)\,dx$.

In (b), the height of the rectangle is smaller than f on $[1, 4]$, hence the area of this rectangle is less than $\int_1^4 f(x)\,dx$.

Finally, in (c) the height of the rectangle is such that the area of the rectangle is *exactly* that of $\int_0^4 f(x)\,dx$. Since rectangles that are "too big", as in (a), and rectangles that are "too little," as in (b), give areas greater/lesser than $\int_1^4 f(x)\,dx$, it makes sense that there is a rectangle, whose top intersects $f(x)$ somewhere on $[1, 4]$, whose area is *exactly* that of the definite integral.

We state this idea formally in a theorem.

Notes:

Theorem 5.4.4   The Mean Value Theorem of Integration

Let f be continuous on $[a, b]$. There exists a value c in $[a, b]$ such that

$$\int_a^b f(x)\, dx = f(c)(b - a).$$

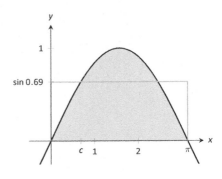

Figure 5.4.6: A graph of $y = \sin x$ on $[0, \pi]$ and the rectangle guaranteed by the Mean Value Theorem.

This is an *existential* statement; c exists, but we do not provide a method of finding it. Theorem 5.4.4 is directly connected to the Mean Value Theorem of Differentiation, given as Theorem 3.2.1; we leave it to the reader to see how.

We demonstrate the principles involved in this version of the Mean Value Theorem in the following example.

Example 5.4.8   Using the Mean Value Theorem

Consider $\int_0^\pi \sin x\, dx$. Find a value c guaranteed by the Mean Value Theorem.

SOLUTION   We first need to evaluate $\int_0^\pi \sin x\, dx$. (This was previously done in Example 5.4.3.)

$$\int_0^\pi \sin x\, dx = -\cos x \Big|_0^\pi = 2.$$

Thus we seek a value c in $[0, \pi]$ such that $\pi \sin c = 2$.

$$\pi \sin c = 2 \;\Rightarrow\; \sin c = 2/\pi \;\Rightarrow\; c = \arcsin(2/\pi) \approx 0.69.$$

In Figure 5.4.6 $\sin x$ is sketched along with a rectangle with height $\sin(0.69)$. The area of the rectangle is the same as the area under $\sin x$ on $[0, \pi]$.

Let f be a function on $[a, b]$ with c such that $f(c)(b-a) = \int_a^b f(x)\, dx$. Consider $\int_a^b \big(f(x) - f(c)\big)\, dx$:

$$\int_a^b \big(f(x) - f(c)\big)\, dx = \int_a^b f(x) - \int_a^b f(c)\, dx$$
$$= f(c)(b - a) - f(c)(b - a)$$
$$= 0.$$

When $f(x)$ is shifted by $-f(c)$, the amount of area under f above the x–axis on $[a, b]$ is the same as the amount of area below the x–axis above f; see Figure 5.4.7 for an illustration of this. In this sense, we can say that $f(c)$ is the *average value* of f on $[a, b]$.

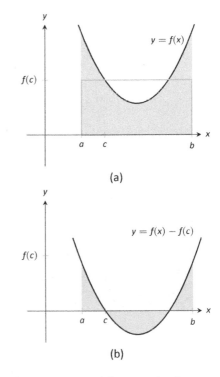

(a)

(b)

Figure 5.4.7: In (a), a graph of $y = f(x)$ and the rectangle guaranteed by the Mean Value Theorem. In (b), $y = f(x)$ is shifted down by $f(c)$; the resulting "area under the curve" is 0.

Notes:

The value $f(c)$ is the average value in another sense. First, recognize that the Mean Value Theorem can be rewritten as

$$f(c) = \frac{1}{b-a} \int_a^b f(x)\, dx,$$

for some value of c in $[a, b]$. Next, partition the interval $[a, b]$ into n equally spaced subintervals, $a = x_1 < x_2 < \ldots < x_{n+1} = b$ and choose any c_i in $[x_i, x_{i+1}]$. The average of the numbers $f(c_1), f(c_2), ..., f(c_n)$ is:

$$\frac{1}{n}\Big(f(c_1) + f(c_2) + \ldots + f(c_n)\Big) = \frac{1}{n}\sum_{i=1}^{n} f(c_i).$$

Multiply this last expression by 1 in the form of $\frac{(b-a)}{(b-a)}$:

$$\frac{1}{n}\sum_{i=1}^{n} f(c_i) = \sum_{i=1}^{n} f(c_i)\frac{1}{n}$$

$$= \sum_{i=1}^{n} f(c_i)\frac{1}{n}\frac{(b-a)}{(b-a)}$$

$$= \frac{1}{b-a}\sum_{i=1}^{n} f(c_i)\frac{b-a}{n}$$

$$= \frac{1}{b-a}\sum_{i=1}^{n} f(c_i)\Delta x \quad \text{(where } \Delta x = (b-a)/n\text{)}$$

Now take the limit as $n \to \infty$:

$$\lim_{n\to\infty} \frac{1}{b-a}\sum_{i=1}^{n} f(c_i)\Delta x \quad = \quad \frac{1}{b-a}\int_a^b f(x)\, dx \quad = \quad f(c).$$

This tells us this: when we evaluate f at n (somewhat) equally spaced points in $[a, b]$, the average value of these samples is $f(c)$ as $n \to \infty$.

This leads us to a definition.

Definition 5.4.1   The Average Value of f on $[a, b]$

Let f be continuous on $[a, b]$. The **average value of f on $[a, b]$** is $f(c)$, where c is a value in $[a, b]$ guaranteed by the Mean Value Theorem. I.e.,

$$\text{Average Value of } f \text{ on } [a, b] = \frac{1}{b-a}\int_a^b f(x)\, dx.$$

Notes:

An application of this definition is given in the following example.

Example 5.4.9   **Finding the average value of a function**
An object moves back and forth along a straight line with a velocity given by $v(t) = (t-1)^2$ on $[0, 3]$, where t is measured in seconds and $v(t)$ is measured in ft/s.

What is the average velocity of the object?

SOLUTION   By our definition, the average velocity is:

$$\frac{1}{3-0}\int_0^3 (t-1)^2\, dt = \frac{1}{3}\int_0^3 \left(t^2 - 2t + 1\right) dt = \frac{1}{3}\left(\frac{1}{3}t^3 - t^2 + t\right)\Bigg|_0^3 = 1 \text{ ft/s.}$$

We can understand the above example through a simpler situation. Suppose you drove 100 miles in 2 hours. What was your average speed? The answer is simple: displacement/time = 100 miles/2 hours = 50 mph.

What was the displacement of the object in Example 5.4.9? We calculate this by integrating its velocity function: $\int_0^3 (t-1)^2\, dt = 3$ ft. Its final position was 3 feet from its initial position after 3 seconds: its average velocity was 1 ft/s.

This section has laid the groundwork for a lot of great mathematics to follow. The most important lesson is this: definite integrals can be evaluated using antiderivatives. Since the previous section established that definite integrals are the limit of Riemann sums, we can later create Riemann sums to approximate values other than "area under the curve," convert the sums to definite integrals, then evaluate these using the Fundamental Theorem of Calculus. This will allow us to compute the work done by a variable force, the volume of certain solids, the arc length of curves, and more.

The downside is this: generally speaking, computing antiderivatives is much more difficult than computing derivatives. The next chapter is devoted to techniques of finding antiderivatives so that a wide variety of definite integrals can be evaluated. Before that, the next section explores techniques of approximating the value of definite integrals beyond using the Left Hand, Right Hand and Midpoint Rules. These techniques are invaluable when antiderivatives cannot be computed, or when the actual function f is unknown and all we know is the value of f at certain x-values.

Notes:

Exercises 5.4

Terms and Concepts

1. How are definite and indefinite integrals related?

2. What constant of integration is most commonly used when evaluating definite integrals?

3. T/F: If f is a continuous function, then $F(x) = \int_a^x f(t)\, dt$ is also a continuous function.

4. The definite integral can be used to find "the area under a curve." Give two other uses for definite integrals.

Problems

In Exercises 5 – 28, evaluate the definite integral.

5. $\int_1^3 (3x^2 - 2x + 1)\, dx$

6. $\int_0^4 (x - 1)^2\, dx$

7. $\int_{-1}^1 (x^3 - x^5)\, dx$

8. $\int_{\pi/2}^{\pi} \cos x\, dx$

9. $\int_0^{\pi/4} \sec^2 x\, dx$

10. $\int_1^e \frac{1}{x}\, dx$

11. $\int_{-1}^1 5^x\, dx$

12. $\int_{-2}^{-1} (4 - 2x^3)\, dx$

13. $\int_0^{\pi} (2\cos x - 2\sin x)\, dx$

14. $\int_1^3 e^x\, dx$

15. $\int_0^4 \sqrt{t}\, dt$

16. $\int_9^{25} \frac{1}{\sqrt{t}}\, dt$

17. $\int_1^8 \sqrt[3]{x}\, dx$

18. $\int_1^2 \frac{1}{x}\, dx$

19. $\int_1^2 \frac{1}{x^2}\, dx$

20. $\int_1^2 \frac{1}{x^3}\, dx$

21. $\int_0^1 x\, dx$

22. $\int_0^1 x^2\, dx$

23. $\int_0^1 x^3\, dx$

24. $\int_0^1 x^{100}\, dx$

25. $\int_{-4}^4 dx$

26. $\int_{-10}^{-5} 3\, dx$

27. $\int_{-2}^2 0\, dx$

28. $\int_{\pi/6}^{\pi/3} \csc x \cot x\, dx$

29. Explain why:

   (a) $\int_{-1}^1 x^n\, dx = 0$, when n is a positive, odd integer, and

   (b) $\int_{-1}^1 x^n\, dx = 2\int_0^1 x^n\, dx$ when n is a positive, even integer.

30. Explain why $\int_a^{a+2\pi} \sin t\, dt = 0$ for all values of a.

In Exercises 31 – 34, find a value c guaranteed by the Mean Value Theorem.

31. $\int_0^2 x^2\, dx$

32. $\int_{-2}^2 x^2\, dx$

33. $\int_0^1 e^x\, dx$

34. $\int_0^{16} \sqrt{x}\, dx$

In Exercises 35 – 40, find the average value of the function on the given interval.

35. $f(x) = \sin x$ on $[0, \pi/2]$

36. $y = \sin x$ on $[0, \pi]$

37. $y = x$ on $[0, 4]$

38. $y = x^2$ on $[0, 4]$

39. $y = x^3$ on $[0, 4]$

40. $g(t) = 1/t$ on $[1, e]$

In Exercises 41 – 46, a velocity function of an object moving along a straight line is given. Find the displacement of the object over the given time interval.

41. $v(t) = -32t + 20$ft/s on $[0, 5]$

42. $v(t) = -32t + 200$ft/s on $[0, 10]$

43. $v(t) = 10$ft/s on $[0, 3]$.

44. $v(t) = 2^t$mph on $[-1, 1]$

45. $v(t) = \cos t$ ft/s on $[0, 3\pi/2]$

46. $v(t) = \sqrt[4]{t}$ ft/s on $[0, 16]$

In Exercises 47 – 50, an acceleration function of an object moving along a straight line is given. Find the change of the object's velocity over the given time interval.

47. $a(t) = -32$ft/s^2 on $[0, 2]$

48. $a(t) = 10$ft/s^2 on $[0, 5]$

49. $a(t) = t$ ft/s^2 on $[0, 2]$

50. $a(t) = \cos t$ ft/s^2 on $[0, \pi]$

In Exercises 51 – 54, sketch the given functions and find the area of the enclosed region.

51. $y = 2x$, $y = 5x$, and $x = 3$.

52. $y = -x + 1$, $y = 3x + 6$, $x = 2$ and $x = -1$.

53. $y = x^2 - 2x + 5$, $y = 5x - 5$.

54. $y = 2x^2 + 2x - 5$, $y = x^2 + 3x + 7$.

In Exercises 55 – 58, find $F'(x)$.

55. $F(x) = \int_2^{x^3 + x} \dfrac{1}{t}\, dt$

56. $F(x) = \int_{x^3}^0 t^3\, dt$

57. $F(x) = \int_x^{x^2} (t + 2)\, dt$

58. $F(x) = \int_{\ln x}^{e^x} \sin t\, dt$

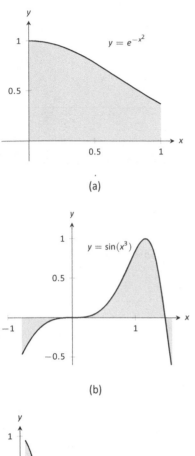

(a)

(b)

(c)

Figure 5.5.1: Graphically representing three definite integrals that cannot be evaluated using antiderivatives.

5.5   Numerical Integration

The Fundamental Theorem of Calculus gives a concrete technique for finding the exact value of a definite integral. That technique is based on computing antiderivatives. Despite the power of this theorem, there are still situations where we must *approximate* the value of the definite integral instead of finding its exact value. The first situation we explore is where we *cannot* compute the antiderivative of the integrand. The second case is when we actually do not know the function in the integrand, but only its value when evaluated at certain points.

An **elementary function** is any function that is a combination of polynomial, n^{th} root, rational, exponential, logarithmic and trigonometric functions. We can compute the derivative of any elementary function, but there are many elementary functions of which we cannot compute an antiderivative. For example, the following functions do not have antiderivatives that we can express with elementary functions:

$$e^{-x^2}, \quad \sin(x^3) \quad \text{and} \quad \frac{\sin x}{x}.$$

The simplest way to refer to the antiderivatives of e^{-x^2} is to simply write $\int e^{-x^2}\,dx$.

This section outlines three common methods of approximating the value of definite integrals. We describe each as a systematic method of approximating area under a curve. By approximating this area accurately, we find an accurate approximation of the corresponding definite integral.

We will apply the methods we learn in this section to the following definite integrals:

$$\int_0^1 e^{-x^2}\,dx, \quad \int_{-\frac{\pi}{4}}^{\frac{\pi}{2}} \sin(x^3)\,dx, \quad \text{and} \quad \int_{0.5}^{4\pi} \frac{\sin(x)}{x}\,dx,$$

as pictured in Figure 5.5.1.

The Left and Right Hand Rule Methods

In Section 5.3 we addressed the problem of evaluating definite integrals by approximating the area under the curve using rectangles. We revisit those ideas here before introducing other methods of approximating definite integrals.

We start with a review of notation. Let f be a continuous function on the interval $[a, b]$. We wish to approximate $\int_a^b f(x)\,dx$. We partition $[a, b]$ into n equally spaced subintervals, each of length $\Delta x = \dfrac{b - a}{n}$. The endpoints of these

Notes:

subintervals are labeled as

$$x_1 = a, \; x_2 = a + \Delta x, \; x_3 = a + 2\Delta x, \; \dots, \; x_i = a + (i-1)\Delta x, \; \dots, \; x_{n+1} = b.$$

Key Idea 5.3.1 states that to use the Left Hand Rule we use the summation $\sum_{i=1}^{n} f(x_i)\Delta x$ and to use the Right Hand Rule we use $\sum_{i=1}^{n} f(x_{i+1})\Delta x$. We review the use of these rules in the context of examples.

Example 5.5.1   **Approximating definite integrals with rectangles**
Approximate $\int_0^1 e^{-x^2}\, dx$ using the Left and Right Hand Rules with 5 equally spaced subintervals.

SOLUTION   We begin by partitioning the interval $[0, 1]$ into 5 equally spaced intervals. We have $\Delta x = \frac{1-0}{5} = 1/5 = 0.2$, so

$$x_1 = 0, \; x_2 = 0.2, \; x_3 = 0.4, \; x_4 = 0.6, \; x_5 = 0.8, \text{ and } x_6 = 1.$$

Using the Left Hand Rule, we have:

$$\begin{aligned}
\sum_{i=1}^{n} f(x_i)\Delta x &= \big(f(x_1) + f(x_2) + f(x_3) + f(x_4) + f(x_5)\big)\Delta x \\
&= \big(f(0) + f(0.2) + f(0.4) + f(0.6) + f(0.8)\big)\Delta x \\
&\approx \big(1 + 0.961 + 0.852 + 0.698 + 0.527\big)(0.2) \\
&\approx 0.808.
\end{aligned}$$

Using the Right Hand Rule, we have:

$$\begin{aligned}
\sum_{i=1}^{n} f(x_{i+1})\Delta x &= \big(f(x_2) + f(x_3) + f(x_4) + f(x_5) + f(x_6)\big)\Delta x \\
&= \big(f(0.2) + f(0.4) + f(0.6) + f(0.8) + f(1)\big)\Delta x \\
&\approx \big(0.961 + 0.852 + 0.698 + 0.527 + 0.368\big)(0.2) \\
&\approx 0.681.
\end{aligned}$$

Figure 5.5.2 shows the rectangles used in each method to approximate the definite integral. These graphs show that in this particular case, the Left Hand Rule is an over approximation and the Right Hand Rule is an under approximation. To get a better approximation, we could use more rectangles, as we did in

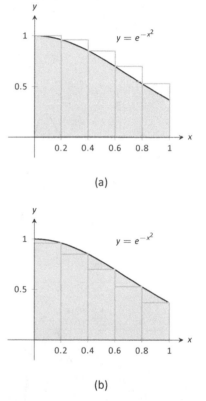

(a)

(b)

Figure 5.5.2: Approximating $\int_0^1 e^{-x^2}\, dx$ in Example 5.5.1.

Notes:

x_i	Exact	Approx.	$\sin(x_i^3)$
x_1	$-\pi/4$	-0.785	-0.466
x_2	$-7\pi/40$	-0.550	-0.165
x_3	$-\pi/10$	-0.314	-0.031
x_4	$-\pi/40$	-0.0785	0
x_5	$\pi/20$	0.157	0.004
x_6	$\pi/8$	0.393	0.061
x_7	$\pi/5$	0.628	0.246
x_8	$11\pi/40$	0.864	0.601
x_9	$7\pi/20$	1.10	0.971
x_{10}	$17\pi/40$	1.34	0.690
x_{11}	$\pi/2$	1.57	-0.670

Figure 5.5.3: Table of values used to approximate $\int_{-\frac{\pi}{4}}^{\frac{\pi}{2}} \sin(x^3)\, dx$ in Example 5.5.2.

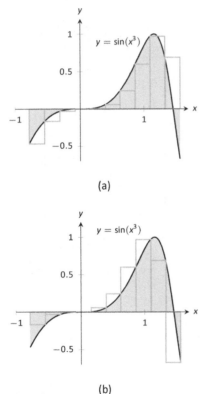

(a)

(b)

Figure 5.5.4: Approximating $\int_{-\frac{\pi}{4}}^{\frac{\pi}{2}} \sin(x^3)\, dx$ in Example 5.5.2.

Section 5.3. We could also average the Left and Right Hand Rule results together, giving

$$\frac{0.808 + 0.681}{2} = 0.7445.$$

The actual answer, accurate to 4 places after the decimal, is 0.7468, showing our average is a good approximation.

Example 5.5.2 **Approximating definite integrals with rectangles**

Approximate $\int_{-\frac{\pi}{4}}^{\frac{\pi}{2}} \sin(x^3)\, dx$ using the Left and Right Hand Rules with 10 equally spaced subintervals.

SOLUTION We begin by finding Δx:

$$\frac{b - a}{n} = \frac{\pi/2 - (-\pi/4)}{10} = \frac{3\pi}{40} \approx 0.236.$$

It is useful to write out the endpoints of the subintervals in a table; in Figure 5.5.3, we give the exact values of the endpoints, their decimal approximations, and decimal approximations of $\sin(x^3)$ evaluated at these points.

Once this table is created, it is straightforward to approximate the definite integral using the Left and Right Hand Rules. (Note: the table itself is easy to create, especially with a standard spreadsheet program on a computer. The last two columns are all that are needed.) The Left Hand Rule sums the first 10 values of $\sin(x_i^3)$ and multiplies the sum by Δx; the Right Hand Rule sums the last 10 values of $\sin(x_i^3)$ and multiplies by Δx. Therefore we have:

Left Hand Rule: $\displaystyle\int_{-\frac{\pi}{4}}^{\frac{\pi}{2}} \sin(x^3)\, dx \approx (1.91)(0.236) = 0.451.$

Right Hand Rule: $\displaystyle\int_{-\frac{\pi}{4}}^{\frac{\pi}{2}} \sin(x^3)\, dx \approx (1.71)(0.236) = 0.404.$

Average of the Left and Right Hand Rules: 0.4275.

The actual answer, accurate to 3 places after the decimal, is 0.460. Our approximations were once again fairly good. The rectangles used in each approximation are shown in Figure 5.5.4. It is clear from the graphs that using more rectangles (and hence, narrower rectangles) should result in a more accurate approximation.

The Trapezoidal Rule

In Example 5.5.1 we approximated the value of $\int_0^1 e^{-x^2}\, dx$ with 5 rectangles of equal width. Figure 5.5.2 shows the rectangles used in the Left and Right

Notes:

Hand Rules. These graphs clearly show that rectangles do not match the shape of the graph all that well, and that accurate approximations will only come by using lots of rectangles.

Instead of using rectangles to approximate the area, we can instead use *trapezoids*. In Figure 5.5.5, we show the region under $f(x) = e^{-x^2}$ on $[0, 1]$ approximated with 5 trapezoids of equal width; the top "corners" of each trapezoid lies on the graph of $f(x)$. It is clear from this figure that these trapezoids more accurately approximate the area under f and hence should give a better approximation of $\int_0^1 e^{-x^2}\, dx$. (In fact, these trapezoids seem to give a *great* approximation of the area!)

The formula for the area of a trapezoid is given in Figure 5.5.6. We approximate $\int_0^1 e^{-x^2}\, dx$ with these trapezoids in the following example.

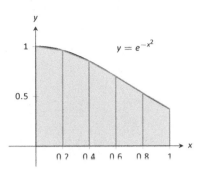

Figure 5.5.5: Approximating $\int_0^1 e^{-x^2}\, dx$ using 5 trapezoids of equal widths.

Example 5.5.3   Approximating definite integrals using trapezoids

Use 5 trapezoids of equal width to approximate $\displaystyle\int_0^1 e^{-x^2}\, dx$.

SOLUTION   To compute the areas of the 5 trapezoids in Figure 5.5.5, it will again be useful to create a table of values as shown in Figure 5.5.7.

The leftmost trapezoid has legs of length 1 and 0.961 and a height of 0.2. Thus, by our formula, the area of the leftmost trapezoid is:

$$\frac{1 + 0.961}{2}(0.2) = 0.1961.$$

Moving right, the next trapezoid has legs of length 0.961 and 0.852 and a height of 0.2. Thus its area is:

$$\frac{0.961 + 0.852}{2}(0.2) = 0.1813.$$

The sum of the areas of all 5 trapezoids is:

$$\frac{1 + 0.961}{2}(0.2) + \frac{0.961 + 0.852}{2}(0.2) + \frac{0.852 + 0.698}{2}(0.2) +$$
$$\frac{0.698 + 0.527}{2}(0.2) + \frac{0.527 + 0.368}{2}(0.2) = 0.7445.$$

We approximate $\displaystyle\int_0^1 e^{-x^2}\, dx \approx 0.7445$.

Figure 5.5.6: The area of a trapezoid.

x_i	$e^{-x_i^2}$
0	1
0.2	0.961
0.4	0.852
0.6	0.698
0.8	0.527
1	0.368

Figure 5.5.7: A table of values of e^{-x^2}.

There are many things to observe in this example. Note how each term in the final summation was multiplied by both 1/2 and by $\Delta x = 0.2$. We can factor these coefficients out, leaving a more concise summation as:

$$\frac{1}{2}(0.2)\Big[(1+0.961)+(0.961+0.852)+(0.852+0.698)+(0.698+0.527)+(0.527+0.368)\Big].$$

Notes:

Now notice that all numbers except for the first and the last are added twice. Therefore we can write the summation even more concisely as

$$\frac{0.2}{2}\Big[1 + 2(0.961 + 0.852 + 0.698 + 0.527) + 0.368\Big].$$

This is the heart of the **Trapezoidal Rule**, wherein a definite integral $\int_a^b f(x)\,dx$ is approximated by using trapezoids of equal widths to approximate the corresponding area under f. Using n equally spaced subintervals with endpoints x_1, $x_2, \ldots, x_{n+1}$, we again have $\Delta x = \dfrac{b-a}{n}$. Thus:

$$\int_a^b f(x)\,dx \approx \sum_{i=1}^n \frac{f(x_i) + f(x_{i+1})}{2}\Delta x$$

$$= \frac{\Delta x}{2}\sum_{i=1}^n \big(f(x_i) + f(x_{i+1})\big)$$

$$= \frac{\Delta x}{2}\Big[f(x_1) + 2\sum_{i=2}^n f(x_i) + f(x_{n+1})\Big].$$

Example 5.5.4 Using the Trapezoidal Rule

Revisit Example 5.5.2 and approximate $\int_{-\frac{\pi}{4}}^{\frac{\pi}{2}} \sin(x^3)\,dx$ using the Trapezoidal Rule and 10 equally spaced subintervals.

SOLUTION We refer back to Figure 5.5.3 for the table of values of $\sin(x^3)$. Recall that $\Delta x = 3\pi/40 \approx 0.236$. Thus we have:

$$\int_{-\frac{\pi}{4}}^{\frac{\pi}{2}} \sin(x^3)\,dx \approx \frac{0.236}{2}\Big[-0.466 + 2\big(-0.165 + (-0.031) + \ldots + 0.69\big) + (-0.67)\Big]$$

$$= 0.4275.$$

Notice how "quickly" the Trapezoidal Rule can be implemented once the table of values is created. This is true for all the methods explored in this section; the real work is creating a table of x_i and $f(x_i)$ values. Once this is completed, approximating the definite integral is not difficult. Again, using technology is wise. Spreadsheets can make quick work of these computations and make using lots of subintervals easy.

Also notice the approximations the Trapezoidal Rule gives. It is the average of the approximations given by the Left and Right Hand Rules! This effectively

Notes:

renders the Left and Right Hand Rules obsolete. They are useful when first learning about definite integrals, but if a real approximation is needed, one is generally better off using the Trapezoidal Rule instead of either the Left or Right Hand Rule.

How can we improve on the Trapezoidal Rule, apart from using more and more trapezoids? The answer is clear once we look back and consider what we have *really* done so far. The Left Hand Rule is not *really* about using rectangles to approximate area. Instead, it approximates a function f with constant functions on small subintervals and then computes the definite integral of these constant functions. The Trapezoidal Rule is really approximating a function f with a linear function on a small subinterval, then computes the definite integral of this linear function. In both of these cases the definite integrals are easy to compute in geometric terms.

So we have a progression: we start by approximating f with a constant function and then with a linear function. What is next? A quadratic function. By approximating the curve of a function with lots of parabolas, we generally get an even better approximation of the definite integral. We call this process **Simpson's Rule**, named after Thomas Simpson (1710-1761), even though others had used this rule as much as 100 years prior.

Simpson's Rule

Given one point, we can create a constant function that goes through that point. Given two points, we can create a linear function that goes through those points. Given three points, we can create a quadratic function that goes through those three points (given that no two have the same x–value).

Consider three points (x_1, y_1), (x_2, y_2) and (x_3, y_3) whose x–values are equally spaced and $x_1 < x_2 < x_3$. Let f be the quadratic function that goes through these three points. It is not hard to show that

$$\int_{x_1}^{x_3} f(x)\, dx = \frac{x_3 - x_1}{6}\left(y_1 + 4y_2 + y_3\right). \tag{5.4}$$

Consider Figure 5.5.8. A function f goes through the 3 points shown and the parabola g that also goes through those points is graphed with a dashed line. Using our equation from above, we know exactly that

$$\int_{1}^{3} g(x)\, dx = \frac{3 - 1}{6}\left(3 + 4(1) + 2\right) = 3.$$

Since g is a good approximation for f on $[1, 3]$, we can state that

$$\int_{1}^{3} f(x)\, dx \approx 3.$$

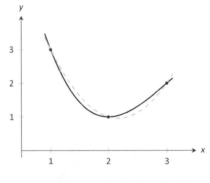

Figure 5.5.8: A graph of a function f and a parabola that approximates it well on $[1, 3]$.

Notes:

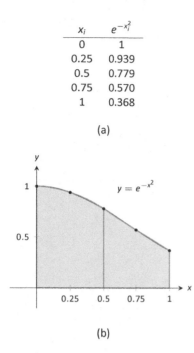

x_i	$e^{-x_i^2}$
0	1
0.25	0.939
0.5	0.779
0.75	0.570
1	0.368

(a)

(b)

Figure 5.5.9: A table of values to approximate $\int_0^1 e^{-x^2}\,dx$, along with a graph of the function.

x_i	$\sin(x_i^3)$
-0.785	-0.466
-0.550	-0.165
-0.314	-0.031
-0.0785	0
0.157	0.004
0.393	0.061
0.628	0.246
0.864	0.601
1.10	0.971
1.34	0.690
1.57	-0.670

Figure 5.5.10: Table of values used to approximate $\int_{-\frac{\pi}{4}}^{\frac{\pi}{2}} \sin(x^3)\,dx$ in Example 5.5.6.

Notice how the interval $[1, 3]$ was split into two subintervals as we needed 3 points. Because of this, whenever we use Simpson's Rule, we need to break the interval into an even number of subintervals.

In general, to approximate $\int_a^b f(x)\,dx$ using Simpson's Rule, subdivide $[a, b]$ into n subintervals, where n is even and each subinterval has width $\Delta x = (b - a)/n$. We approximate f with $n/2$ parabolic curves, using Equation (5.4) to compute the area under these parabolas. Adding up these areas gives the formula:

$$\int_a^b f(x)\,dx \approx \frac{\Delta x}{3}\Big[f(x_1)+4f(x_2)+2f(x_3)+4f(x_4)+\ldots+2f(x_{n-1})+4f(x_n)+f(x_{n+1})\Big].$$

Note how the coefficients of the terms in the summation have the pattern 1, 4, 2, 4, 2, 4, . . ., 2, 4, 1.

Let's demonstrate Simpson's Rule with a concrete example.

Example 5.5.5   Using Simpson's Rule

Approximate $\int_0^1 e^{-x^2}\,dx$ using Simpson's Rule and 4 equally spaced subintervals.

SOLUTION   We begin by making a table of values as we have in the past, as shown in Figure 5.5.9(a). Simpson's Rule states that

$$\int_0^1 e^{-x^2}\,dx \approx \frac{0.25}{3}\Big[1 + 4(0.939) + 2(0.779) + 4(0.570) + 0.368\Big] = 0.74683\overline{3}.$$

Recall in Example 5.5.1 we stated that the correct answer, accurate to 4 places after the decimal, was 0.7468. Our approximation with Simpson's Rule, with 4 subintervals, is better than our approximation with the Trapezoidal Rule using 5!

Figure 5.5.9(b) shows $f(x) = e^{-x^2}$ along with its approximating parabolas, demonstrating how good our approximation is. The approximating curves are nearly indistinguishable from the actual function.

Example 5.5.6   Using Simpson's Rule

Approximate $\int_{-\frac{\pi}{4}}^{\frac{\pi}{2}} \sin(x^3)\,dx$ using Simpson's Rule and 10 equally spaced intervals.

SOLUTION   Figure 5.5.10 shows the table of values that we used in the past for this problem, shown here again for convenience. Again, $\Delta x = (\pi/2 + \pi/4)/10 \approx 0.236$.

Notes:

Simpson's Rule states that

$$\int_{-\frac{\pi}{4}}^{\frac{\pi}{2}} \sin(x^3)\, dx \approx \frac{0.236}{3}\Big[(-0.466) + 4(-0.165) + 2(-0.031) + \ldots$$

$$\ldots + 2(0.971) + 4(0.69) + (-0.67)\Big]$$

$$= 0.4701$$

Recall that the actual value, accurate to 3 decimal places, is 0.460. Our approximation is within one 1/100th of the correct value. The graph in Figure 5.5.11 shows how closely the parabolas match the shape of the graph.

Summary and Error Analysis

We summarize the key concepts of this section thus far in the following Key Idea.

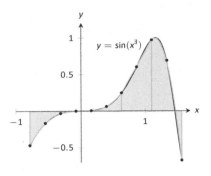

Figure 5.5.11: Approximating $\int_{-\frac{\pi}{4}}^{\frac{\pi}{2}} \sin(x^3)\, dx$ in Example 5.5.6 with Simpson's Rule and 10 equally spaced intervals.

Key Idea 5.5.1   Numerical Integration

Let f be a continuous function on $[a, b]$, let n be a positive integer, and let $\Delta x = \dfrac{b-a}{n}$.

Set $x_1 = a$, $x_2 = a + \Delta x$, ..., $x_i = a + (i-1)\Delta x$, $x_{n+1} = b$.

Consider $\displaystyle\int_a^b f(x)\, dx$.

Left Hand Rule: $\displaystyle\int_a^b f(x)\, dx \approx \Delta x\Big[f(x_1) + f(x_2) + \ldots + f(x_n)\Big]$.

Right Hand Rule: $\displaystyle\int_a^b f(x)\, dx \approx \Delta x\Big[f(x_2) + f(x_3) + \ldots + f(x_{n+1})\Big]$.

Trapezoidal Rule: $\displaystyle\int_a^b f(x)\, dx \approx \dfrac{\Delta x}{2}\Big[f(x_1) + 2f(x_2) + 2f(x_3) + \ldots + 2f(x_n) + f(x_{n+1})\Big]$.

Simpson's Rule: $\displaystyle\int_a^b f(x)\, dx \approx \dfrac{\Delta x}{3}\Big[f(x_1) + 4f(x_2) + 2f(x_3) + \ldots + 4f(x_n) + f(x_{n+1})\Big]$ (n even).

In our examples, we approximated the value of a definite integral using a given method then compared it to the "right" answer. This should have raised several questions in the reader's mind, such as:

1. How was the "right" answer computed?

2. If the right answer can be found, what is the point of approximating?

3. If there is value to approximating, how are we supposed to know if the approximation is any good?

Notes:

These are good questions, and their answers are educational. In the examples, *the* right answer was never computed. Rather, an approximation accurate to a certain number of places after the decimal was given. In Example 5.5.1, we do not know the *exact* answer, but we know it starts with 0.7468. These more accurate approximations were computed using numerical integration but with more precision (i.e., more subintervals and the help of a computer).

Since the exact answer cannot be found, approximation still has its place. How are we to tell if the approximation is any good?

"Trial and error" provides one way. Using technology, make an approximation with, say, 10, 100, and 200 subintervals. This likely will not take much time at all, and a trend should emerge. If a trend does not emerge, try using yet more subintervals. Keep in mind that trial and error is never foolproof; you might stumble upon a problem in which a trend will not emerge.

A second method is to use Error Analysis. While the details are beyond the scope of this text, there are some formulas that give *bounds* for how good your approximation will be. For instance, the formula might state that the approximation is within 0.1 of the correct answer. If the approximation is 1.58, then one knows that the correct answer is between 1.48 and 1.68. By using lots of subintervals, one can get an approximation as accurate as one likes. Theorem 5.5.1 states what these bounds are.

Theorem 5.5.1   Error Bounds in the Trapezoidal Rule and Simpson's Rule

1. Let E_T be the error in approximating $\int_a^b f(x)\,dx$ using the Trapezoidal Rule with n subintervals.

   If f has a continuous 2nd derivative on $[a, b]$ and M is any upper bound of $|f''(x)|$ on $[a, b]$, then

   $$E_T \leq \frac{(b-a)^3}{12n^2}M.$$

2. Let E_S be the error in approximating $\int_a^b f(x)\,dx$ using Simpson's Rule with n subintervals.

   If f has a continuous 4th derivative on $[a, b]$ and M is any upper bound of $|f^{(4)}|$ on $[a, b]$, then

   $$E_S \leq \frac{(b-a)^5}{180n^4}M.$$

Notes:

There are some key things to note about this theorem.

1. The larger the interval, the larger the error. This should make sense intuitively.

2. The error shrinks as more subintervals are used (i.e., as n gets larger).

3. The error in Simpson's Rule has a term relating to the 4th derivative of f. Consider a cubic polynomial: it's 4th derivative is 0. Therefore, the error in approximating the definite integral of a cubic polynomial with Simpson's Rule is 0 – Simpson's Rule computes the exact answer!

We revisit Examples 5.5.3 and 5.5.5 and compute the error bounds using Theorem 5.5.1 in the following example.

Example 5.5.7   Computing error bounds

Find the error bounds when approximating $\displaystyle\int_0^1 e^{-x^2}\, dx$ using the Trapezoidal Rule and 5 subintervals, and using Simpson's Rule with 4 subintervals.

SOLUTION
Trapezoidal Rule with $n = 5$:

We start by computing the 2nd derivative of $f(x) = e^{-x^2}$:

$$f''(x) = e^{-x^2}(4x^2 - 2).$$

Figure 5.5.12 shows a graph of $f''(x)$ on $[0, 1]$. It is clear that the largest value of f'', in absolute value, is 2. Thus we let $M = 2$ and apply the error formula from Theorem 5.5.1.

$$E_T = \frac{(1-0)^3}{12 \cdot 5^2} \cdot 2 = 0.00\overline{6}.$$

Our error estimation formula states that our approximation of 0.7445 found in Example 5.5.3 is within 0.0067 of the correct answer, hence we know that

$$0.7445 - 0.0067 = .7378 \leq \int_0^1 e^{-x^2}\, dx \leq 0.7512 = 0.7445 + 0.0067.$$

We had earlier computed the exact answer, correct to 4 decimal places, to be 0.7468, affirming the validity of Theorem 5.5.1.

Simpson's Rule with $n = 4$:

We start by computing the 4th derivative of $f(x) = e^{-x^2}$:

$$f^{(4)}(x) = e^{-x^2}(16x^4 - 48x^2 + 12).$$

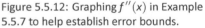

$y = e^{-x^2}(4x^2 - 2)$

Figure 5.5.12: Graphing $f''(x)$ in Example 5.5.7 to help establish error bounds.

Notes:

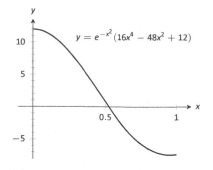

Figure 5.5.13: Graphing $f^{(4)}(x)$ in Example 5.5.7 to help establish error bounds.

Time	Speed (mph)
0	0
1	25
2	22
3	19
4	39
5	0
6	43
7	59
8	54
9	51
10	43
11	35
12	40
13	43
14	30
15	0
16	0
17	28
18	40
19	42
20	40
21	39
22	40
23	23
24	0

Figure 5.5.14: Speed data collected at 30 second intervals for Example 5.5.8.

Figure 5.5.13 shows a graph of $f^{(4)}(x)$ on $[0, 1]$. It is clear that the largest value of $f^{(4)}$, in absolute value, is 12. Thus we let $M = 12$ and apply the error formula from Theorem 5.5.1.

$$E_s = \frac{(1-0)^5}{180 \cdot 4^4} \cdot 12 = 0.00026.$$

Our error estimation formula states that our approximation of $0.7468\overline{3}$ found in Example 5.5.5 is within 0.00026 of the correct answer, hence we know that

$$0.74683 - 0.00026 = .74657 \leq \int_0^1 e^{-x^2}\, dx \leq 0.74709 = 0.74683 + 0.00026.$$

Once again we affirm the validity of Theorem 5.5.1.

At the beginning of this section we mentioned two main situations where numerical integration was desirable. We have considered the case where an antiderivative of the integrand cannot be computed. We now investigate the situation where the integrand is not known. This is, in fact, the most widely used application of Numerical Integration methods. "Most of the time" we observe behavior but do not know "the" function that describes it. We instead collect data about the behavior and make approximations based on this data. We demonstrate this in an example.

Example 5.5.8   Approximating distance traveled

One of the authors drove his daughter home from school while she recorded their speed every 30 seconds. The data is given in Figure 5.5.14. Approximate the distance they traveled.

SOLUTION   Recall that by integrating a speed function we get distance traveled. We have information about $v(t)$; we will use Simpson's Rule to approximate $\int_a^b v(t)\, dt$.

The most difficult aspect of this problem is converting the given data into the form we need it to be in. The speed is measured in miles per hour, whereas the time is measured in 30 second increments.

We need to compute $\Delta x = (b - a)/n$. Clearly, $n = 24$. What are a and b? Since we start at time $t = 0$, we have that $a = 0$. The final recorded time came after 24 periods of 30 seconds, which is 12 minutes or 1/5 of an hour. Thus we have

$$\Delta x = \frac{b - a}{n} = \frac{1/5 - 0}{24} = \frac{1}{120}; \quad \frac{\Delta x}{3} = \frac{1}{360}.$$

Notes:

Thus the distance traveled is approximately:

$$\int_0^{0.2} v(t)\,dt \approx \frac{1}{360}\Big[f(x_1) + 4f(x_2) + 2f(x_3) + \cdots + 4f(x_n) + f(x_{n+1})\Big]$$

$$= \frac{1}{360}\Big[0 + 4\cdot 25 + 2 \cdot 22 + \cdots + 2\cdot 40 + 4\cdot 23 + 0\Big]$$

$$\approx 6.2167 \text{ miles.}$$

We approximate the author drove 6.2 miles. (Because we are sure the reader wants to know, the author's odometer recorded the distance as about 6.05 miles.)

We started this chapter learning about antiderivatives and indefinite integrals. We then seemed to change focus by looking at areas between the graph of a function and the *x*-axis. We defined these areas as the definite integral of the function, using a notation very similar to the notation of the indefinite integral. The Fundamental Theorem of Calculus tied these two seemingly separate concepts together: we can find areas under a curve, i.e., we can evaluate a definite integral, using antiderivatives.

We ended the chapter by noting that antiderivatives are sometimes more than difficult to find: they are impossible. Therefore we developed numerical techniques that gave us good approximations of definite integrals.

We used the definite integral to compute areas, and also to compute displacements and distances traveled. There is far more we can do than that. In Chapter 7 we'll see more applications of the definite integral. Before that, in Chapter 6 we'll learn advanced techniques of integration, analogous to learning rules like the Product, Quotient and Chain Rules of differentiation.

Notes:

Exercises 5.5

Terms and Concepts

1. T/F: Simpson's Rule is a method of approximating antiderivatives.

2. What are the two basic situations where approximating the value of a definite integral is necessary?

3. Why are the Left and Right Hand Rules rarely used?

4. Simpson's Rule is based on approximating portions of a function with what type of function?

Problems

In Exercises 5 – 12, a definite integral is given.

(a) Approximate the definite integral with the Trapezoidal Rule and $n = 4$.

(b) Approximate the definite integral with Simpson's Rule and $n = 4$.

(c) Find the exact value of the integral.

5. $\int_{-1}^{1} x^2 \, dx$

6. $\int_{0}^{10} 5x \, dx$

7. $\int_{0}^{\pi} \sin x \, dx$

8. $\int_{0}^{4} \sqrt{x} \, dx$

9. $\int_{0}^{3} (x^3 + 2x^2 - 5x + 7) \, dx$

10. $\int_{0}^{1} x^4 \, dx$

11. $\int_{0}^{2\pi} \cos x \, dx$

12. $\int_{-3}^{3} \sqrt{9 - x^2} \, dx$

In Exercises 13 – 20, approximate the definite integral with the Trapezoidal Rule and Simpson's Rule, with $n = 6$.

13. $\int_{0}^{1} \cos (x^2) \, dx$

14. $\int_{-1}^{1} e^{x^2} \, dx$

15. $\int_{0}^{5} \sqrt{x^2 + 1} \, dx$

16. $\int_{0}^{\pi} x \sin x \, dx$

17. $\int_{0}^{\pi/2} \sqrt{\cos x} \, dx$

18. $\int_{1}^{4} \ln x \, dx$

19. $\int_{-1}^{1} \frac{1}{\sin x + 2} \, dx$

20. $\int_{0}^{6} \frac{1}{\sin x + 2} \, dx$

In Exercises 21 – 24, find n such that the error in approximating the given definite integral is less than 0.0001 when using:

(a) the Trapezoidal Rule

(b) Simpson's Rule

21. $\int_{0}^{\pi} \sin x \, dx$

22. $\int_{1}^{4} \frac{1}{\sqrt{x}} \, dx$

23. $\int_{0}^{\pi} \cos (x^2) \, dx$

24. $\int_{0}^{5} x^4 \, dx$

In Exercises 25 – 26, a region is given. Find the area of the region using Simpson's Rule:

(a) where the measurements are in centimeters, taken in 1 cm increments, and

(b) where the measurements are in hundreds of yards, taken in 100 yd increments.

25.

26.

6: TECHNIQUES OF ANTIDIFFERENTIATION

The previous chapter introduced the antiderivative and connected it to signed areas under a curve through the Fundamental Theorem of Calculus. The next chapter explores more applications of definite integrals than just area. As evaluating definite integrals will become important, we will want to find antiderivatives of a variety of functions.

This chapter is devoted to exploring techniques of antidifferentiation. While not every function has an antiderivative in terms of elementary functions (a concept introduced in the section on Numerical Integration), we can still find antiderivatives of a wide variety of functions.

6.1 Substitution

We motivate this section with an example. Let $f(x) = (x^2 + 3x - 5)^{10}$. We can compute $f'(x)$ using the Chain Rule. It is:

$$f'(x) = 10(x^2 + 3x - 5)^9 \cdot (2x + 3) = (20x + 30)(x^2 + 3x - 5)^9.$$

Now consider this: What is $\int (20x + 30)(x^2 + 3x - 5)^9 \, dx$? We have the answer in front of us;

$$\int (20x + 30)(x^2 + 3x - 5)^9 \, dx = (x^2 + 3x - 5)^{10} + C.$$

How would we have evaluated this indefinite integral without starting with $f(x)$ as we did?

This section explores *integration by substitution*. It allows us to "undo the Chain Rule." Substitution allows us to evaluate the above integral without knowing the original function first.

The underlying principle is to rewrite a "complicated" integral of the form $\int f(x) \, dx$ as a not–so–complicated integral $\int h(u) \, du$. We'll formally establish later how this is done. First, consider again our introductory indefinite integral, $\int (20x + 30)(x^2 + 3x - 5)^9 \, dx$. Arguably the most "complicated" part of the integrand is $(x^2 + 3x - 5)^9$. We wish to make this simpler; we do so through a substitution. Let $u = x^2 + 3x - 5$. Thus

$$(x^2 + 3x - 5)^9 = u^9.$$

We have established u as a function of x, so now consider the differential of u:

$$du = (2x + 3)dx.$$

Keep in mind that $(2x+3)$ and dx are multiplied; the dx is not "just sitting there." Return to the original integral and do some substitutions through algebra:

$$\int (20x + 30)(x^2 + 3x - 5)^9\, dx = \int 10(2x + 3)(x^2 + 3x - 5)^9\, dx$$

$$= \int 10\underbrace{(x^2 + 3x - 5)}_{u}{}^9\, \underbrace{(2x + 3)\, dx}_{du}$$

$$= \int 10u^9\, du$$

$$= u^{10} + C \quad \text{(replace u with $x^2 + 3x - 5$)}$$

$$= (x^2 + 3x - 5)^{10} + C$$

One might well look at this and think "I (sort of) followed how that worked, but I could never come up with that on my own," but the process is learnable. This section contains numerous examples through which the reader will gain understanding and mathematical maturity enabling them to regard substitution as a natural tool when evaluating integrals.

We stated before that integration by substitution "undoes" the Chain Rule. Specifically, let $F(x)$ and $g(x)$ be differentiable functions and consider the derivative of their composition:

$$\frac{d}{dx}\Big(F(g(x))\Big) = F'(g(x))g'(x).$$

Thus

$$\int F'(g(x))g'(x)\, dx = F(g(x)) + C.$$

Integration by substitution works by recognizing the "inside" function $g(x)$ and replacing it with a variable. By setting $u = g(x)$, we can rewrite the derivative as

$$\frac{d}{dx}\Big(F(u)\Big) = F'(u)u'.$$

Since $du = g'(x)dx$, we can rewrite the above integral as

$$\int F'(g(x))g'(x)\, dx = \int F'(u)du = F(u) + C = F(g(x)) + C.$$

This concept is important so we restate it in the context of a theorem.

Notes:

Theorem 6.1.1 **Integration by Substitution**

Let F and g be differentiable functions, where the range of g is an interval I contained in the domain of F. Then

$$\int F'(g(x))g'(x)\,dx = F(g(x)) + C.$$

If $u = g(x)$, then $du = g'(x)dx$ and

$$\int F'(g(x))g'(x)\,dx = \int F'(u)\,du = F(u) + C = F(g(x)) + C.$$

The point of substitution is to make the integration step easy. Indeed, the step $\int F'(u)\,du = F(u) + C$ looks easy, as the antiderivative of the derivative of F is just F, plus a constant. The "work" involved is making the proper substitution. There is not a step–by–step process that one can memorize; rather, experience will be one's guide. To gain experience, we now embark on many examples.

Example 6.1.1 **Integrating by substitution**

Evaluate $\int x\sin(x^2 + 5)\,dx$.

SOLUTION   Knowing that substitution is related to the Chain Rule, we choose to let u be the "inside" function of $\sin(x^2 + 5)$. (This is not *always* a good choice, but it is often the best place to start.)

Let $u = x^2 + 5$, hence $du = 2x\,dx$. The integrand has an $x\,dx$ term, but not a $2x\,dx$ term. (Recall that multiplication is commutative, so the x does not physically have to be next to dx for there to be an $x\,dx$ term.) We can divide both sides of the du expression by 2:

$$du = 2x\,dx \quad \Rightarrow \quad \frac{1}{2}du = x\,dx.$$

We can now substitute.

$$\int x\sin(x^2 + 5)\,dx = \int \sin(\underbrace{x^2 + 5}_{u})\,\underbrace{x\,dx}_{\frac{1}{2}du}$$

$$= \int \frac{1}{2}\sin u\,du$$

Notes:

$$= -\frac{1}{2}\cos u + C \quad \text{(now replace u with $x^2 + 5$)}$$

$$= -\frac{1}{2}\cos(x^2 + 5) + C.$$

Thus $\int x\sin(x^2 + 5)\,dx = -\frac{1}{2}\cos(x^2 + 5) + C$. We can check our work by evaluating the derivative of the right hand side.

Example 6.1.2   Integrating by substitution

Evaluate $\int \cos(5x)\,dx$.

SOLUTION   Again let u replace the "inside" function. Letting $u = 5x$, we have $du = 5dx$. Since our integrand does not have a $5dx$ term, we can divide the previous equation by 5 to obtain $\frac{1}{5}du = dx$. We can now substitute.

$$\int \cos(5x)\,dx = \int \cos(\underbrace{5x}_{u})\underbrace{dx}_{\frac{1}{5}du}$$

$$= \int \frac{1}{5}\cos u\,du$$

$$= \frac{1}{5}\sin u + C$$

$$= \frac{1}{5}\sin(5x) + C.$$

We can again check our work through differentiation.

The previous example exhibited a common, and simple, type of substitution. The "inside" function was a linear function (in this case, $y = 5x$). When the inside function is linear, the resulting integration is very predictable, outlined here.

Key Idea 6.1.1   Substitution With A Linear Function

Consider $\int F'(ax + b)\,dx$, where $a \neq 0$ and b are constants. Letting $u = ax + b$ gives $du = a \cdot dx$, leading to the result

$$\int F'(ax + b)\,dx = \frac{1}{a}F(ax + b) + C.$$

Thus $\int \sin(7x - 4)\,dx = -\frac{1}{7}\cos(7x - 4) + C$. Our next example can use Key Idea 6.1.1, but we will only employ it after going through all of the steps.

Notes:

Example 6.1.3 **Integrating by substituting a linear function**

Evaluate $\int \dfrac{7}{-3x+1}\,dx.$

SOLUTION View the integrand as the composition of functions $f(g(x))$, where $f(x) = 7/x$ and $g(x) = -3x+1$. Employing our understanding of substitution, we let $u = -3x+1$, the inside function. Thus $du = -3dx$. The integrand lacks a -3; hence divide the previous equation by -3 to obtain $-du/3 = dx$. We can now evaluate the integral through substitution.

$$\int \frac{7}{-3x+1}\,dx = \int \frac{7}{u}\frac{du}{-3}$$
$$= \frac{-7}{3}\int \frac{du}{u}$$
$$= \frac{-7}{3}\ln|u| + C$$
$$= -\frac{7}{3}\ln|-3x+1| + C.$$

Using Key Idea 6.1.1 is faster, recognizing that u is linear and $a = -3$. One may want to continue writing out all the steps until they are comfortable with this particular shortcut.

Not all integrals that benefit from substitution have a clear "inside" function. Several of the following examples will demonstrate ways in which this occurs.

Example 6.1.4 **Integrating by substitution**

Evaluate $\int \sin x \cos x\,dx.$

SOLUTION There is not a composition of function here to exploit; rather, just a product of functions. Do not be afraid to experiment; when given an integral to evaluate, it is often beneficial to think "If I let u be *this*, then du must be *that* ..." and see if this helps simplify the integral at all.

In this example, let's set $u = \sin x$. Then $du = \cos x\,dx$, which we have as part of the integrand! The substitution becomes very straightforward:

$$\int \sin x \cos x\,dx = \int u\,du$$
$$= \frac{1}{2}u^2 + C$$
$$= \frac{1}{2}\sin^2 x + C.$$

Notes:

One would do well to ask "What would happen if we let $u = \cos x$?" The result is just as easy to find, yet looks very different. The challenge to the reader is to evaluate the integral letting $u = \cos x$ and discover why the answer is the same, yet looks different.

Our examples so far have required "basic substitution." The next example demonstrates how substitutions can be made that often strike the new learner as being "nonstandard."

Example 6.1.5   **Integrating by substitution**

Evaluate $\displaystyle\int x\sqrt{x+3}\,dx$.

SOLUTION   Recognizing the composition of functions, set $u = x + 3$. Then $du = dx$, giving what seems initially to be a simple substitution. But at this stage, we have:

$$\int x\sqrt{x+3}\,dx = \int x\sqrt{u}\,du.$$

We cannot evaluate an integral that has both an x and an u in it. We need to convert the x to an expression involving just u.

Since we set $u = x + 3$, we can also state that $u - 3 = x$. Thus we can replace x in the integrand with $u - 3$. It will also be helpful to rewrite $\sqrt{u}$ as $u^{\frac{1}{2}}$.

$$\int x\sqrt{x+3}\,dx = \int (u-3)u^{\frac{1}{2}}\,du$$

$$= \int \left(u^{\frac{3}{2}} - 3u^{\frac{1}{2}}\right)du$$

$$= \frac{2}{5}u^{\frac{5}{2}} - 2u^{\frac{3}{2}} + C$$

$$= \frac{2}{5}(x+3)^{\frac{5}{2}} - 2(x+3)^{\frac{3}{2}} + C.$$

Checking your work is always a good idea. In this particular case, some algebra will be needed to make one's answer match the integrand in the original problem.

Example 6.1.6   **Integrating by substitution**

Evaluate $\displaystyle\int \frac{1}{x\ln x}\,dx$.

SOLUTION   This is another example where there does not seem to be an obvious composition of functions. The line of thinking used in Example 6.1.5 is useful here: choose something for u and consider what this implies du must

Notes:

be. If u can be chosen such that du also appears in the integrand, then we have chosen well.

Choosing $u = 1/x$ makes $du = -1/x^2 \, dx$; that does not seem helpful. However, setting $u = \ln x$ makes $du = 1/x \, dx$, which is part of the integrand. Thus:

$$\int \frac{1}{x \ln x} \, dx = \int \underbrace{\frac{1}{\ln x}}_{u} \underbrace{\frac{1}{x} \, dx}_{du}$$

$$= \int \frac{1}{u} \, du$$

$$= \ln|u| + C$$

$$= \ln|\ln x| + C.$$

The final answer is interesting; the natural log of the natural log. Take the derivative to confirm this answer is indeed correct.

Integrals Involving Trigonometric Functions

Section 6.3 delves deeper into integrals of a variety of trigonometric functions; here we use substitution to establish a foundation that we will build upon.

The next three examples will help fill in some missing pieces of our antiderivative knowledge. We know the antiderivatives of the sine and cosine functions; what about the other standard functions tangent, cotangent, secant and cosecant? We discover these next.

Example 6.1.7   **Integration by substitution: antiderivatives of** $\tan x$
Evaluate $\displaystyle\int \tan x \, dx$.

SOLUTION   The previous paragraph established that we did not know the antiderivatives of tangent, hence we must assume that we have learned something in this section that can help us evaluate this indefinite integral.

Rewrite $\tan x$ as $\sin x / \cos x$. While the presence of a composition of functions may not be immediately obvious, recognize that $\cos x$ is "inside" the $1/x$ function. Therefore, we see if setting $u = \cos x$ returns usable results. We have

Notes:

that $du = -\sin x\, dx$, hence $-du = \sin x\, dx$. We can integrate:

$$
\begin{aligned}
\int \tan x\, dx &= \int \frac{\sin x}{\cos x}\, dx \\
&= \int \underbrace{\frac{1}{\cos x}}_{u}\, \underbrace{\sin x\, dx}_{-du} \\
&= \int \frac{-1}{u}\, du \\
&= -\ln|u| + C \\
&= -\ln|\cos x| + C.
\end{aligned}
$$

Some texts prefer to bring the -1 inside the logarithm as a power of $\cos x$, as in:

$$
\begin{aligned}
-\ln|\cos x| + C &= \ln\left|(\cos x)^{-1}\right| + C \\
&= \ln\left|\frac{1}{\cos x}\right| + C \\
&= \ln|\sec x| + C.
\end{aligned}
$$

Thus the result they give is $\int \tan x\, dx = \ln|\sec x| + C$. These two answers are equivalent.

Example 6.1.8   Integrating by substitution: antiderivatives of $\sec x$

Evaluate $\displaystyle\int \sec x\, dx$.

SOLUTION   This example employs a wonderful trick: multiply the integrand by "1" so that we see how to integrate more clearly. In this case, we write "1" as

$$
1 = \frac{\sec x + \tan x}{\sec x + \tan x}.
$$

This may seem like it came out of left field, but it works beautifully. Consider:

$$
\begin{aligned}
\int \sec x\, dx &= \int \sec x \cdot \frac{\sec x + \tan x}{\sec x + \tan x}\, dx \\
&= \int \frac{\sec^2 x + \sec x \tan x}{\sec x + \tan x}\, dx.
\end{aligned}
$$

Notes:

Now let $u = \sec x + \tan x$; this means $du = (\sec x \tan x + \sec^2 x)\, dx$, which is our numerator. Thus:

$$= \int \frac{du}{u}$$
$$= \ln|u| + C$$
$$= \ln|\sec x + \tan x| + C.$$

We can use similar techniques to those used in Examples 6.1.7 and 6.1.8 to find antiderivatives of $\cot x$ and $\csc x$ (which the reader can explore in the exercises.) We summarize our results here.

Theorem 6.1.2   Antiderivatives of Trigonometric Functions

1. $\int \sin x\, dx = -\cos x + C$ 4. $\int \csc x\, dx = -\ln|\csc x + \cot x| + C$

2. $\int \cos x\, dx = \sin x + C$ 5. $\int \sec x\, dx = \ln|\sec x + \tan x| + C$

3. $\int \tan x\, dx = -\ln|\cos x| + C$ 6. $\int \cot x\, dx = \ln|\sin x| + C$

We explore one more common trigonometric integral.

Example 6.1.9   Integration by substitution: powers of $\cos x$ and $\sin x$
Evaluate $\int \cos^2 x\, dx$.

SOLUTION   We have a composition of functions as $\cos^2 x = (\cos x)^2$. However, setting $u = \cos x$ means $du = -\sin x\, dx$, which we do not have in the integral. Another technique is needed.

The process we'll employ is to use a Power Reducing formula for $\cos^2 x$ (perhaps consult the back of this text for this formula), which states

$$\cos^2 x = \frac{1 + \cos(2x)}{2}.$$

The right hand side of this equation is not difficult to integrate. We have:

$$\int \cos^2 x\, dx = \int \frac{1 + \cos(2x)}{2}\, dx$$
$$= \int \left(\frac{1}{2} + \frac{1}{2}\cos(2x)\right) dx.$$

Notes:

Now use Key Idea 6.1.1:

$$= \frac{1}{2}x + \frac{1}{2}\frac{\sin(2x)}{2} + C$$

$$= \frac{1}{2}x + \frac{\sin(2x)}{4} + C.$$

We'll make significant use of this power–reducing technique in future sections.

Simplifying the Integrand

It is common to be reluctant to manipulate the integrand of an integral; at first, our grasp of integration is tenuous and one may think that working with the integrand will improperly change the results. Integration by substitution works using a different logic: as long as *equality* is maintained, the integrand can be manipulated so that its *form* is easier to deal with. The next two examples demonstrate common ways in which using algebra first makes the integration easier to perform.

Example 6.1.10   Integration by substitution: simplifying first
Evaluate $\displaystyle\int \frac{x^3 + 4x^2 + 8x + 5}{x^2 + 2x + 1}\,dx.$

SOLUTION   One may try to start by setting u equal to either the numerator or denominator; in each instance, the result is not workable.

When dealing with rational functions (i.e., quotients made up of polynomial functions), it is an almost universal rule that everything works better when the degree of the numerator is less than the degree of the denominator. Hence we use polynomial division.

We skip the specifics of the steps, but note that when $x^2 + 2x + 1$ is divided into $x^3 + 4x^2 + 8x + 5$, it goes in $x + 2$ times with a remainder of $3x + 3$. Thus

$$\frac{x^3 + 4x^2 + 8x + 5}{x^2 + 2x + 1} = x + 2 + \frac{3x + 3}{x^2 + 2x + 1}.$$

Integrating $x + 2$ is simple. The fraction can be integrated by setting $u = x^2 + 2x + 1$, giving $du = (2x + 2)\,dx$. This is very similar to the numerator. Note that

Notes:

$du/2 = (x + 1)\, dx$ and then consider the following:

$$\int \frac{x^3 + 4x^2 + 8x + 5}{x^2 + 2x + 1}\, dx = \int \left(x + 2 + \frac{3x + 3}{x^2 + 2x + 1} \right) dx$$

$$= \int (x + 2)\, dx + \int \frac{3(x + 1)}{x^2 + 2x + 1}\, dx$$

$$= \frac{1}{2}x^2 + 2x + C_1 + \int \frac{3}{u}\frac{du}{2}$$

$$= \frac{1}{2}x^2 + 2x + C_1 + \frac{3}{2}\ln|u| + C_2$$

$$= \frac{1}{2}x^2 + 2x + \frac{3}{2}\ln|x^2 + 2x + 1| + C.$$

In some ways, we "lucked out" in that after dividing, substitution was able to be done. In later sections we'll develop techniques for handling rational functions where substitution is not directly feasible.

Example 6.1.11 Integration by alternate methods

Evaluate $\displaystyle\int \frac{x^2 + 2x + 3}{\sqrt{x}}\, dx$ with, and without, substitution.

SOLUTION We already know how to integrate this particular example. Rewrite $\sqrt{x}$ as $x^{\frac{1}{2}}$ and simplify the fraction:

$$\frac{x^2 + 2x + 3}{x^{1/2}} = x^{\frac{3}{2}} + 2x^{\frac{1}{2}} + 3x^{-\frac{1}{2}}.$$

We can now integrate using the Power Rule:

$$\int \frac{x^2 + 2x + 3}{x^{1/2}}\, dx = \int \left(x^{\frac{3}{2}} + 2x^{\frac{1}{2}} + 3x^{-\frac{1}{2}} \right) dx$$

$$= \frac{2}{5}x^{\frac{5}{2}} + \frac{4}{3}x^{\frac{3}{2}} + 6x^{\frac{1}{2}} + C$$

This is a perfectly fine approach. We demonstrate how this can also be solved using substitution as its implementation is rather clever.

Let $u = \sqrt{x} = x^{\frac{1}{2}}$; therefore

$$du = \frac{1}{2}x^{-\frac{1}{2}}dx = \frac{1}{2\sqrt{x}}\, dx \quad \Rightarrow \quad 2du = \frac{1}{\sqrt{x}}\, dx.$$

This gives us $\displaystyle\int \frac{x^2 + 2x + 3}{\sqrt{x}}\, dx = \int (x^2 + 2x + 3) \cdot 2\, du.$ What are we to do with the other x terms? Since $u = x^{\frac{1}{2}}$, $u^2 = x$, etc. We can then replace x^2 and

Notes:

x with appropriate powers of u. We thus have

$$\int \frac{x^2 + 2x + 3}{\sqrt{x}}\, dx = \int (x^2 + 2x + 3) \cdot 2\, du$$

$$= \int 2(u^4 + 2u^2 + 3)\, du$$

$$= \frac{2}{5}u^5 + \frac{4}{3}u^3 + 6u + C$$

$$= \frac{2}{5}x^{\frac{5}{2}} + \frac{4}{3}x^{\frac{3}{2}} + 6x^{\frac{1}{2}} + C,$$

which is obviously the same answer we obtained before. In this situation, substitution is arguably more work than our other method. The fantastic thing is that it works. It demonstrates how flexible integration is.

Substitution and Inverse Trigonometric Functions

When studying derivatives of inverse functions, we learned that

$$\frac{d}{dx}\left(\tan^{-1} x\right) = \frac{1}{1 + x^2}.$$

Applying the Chain Rule to this is not difficult; for instance,

$$\frac{d}{dx}\left(\tan^{-1} 5x\right) = \frac{5}{1 + 25x^2}.$$

We now explore how Substitution can be used to "undo" certain derivatives that are the result of the Chain Rule applied to Inverse Trigonometric functions. We begin with an example.

Example 6.1.12　Integrating by substitution: inverse trigonometric functions
Evaluate $\displaystyle\int \frac{1}{25 + x^2}\, dx$.

SOLUTION　The integrand looks similar to the derivative of the arctangent function. Note:

$$\frac{1}{25 + x^2} = \frac{1}{25\left(1 + \frac{x^2}{25}\right)}$$

$$= \frac{1}{25\left(1 + \left(\frac{x}{5}\right)^2\right)}$$

$$= \frac{1}{25}\frac{1}{1 + \left(\frac{x}{5}\right)^2}.$$

Notes:

Thus

$$\int \frac{1}{25 + x^2}\, dx = \frac{1}{25} \int \frac{1}{1 + \left(\frac{x}{5}\right)^2}\, dx.$$

This can be integrated using Substitution. Set $u = x/5$, hence $du = dx/5$ or $dx = 5du$. Thus

$$\int \frac{1}{25 + x^2}\, dx = \frac{1}{25} \int \frac{1}{1 + \left(\frac{x}{5}\right)^2}\, dx$$

$$= \frac{1}{5} \int \frac{1}{1 + u^2}\, du$$

$$= \frac{1}{5} \tan^{-1} u + C$$

$$= \frac{1}{5} \tan^{-1}\left(\frac{x}{5}\right) + C$$

Example 6.1.12 demonstrates a general technique that can be applied to other integrands that result in inverse trigonometric functions. The results are summarized here.

Theorem 6.1.3   Integrals Involving Inverse Trigonometric Functions

Let $a > 0$.

1. $\displaystyle\int \frac{1}{a^2 + x^2}\, dx = \frac{1}{a}\tan^{-1}\left(\frac{x}{a}\right) + C$

2. $\displaystyle\int \frac{1}{\sqrt{a^2 - x^2}}\, dx = \sin^{-1}\left(\frac{x}{a}\right) + C$

3. $\displaystyle\int \frac{1}{x\sqrt{x^2 - a^2}}\, dx = \frac{1}{a}\sec^{-1}\left(\frac{|x|}{a}\right) + C$

Let's practice using Theorem 6.1.3.

Example 6.1.13   Integrating by substitution: inverse trigonometric functions
Evaluate the given indefinite integrals.

1. $\displaystyle\int \frac{1}{9 + x^2}\, dx,$ 2. $\displaystyle\int \frac{1}{x\sqrt{x^2 - \frac{1}{100}}}\, dx$ 3. $\displaystyle\int \frac{1}{\sqrt{5 - x^2}}\, dx.$

Notes:

SOLUTION Each can be answered using a straightforward application of Theorem 6.1.3.

1. $\displaystyle\int \frac{1}{9+x^2}\,dx = \frac{1}{3}\tan^{-1}\frac{x}{3}+C$, as $a=3$.

2. $\displaystyle\int \frac{1}{x\sqrt{x^2-\frac{1}{100}}}\,dx = 10\sec^{-1}10x+C$, as $a=\frac{1}{10}$.

3. $\displaystyle\int \frac{1}{\sqrt{5-x^2}} = \sin^{-1}\frac{x}{\sqrt{5}}+C$, as $a=\sqrt{5}$.

Most applications of Theorem 6.1.3 are not as straightforward. The next examples show some common integrals that can still be approached with this theorem.

Example 6.1.14   Integrating by substitution: completing the square

Evaluate $\displaystyle\int \frac{1}{x^2-4x+13}\,dx$.

SOLUTION Initially, this integral seems to have nothing in common with the integrals in Theorem 6.1.3. As it lacks a square root, it almost certainly is not related to arcsine or arcsecant. It is, however, related to the arctangent function.

We see this by *completing the square* in the denominator. We give a brief reminder of the process here.

Start with a quadratic with a leading coefficient of 1. It will have the form of x^2+bx+c. Take 1/2 of b, square it, and add/subtract it back into the expression. I.e.,

$$x^2+bx+c = \underbrace{x^2+bx+\frac{b^2}{4}}_{(x+b/2)^2}-\frac{b^2}{4}+c$$

$$= \left(x+\frac{b}{2}\right)^2+c-\frac{b^2}{4}$$

In our example, we take half of -4 and square it, getting 4. We add/subtract it into the denominator as follows:

$$\frac{1}{x^2-4x+13} = \frac{1}{\underbrace{x^2-4x+4}_{(x-2)^2}-4+13}$$

$$= \frac{1}{(x-2)^2+9}$$

Notes:

We can now integrate this using the arctangent rule. Technically, we need to substitute first with $u = x - 2$, but we can employ Key Idea 6.1.1 instead. Thus we have

$$\int \frac{1}{x^2 - 4x + 13}\, dx = \int \frac{1}{(x-2)^2 + 9}\, dx = \frac{1}{3} \tan^{-1} \frac{x-2}{3} + C.$$

Example 6.1.15   Integrals requiring multiple methods

Evaluate $\displaystyle\int \frac{4-x}{\sqrt{16-x^2}}\, dx.$

SOLUTION   This integral requires two different methods to evaluate it. We get to those methods by splitting up the integral:

$$\int \frac{4-x}{\sqrt{16-x^2}}\, dx = \int \frac{4}{\sqrt{16-x^2}}\, dx - \int \frac{x}{\sqrt{16-x^2}}\, dx.$$

The first integral is handled using a straightforward application of Theorem 6.1.3; the second integral is handled by substitution, with $u = 16 - x^2$. We handle each separately.

$$\int \frac{4}{\sqrt{16-x^2}}\, dx = 4\sin^{-1} \frac{x}{4} + C.$$

$\displaystyle\int \frac{x}{\sqrt{16-x^2}}\, dx$: Set $u = 16 - x^2$, so $du = -2x\,dx$ and $x\,dx = -du/2$. We have

$$\int \frac{x}{\sqrt{16-x^2}}\, dx = \int \frac{-du/2}{\sqrt{u}}$$
$$= -\frac{1}{2} \int \frac{1}{\sqrt{u}}\, du$$
$$= -\sqrt{u} + C$$
$$= -\sqrt{16-x^2} + C.$$

Combining these together, we have

$$\int \frac{4-x}{\sqrt{16-x^2}}\, dx = 4\sin^{-1} \frac{x}{4} + \sqrt{16-x^2} + C.$$

Substitution and Definite Integration

This section has focused on evaluating indefinite integrals as we are learning a new technique for finding antiderivatives. However, much of the time integration is used in the context of a definite integral. Definite integrals that require substitution can be calculated using the following workflow:

Notes:

1. Start with a definite integral $\int_a^b f(x)\,dx$ that requires substitution.

2. Ignore the bounds; use substitution to evaluate $\int f(x)\,dx$ and find an antiderivative $F(x)$.

3. Evaluate $F(x)$ at the bounds; that is, evaluate $F(x)\Big|_a^b = F(b) - F(a)$.

This workflow works fine, but substitution offers an alternative that is powerful and amazing (and a little time saving).

At its heart, (using the notation of Theorem 6.1.1) substitution converts integrals of the form $\int F'(g(x))g'(x)\,dx$ into an integral of the form $\int F'(u)\,du$ with the substitution of $u = g(x)$. The following theorem states how the bounds of a definite integral can be changed as the substitution is performed.

Theorem 6.1.4   Substitution with Definite Integrals

Let F and g be differentiable functions, where the range of g is an interval I that is contained in the domain of F. Then

$$\int_a^b F'\big(g(x)\big)g'(x)\,dx = \int_{g(a)}^{g(b)} F'(u)\,du.$$

In effect, Theorem 6.1.4 states that once you convert to integrating with respect to u, you do not need to switch back to evaluating with respect to x. A few examples will help one understand.

Example 6.1.16   Definite integrals and substitution: changing the bounds

Evaluate $\int_0^2 \cos(3x - 1)\,dx$ using Theorem 6.1.4.

SOLUTION   Observing the composition of functions, let $u = 3x - 1$, hence $du = 3dx$. As $3dx$ does not appear in the integrand, divide the latter equation by 3 to get $du/3 = dx$.

By setting $u = 3x - 1$, we are implicitly stating that $g(x) = 3x - 1$. Theorem 6.1.4 states that the new lower bound is $g(0) = -1$; the new upper bound is

---

Notes:

$g(2) = 5$. We now evaluate the definite integral:

$$\int_0^2 \cos(3x - 1)\, dx = \int_{-1}^5 \cos u \frac{du}{3}$$

$$= \frac{1}{3} \sin u \Big|_{-1}^5$$

$$= \frac{1}{3}\big(\sin 5 - \sin(-1)\big) \approx -0.039.$$

Notice how once we converted the integral to be in terms of u, we never went back to using x.

The graphs in Figure 6.1.1 tell more of the story. In (a) the area defined by the original integrand is shaded, whereas in (b) the area defined by the new integrand is shaded. In this particular situation, the areas look very similar; the new region is "shorter" but "wider," giving the same area.

Example 6.1.17   Definite integrals and substitution: changing the bounds

Evaluate $\displaystyle\int_0^{\pi/2} \sin x \cos x\, dx$ using Theorem 6.1.4.

    SOLUTION   We saw the corresponding indefinite integral in Example 6.1.4. In that example we set $u = \sin x$ but stated that we could have let $u = \cos x$. For variety, we do the latter here.

Let $u = g(x) = \cos x$, giving $du = -\sin x\, dx$ and hence $\sin x\, dx = -du$. The new upper bound is $g(\pi/2) = 0$; the new lower bound is $g(0) = 1$. Note how the lower bound is actually larger than the upper bound now. We have

$$\int_0^{\pi/2} \sin x \cos x\, dx = \int_1^0 -u\, du \quad \text{(switch bounds \& change sign)}$$

$$= \int_0^1 u\, du$$

$$= \frac{1}{2}u^2 \Big|_0^1 = 1/2.$$

In Figure 6.1.2 we have again graphed the two regions defined by our definite integrals. Unlike the previous example, they bear no resemblance to each other. However, Theorem 6.1.4 guarantees that they have the same area.

    Integration by substitution is a powerful and useful integration technique. The next section introduces another technique, called Integration by Parts. As substitution "undoes" the Chain Rule, integration by parts "undoes" the Product Rule. Together, these two techniques provide a strong foundation on which most other integration techniques are based.

Notes:

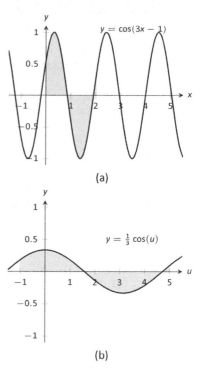

Figure 6.1.1: Graphing the areas defined by the definite integrals of Example 6.1.16.

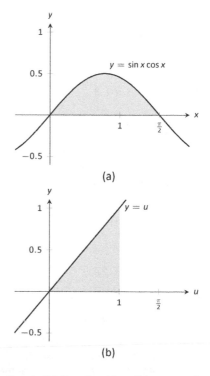

Figure 6.1.2: Graphing the areas defined by the definite integrals of Example 6.1.17.

Exercises 6.1

Terms and Concepts

1. Substitution "undoes" what derivative rule?

2. T/F: One can use algebra to rewrite the integrand of an integral to make it easier to evaluate.

Problems

In Exercises 3 – 14, evaluate the indefinite integral to develop an understanding of Substitution.

3. $\int 3x^2 \left(x^3 - 5\right)^7 dx$

4. $\int (2x - 5) \left(x^2 - 5x + 7\right)^3 dx$

5. $\int x \left(x^2 + 1\right)^8 dx$

6. $\int (12x + 14) \left(3x^2 + 7x - 1\right)^5 dx$

7. $\int \dfrac{1}{2x + 7} dx$

8. $\int \dfrac{1}{\sqrt{2x + 3}} dx$

9. $\int \dfrac{x}{\sqrt{x + 3}} dx$

10. $\int \dfrac{x^3 - x}{\sqrt{x}} dx$

11. $\int \dfrac{e^{\sqrt{x}}}{\sqrt{x}} dx$

12. $\int \dfrac{x^4}{\sqrt{x^5 + 1}} dx$

13. $\int \dfrac{\frac{1}{x} + 1}{x^2} dx$

14. $\int \dfrac{\ln(x)}{x} dx$

In Exercises 15 – 24, use Substitution to evaluate the indefinite integral involving trigonometric functions.

15. $\int \sin^2(x) \cos(x) dx$

16. $\int \cos^3(x) \sin(x) dx$

17. $\int \cos(3 - 6x) dx$

18. $\int \sec^2(4 - x) dx$

19. $\int \sec(2x) dx$

20. $\int \tan^2(x) \sec^2(x) dx$

21. $\int x \cos\left(x^2\right) dx$

22. $\int \tan^2(x) dx$

23. $\int \cot x \, dx$. Do not just refer to Theorem 6.1.2 for the answer; justify it through Substitution.

24. $\int \csc x \, dx$. Do not just refer to Theorem 6.1.2 for the answer; justify it through Substitution.

In Exercises 25 – 32, use Substitution to evaluate the indefinite integral involving exponential functions.

25. $\int e^{3x-1} dx$

26. $\int e^{x^3} x^2 dx$

27. $\int e^{x^2 - 2x + 1}(x - 1) dx$

28. $\int \dfrac{e^x + 1}{e^x} dx$

29. $\int \dfrac{e^x}{e^x + 1} dx$

30. $\int \dfrac{e^x - e^{-x}}{e^{2x}} dx$

31. $\int 3^{3x} dx$

32. $\int 4^{2x} dx$

In Exercises 33 – 36, use Substitution to evaluate the indefinite integral involving logarithmic functions.

33. $\int \dfrac{\ln x}{x} dx$

34. $\int \dfrac{\left(\ln x\right)^2}{x} dx$

35. $\displaystyle\int \frac{\ln\left(x^3\right)}{x}\,dx$

36. $\displaystyle\int \frac{1}{x\ln\left(x^2\right)}\,dx$

In Exercises 37 – 42, use Substitution to evaluate the indefinite integral involving rational functions.

37. $\displaystyle\int \frac{x^2 + 3x + 1}{x}\,dx$

38. $\displaystyle\int \frac{x^3 + x^2 + x + 1}{x}\,dx$

39. $\displaystyle\int \frac{x^3 - 1}{x + 1}\,dx$

40. $\displaystyle\int \frac{x^2 + 2x - 5}{x - 3}\,dx$

41. $\displaystyle\int \frac{3x^2 - 5x + 7}{x + 1}\,dx$

42. $\displaystyle\int \frac{x^2 + 2x + 1}{x^3 + 3x^2 + 3x}\,dx$

In Exercises 43 – 52, use Substitution to evaluate the indefinite integral involving inverse trigonometric functions.

43. $\displaystyle\int \frac{7}{x^2 + 7}\,dx$

44. $\displaystyle\int \frac{3}{\sqrt{9 - x^2}}\,dx$

45. $\displaystyle\int \frac{14}{\sqrt{5 - x^2}}\,dx$

46. $\displaystyle\int \frac{2}{x\sqrt{x^2 - 9}}\,dx$

47. $\displaystyle\int \frac{5}{\sqrt{x^4 - 16x^2}}\,dx$

48. $\displaystyle\int \frac{x}{\sqrt{1 - x^4}}\,dx$

49. $\displaystyle\int \frac{1}{x^2 - 2x + 8}\,dx$

50. $\displaystyle\int \frac{2}{\sqrt{-x^2 + 6x + 7}}\,dx$

51. $\displaystyle\int \frac{3}{\sqrt{-x^2 + 8x + 9}}\,dx$

52. $\displaystyle\int \frac{5}{x^2 + 6x + 34}\,dx$

In Exercises 53 – 78, evaluate the indefinite integral.

53. $\displaystyle\int \frac{x^2}{\left(x^3 + 3\right)^2}\,dx$

54. $\displaystyle\int \left(3x^2 + 2x\right)\left(5x^3 + 5x^2 + 2\right)^8\,dx$

55. $\displaystyle\int \frac{x}{\sqrt{1 - x^2}}\,dx$

56. $\displaystyle\int x^2 \csc^2\left(x^3 + 1\right)\,dx$

57. $\displaystyle\int \sin(x)\sqrt{\cos(x)}\,dx$

58. $\displaystyle\int \sin\left(5x + 1\right)\,dx$

59. $\displaystyle\int \frac{1}{x - 5}\,dx$

60. $\displaystyle\int \frac{7}{3x + 2}\,dx$

61. $\displaystyle\int \frac{3x^3 + 4x^2 + 2x - 22}{x^2 + 3x + 5}\,dx$

62. $\displaystyle\int \frac{2x + 7}{x^2 + 7x + 3}\,dx$

63. $\displaystyle\int \frac{9(2x + 3)}{3x^2 + 9x + 7}\,dx$

64. $\displaystyle\int \frac{-x^3 + 14x^2 - 46x - 7}{x^2 - 7x + 1}\,dx$

65. $\displaystyle\int \frac{x}{x^4 + 81}\,dx$

66. $\displaystyle\int \frac{2}{4x^2 + 1}\,dx$

67. $\displaystyle\int \frac{1}{x\sqrt{4x^2 - 1}}\,dx$

68. $\displaystyle\int \frac{1}{\sqrt{16 - 9x^2}}\,dx$

69. $\displaystyle\int \frac{3x - 2}{x^2 - 2x + 10}\,dx$

70. $\displaystyle\int \frac{7 - 2x}{x^2 + 12x + 61}\,dx$

71. $\displaystyle\int \frac{x^2 + 5x - 2}{x^2 - 10x + 32}\,dx$

72. $\displaystyle\int \frac{x^3}{x^2 + 9}\,dx$

73. $\displaystyle\int \frac{x^3 - x}{x^2 + 4x + 9}\,dx$

74. $\displaystyle\int \frac{\sin(x)}{\cos^2(x) + 1}\,dx$

75. $\displaystyle\int \frac{\cos(x)}{\sin^2(x) + 1}\,dx$

76. $\displaystyle\int \frac{\cos(x)}{1 - \sin^2(x)}\,dx$

77. $\displaystyle\int \frac{3x - 3}{\sqrt{x^2 - 2x - 6}}\,dx$

78. $\displaystyle\int \frac{x - 3}{\sqrt{x^2 - 6x + 8}}\,dx$

In Exercises 79 – 86, evaluate the definite integral.

79. $\displaystyle\int_1^3 \frac{1}{x - 5}\,dx$

80. $\displaystyle\int_2^6 x\sqrt{x - 2}\,dx$

81. $\displaystyle\int_{-\pi/2}^{\pi/2} \sin^2 x \cos x\,dx$

82. $\displaystyle\int_0^1 2x(1 - x^2)^4\,dx$

83. $\displaystyle\int_{-2}^{-1} (x + 1)e^{x^2 + 2x + 1}\,dx$

84. $\displaystyle\int_{-1}^1 \frac{1}{1 + x^2}\,dx$

85. $\displaystyle\int_2^4 \frac{1}{x^2 - 6x + 10}\,dx$

86. $\displaystyle\int_1^{\sqrt{3}} \frac{1}{\sqrt{4 - x^2}}\,dx$

6.2 Integration by Parts

Here's a simple integral that we can't yet evaluate:

$$\int x \cos x \, dx.$$

It's a simple matter to take the derivative of the integrand using the Product Rule, but there is no Product Rule for integrals. However, this section introduces *Integration by Parts*, a method of integration that is based on the Product Rule for derivatives. It will enable us to evaluate this integral.

The Product Rule says that if u and v are functions of x, then $(uv)' = u'v + uv'$. For simplicity, we've written u for $u(x)$ and v for $v(x)$. Suppose we integrate both sides with respect to x. This gives

$$\int (uv)' \, dx = \int (u'v + uv') \, dx.$$

By the Fundamental Theorem of Calculus, the left side integrates to uv. The right side can be broken up into two integrals, and we have

$$uv = \int u'v \, dx + \int uv' \, dx.$$

Solving for the second integral we have

$$\int uv' \, dx = uv - \int u'v \, dx.$$

Using differential notation, we can write $du = u'(x)dx$ and $dv = v'(x)dx$ and the expression above can be written as follows:

$$\int u \, dv = uv - \int v \, du.$$

This is the Integration by Parts formula. For reference purposes, we state this in a theorem.

Theorem 6.2.1 Integration by Parts

Let u and v be differentiable functions of x on an interval I containing a and b. Then

$$\int u \, dv = uv - \int v \, du,$$

and

$$\int_{x=a}^{x=b} u \, dv = uv \Big|_{a}^{b} - \int_{x=a}^{x=b} v \, du.$$

Notes:

Let's try an example to understand our new technique.

Example 6.2.1   Integrating using Integration by Parts

Evaluate $\int x \cos x \, dx$.

SOLUTION   The key to Integration by Parts is to identify part of the integrand as "*u*" and part as "*dv*." Regular practice will help one make good identifications, and later we will introduce some principles that help. For now, let $u = x$ and $dv = \cos x \, dx$.

It is generally useful to make a small table of these values as done below. Right now we only know u and dv as shown on the left of Figure 6.2.1; on the right we fill in the rest of what we need. If $u = x$, then $du = dx$. Since $dv = \cos x \, dx$, v is an antiderivative of $\cos x$. We choose $v = \sin x$.

$$
\begin{array}{cc}
u = x & v = ? \\
du = ? & dv = \cos x \, dx
\end{array}
\quad \Rightarrow \quad
\begin{array}{cc}
u = x & v = \sin x \\
du = dx & dv = \cos x \, dx
\end{array}
$$

Figure 6.2.1: Setting up Integration by Parts.

Now substitute all of this into the Integration by Parts formula, giving

$$\int x \cos x \, dx = x \sin x - \int \sin x \, dx.$$

We can then integrate $\sin x$ to get $-\cos x + C$ and overall our answer is

$$\int x \cos x \, dx = x \sin x + \cos x + C.$$

Note how the antiderivative contains a product, $x \sin x$. This product is what makes Integration by Parts necessary.

The example above demonstrates how Integration by Parts works in general. We try to identify u and dv in the integral we are given, and the key is that we usually want to choose u and dv so that du is simpler than u and v is hopefully not too much more complicated than dv. This will mean that the integral on the right side of the Integration by Parts formula, $\int v \, du$ will be simpler to integrate than the original integral $\int u \, dv$.

In the example above, we chose $u = x$ and $dv = \cos x \, dx$. Then $du = dx$ was simpler than u and $v = \sin x$ is no more complicated than dv. Therefore, instead of integrating $x \cos x \, dx$, we could integrate $\sin x \, dx$, which we knew how to do.

A useful mnemonic for helping to determine u is "LIATE," where

L = **L**ogarithmic, I = **I**nverse Trig., A = **A**lgebraic (polynomials),
T = **T**rigonometric, and E = **E**xponential.

Notes:

If the integrand contains both a logarithmic and an algebraic term, in general letting u be the logarithmic term works best, as indicated by L coming before A in LIATE.

We now consider another example.

Example 6.2.2   **Integrating using Integration by Parts**

Evaluate $\int xe^x\, dx$.

    SOLUTION   The integrand contains an **A**lgebraic term (x) and an **E**xponential term (e^x). Our mnemonic suggests letting u be the algebraic term, so we choose $u = x$ and $dv = e^x\, dx$. Then $du = dx$ and $v = e^x$ as indicated by the tables below.

$$
\begin{array}{ll}
u = x & v = ? \\
du = ? & dv = e^x\, dx
\end{array}
\quad \Rightarrow \quad
\begin{array}{ll}
u = x & v = e^x \\
du = dx & dv = e^x\, dx
\end{array}
$$

Figure 6.2.2: Setting up Integration by Parts.

We see du is simpler than u, while there is no change in going from dv to v. This is good. The Integration by Parts formula gives

$$\int xe^x\, dx = xe^x - \int e^x\, dx.$$

The integral on the right is simple; our final answer is

$$\int xe^x\, dx = xe^x - e^x + C.$$

Note again how the antiderivatives contain a product term.

Example 6.2.3   **Integrating using Integration by Parts**

Evaluate $\int x^2 \cos x\, dx$.

    SOLUTION   The mnemonic suggests letting $u = x^2$ instead of the trigonometric function, hence $dv = \cos x\, dx$. Then $du = 2x\, dx$ and $v = \sin x$ as shown below.

$$
\begin{array}{ll}
u = x^2 & v = ? \\
du = ? & dv = \cos x\, dx
\end{array}
\quad \Rightarrow \quad
\begin{array}{ll}
u = x^2 & v = \sin x \\
du = 2x\, dx & dv = \cos x\, dx
\end{array}
$$

Figure 6.2.3: Setting up Integration by Parts.

Notes:

The Integration by Parts formula gives

$$\int x^2 \cos x \, dx = x^2 \sin x - \int 2x \sin x \, dx.$$

At this point, the integral on the right is indeed simpler than the one we started with, but to evaluate it, we need to do Integration by Parts again. Here we choose $u = 2x$ and $dv = \sin x$ and fill in the rest below.

$$u = 2x \qquad v = ?$$
$$du = ? \qquad dv = \sin x \, dx \qquad \Rightarrow \qquad u = 2x \qquad v = -\cos x$$
$$du = 2\,dx \qquad dv = \sin x \, dx$$

Figure 6.2.4: Setting up Integration by Parts (again).

$$\int x^2 \cos x \, dx = x^2 \sin x - \left(-2x \cos x - \int -2 \cos x \, dx \right).$$

The integral all the way on the right is now something we can evaluate. It evaluates to $-2 \sin x$. Then going through and simplifying, being careful to keep all the signs straight, our answer is

$$\int x^2 \cos x \, dx = x^2 \sin x + 2x \cos x - 2 \sin x + C.$$

Example 6.2.4   Integrating using Integration by Parts

Evaluate $\int e^x \cos x \, dx$.

SOLUTION   This is a classic problem. Our mnemonic suggests letting u be the trigonometric function instead of the exponential. In this particular example, one can let u be either $\cos x$ or e^x; to demonstrate that we do not have to follow LIATE, we choose $u = e^x$ and hence $dv = \cos x \, dx$. Then $du = e^x \, dx$ and $v = \sin x$ as shown below.

$$u = e^x \qquad v = ?$$
$$du = ? \qquad dv = \cos x \, dx \qquad \Rightarrow \qquad u = e^x \qquad v = \sin x$$
$$du = e^x \, dx \qquad dv = \cos x \, dx$$

Figure 6.2.5: Setting up Integration by Parts.

Notice that du is no simpler than u, going against our general rule (but bear with us). The Integration by Parts formula yields

$$\int e^x \cos x \, dx = e^x \sin x - \int e^x \sin x \, dx.$$

Notes:

The integral on the right is not much different than the one we started with, so it seems like we have gotten nowhere. Let's keep working and apply Integration by Parts to the new integral, using $u = e^x$ and $dv = \sin x\, dx$. This leads us to the following:

$$
\begin{array}{llll}
u = e^x & v = ? & & u = e^x & v = -\cos x \\
du = ? & dv = \sin x\, dx & \Rightarrow & du = e^x\, dx & dv = \sin x\, dx
\end{array}
$$

Figure 6.2.6: Setting up Integration by Parts (again).

The Integration by Parts formula then gives:

$$
\int e^x \cos x\, dx = e^x \sin x - \left(-e^x \cos x - \int -e^x \cos x\, dx \right)
$$

$$
= e^x \sin x + e^x \cos x - \int e^x \cos x\, dx.
$$

It seems we are back right where we started, as the right hand side contains $\int e^x \cos x\, dx$. But this is actually a good thing.

Add $\int e^x \cos x\, dx$ to both sides. This gives

$$
2\int e^x \cos x\, dx = e^x \sin x + e^x \cos x
$$

Now divide both sides by 2:

$$
\int e^x \cos x\, dx = \frac{1}{2}\left(e^x \sin x + e^x \cos x \right).
$$

Simplifying a little and adding the constant of integration, our answer is thus

$$
\int e^x \cos x\, dx = \frac{1}{2}e^x \left(\sin x + \cos x \right) + C.
$$

Example 6.2.5 **Integrating using Integration by Parts: antiderivative of** $\ln x$

Evaluate $\int \ln x\, dx$.

SOLUTION One may have noticed that we have rules for integrating the familiar trigonometric functions and e^x, but we have not yet given a rule for integrating $\ln x$. That is because $\ln x$ can't easily be integrated with any of the rules we have learned up to this point. But we can find its antiderivative by a

Notes:

clever application of Integration by Parts. Set $u = \ln x$ and $dv = dx$. This is a good, sneaky trick to learn as it can help in other situations. This determines $du = (1/x)\,dx$ and $v = x$ as shown below.

$$
\begin{array}{ll}
u = \ln x & v = ? \\
du = ? & dv = dx
\end{array}
\qquad \Rightarrow \qquad
\begin{array}{ll}
u = \ln x & v = x \\
du = 1/x\,dx & dv = dx
\end{array}
$$

Figure 6.2.7: Setting up Integration by Parts.

Putting this all together in the Integration by Parts formula, things work out very nicely:

$$\int \ln x\,dx = x\ln x - \int x\,\frac{1}{x}\,dx.$$

The new integral simplifies to $\int 1\,dx$, which is about as simple as things get. Its integral is $x + C$ and our answer is

$$\int \ln x\,dx = x\ln x - x + C.$$

Example 6.2.6   **Integrating using Int. by Parts: antiderivative of** $\arctan x$

Evaluate $\displaystyle\int \arctan x\,dx$.

SOLUTION   The same sneaky trick we used above works here. Let $u = \arctan x$ and $dv = dx$. Then $du = 1/(1 + x^2)\,dx$ and $v = x$. The Integration by Parts formula gives

$$\int \arctan x\,dx = x\arctan x - \int \frac{x}{1 + x^2}\,dx.$$

The integral on the right can be solved by substitution. Taking $u = 1 + x^2$, we get $du = 2x\,dx$. The integral then becomes

$$\int \arctan x\,dx = x\arctan x - \frac{1}{2}\int \frac{1}{u}\,du.$$

The integral on the right evaluates to $\ln|u| + C$, which becomes $\ln(1 + x^2) + C$ (we can drop the absolute values as $1 + x^2$ is always postive). Therefore, the answer is

$$\int \arctan x\,dx = x\arctan x - \frac{1}{2}\ln(1 + x^2) + C.$$

Notes:

Substitution Before Integration

When taking derivatives, it was common to employ multiple rules (such as using both the Quotient and the Chain Rules). It should then come as no surprise that some integrals are best evaluated by combining integration techniques. In particular, here we illustrate making an "unusual" substitution first before using Integration by Parts.

Example 6.2.7 **Integration by Parts after substitution**

Evaluate $\int \cos(\ln x)\,dx$.

SOLUTION The integrand contains a composition of functions, leading us to think Substitution would be beneficial. Letting $u = \ln x$, we have $du = 1/x\,dx$. This seems problematic, as we do not have a $1/x$ in the integrand. But consider:

$$du = \frac{1}{x}\,dx \Rightarrow x \cdot du = dx.$$

Since $u = \ln x$, we can use inverse functions and conclude that $x = e^u$. Therefore we have that

$$dx = x \cdot du$$
$$= e^u\,du.$$

We can thus replace $\ln x$ with u and dx with $e^u\,du$. Thus we rewrite our integral as

$$\int \cos(\ln x)\,dx = \int e^u \cos u\,du.$$

We evaluated this integral in Example 6.2.4. Using the result there, we have:

$$\int \cos(\ln x)\,dx = \int e^u \cos u\,du$$
$$= \frac{1}{2}e^u\big(\sin u + \cos u\big) + C$$
$$= \frac{1}{2}e^{\ln x}\big(\sin(\ln x) + \cos(\ln x)\big) + C$$
$$= \frac{1}{2}x\big(\sin(\ln x) + \cos(\ln x)\big) + C.$$

Definite Integrals and Integration By Parts

So far we have focused only on evaluating indefinite integrals. Of course, we can use Integration by Parts to evaluate definite integrals as well, as Theorem

Notes:

6.2.1 states. We do so in the next example.

Example 6.2.8   Definite integration using Integration by Parts

Evaluate $\int_1^2 x^2 \ln x \, dx$.

SOLUTION   Our mnemonic suggests letting $u = \ln x$, hence $dv = x^2 \, dx$. We then get $du = (1/x) \, dx$ and $v = x^3/3$ as shown below.

$$u = \ln x \qquad v = \, ? \qquad\qquad\qquad u = \ln x \qquad v = x^3/3$$
$$du = \, ? \qquad dv = x^2 \, dx \qquad \Rightarrow \qquad du = 1/x \, dx \qquad dv = x^2 \, dx$$

Figure 6.2.8: Setting up Integration by Parts.

The Integration by Parts formula then gives

$$\int_1^2 x^2 \ln x \, dx = \frac{x^3}{3} \ln x \Big|_1^2 - \int_1^2 \frac{x^3}{3} \frac{1}{x} \, dx$$

$$= \frac{x^3}{3} \ln x \Big|_1^2 - \int_1^2 \frac{x^2}{3} \, dx$$

$$= \frac{x^3}{3} \ln x \Big|_1^2 - \frac{x^3}{9} \Big|_1^2$$

$$= \left(\frac{x^3}{3} \ln x - \frac{x^3}{9} \right) \Big|_1^2$$

$$= \left(\frac{8}{3} \ln 2 - \frac{8}{9} \right) - \left(\frac{1}{3} \ln 1 - \frac{1}{9} \right)$$

$$= \frac{8}{3} \ln 2 - \frac{7}{9}$$

$$\approx 1.07.$$

In general, Integration by Parts is useful for integrating certain products of functions, like $\int xe^x \, dx$ or $\int x^3 \sin x \, dx$. It is also useful for integrals involving logarithms and inverse trigonometric functions.

As stated before, integration is generally more difficult than derivation. We are developing tools for handling a large array of integrals, and experience will tell us when one tool is preferable/necessary over another. For instance, consider the three similar–looking integrals

$$\int xe^x \, dx, \qquad \int xe^{x^2} \, dx \qquad \text{and} \qquad \int xe^{x^3} \, dx.$$

Notes:

While the first is calculated easily with Integration by Parts, the second is best approached with Substitution. Taking things one step further, the third integral has no answer in terms of elementary functions, so none of the methods we learn in calculus will get us the exact answer.

Integration by Parts is a very useful method, second only to Substitution. In the following sections of this chapter, we continue to learn other integration techniques. The next section focuses on handling integrals containing trigonometric functions.

Notes:

Exercises 6.2

Terms and Concepts

1. T/F: Integration by Parts is useful in evaluating integrands that contain products of functions.

2. T/F: Integration by Parts can be thought of as the "opposite of the Chain Rule."

3. For what is "LIATE" useful?

4. T/F: If the integral that results from Integration by Parts appears to also need Integration by Parts, then a mistake was made in the orginal choice of "u".

Problems

In Exercises 5 – 34, evaluate the given indefinite integral.

5. $\int x \sin x \, dx$

6. $\int xe^{-x} \, dx$

7. $\int x^2 \sin x \, dx$

8. $\int x^3 \sin x \, dx$

9. $\int xe^{x^2} \, dx$

10. $\int x^3 e^x \, dx$

11. $\int xe^{-2x} \, dx$

12. $\int e^x \sin x \, dx$

13. $\int e^{2x} \cos x \, dx$

14. $\int e^{2x} \sin(3x) \, dx$

15. $\int e^{5x} \cos(5x) \, dx$

16. $\int \sin x \cos x \, dx$

17. $\int \sin^{-1} x \, dx$

18. $\int \tan^{-1}(2x) \, dx$

19. $\int x \tan^{-1} x \, dx$

20. $\int \sin^{-1} x \, dx$

21. $\int x \ln x \, dx$

22. $\int (x-2) \ln x \, dx$

23. $\int x \ln(x-1) \, dx$

24. $\int x \ln(x^2) \, dx$

25. $\int x^2 \ln x \, dx$

26. $\int (\ln x)^2 \, dx$

27. $\int (\ln(x+1))^2 \, dx$

28. $\int x \sec^2 x \, dx$

29. $\int x \csc^2 x \, dx$

30. $\int x\sqrt{x-2} \, dx$

31. $\int x\sqrt{x^2-2} \, dx$

32. $\int \sec x \tan x \, dx$

33. $\int x \sec x \tan x \, dx$

34. $\int x \csc x \cot x \, dx$

In Exercises 35 – 40, evaluate the indefinite integral after first making a substitution.

35. $\int \sin(\ln x) \, dx$

36. $\int e^{2x} \cos(e^x) \, dx$

37. $\displaystyle\int \sin(\sqrt{x})\, dx$

38. $\displaystyle\int \ln(\sqrt{x})\, dx$

39. $\displaystyle\int e^{\sqrt{x}}\, dx$

40. $\displaystyle\int e^{\ln x}\, dx$

In Exercises 41 – 49, evaluate the definite integral. Note: the corresponding indefinite integrals appear in Exercises 5 – 13.

41. $\displaystyle\int_0^\pi x \sin x\, dx$

42. $\displaystyle\int_{-1}^1 xe^{-x}\, dx$

43. $\displaystyle\int_{-\pi/4}^{\pi/4} x^2 \sin x\, dx$

44. $\displaystyle\int_{-\pi/2}^{\pi/2} x^3 \sin x\, dx$

45. $\displaystyle\int_0^{\sqrt{\ln 2}} xe^{x^2}\, dx$

46. $\displaystyle\int_0^1 x^3 e^x\, dx$

47. $\displaystyle\int_1^2 xe^{-2x}\, dx$

48. $\displaystyle\int_0^\pi e^x \sin x\, dx$

49. $\displaystyle\int_{-\pi/2}^{\pi/2} e^{2x} \cos x\, dx$

6.3   Trigonometric Integrals

Functions involving trigonometric functions are useful as they are good at describing periodic behavior. This section describes several techniques for finding antiderivatives of certain combinations of trigonometric functions.

Integrals of the form $\int \sin^m x \cos^n x \, dx$

In learning the technique of Substitution, we saw the integral $\int \sin x \cos x \, dx$ in Example 6.1.4. The integration was not difficult, and one could easily evaluate the indefinite integral by letting $u = \sin x$ or by letting $u = \cos x$. This integral is easy since the power of both sine and cosine is 1.

We generalize this integral and consider integrals of the form $\int \sin^m x \cos^n x \, dx$, where m, n are nonnegative integers. Our strategy for evaluating these integrals is to use the identity $\cos^2 x + \sin^2 x = 1$ to convert high powers of one trigonometric function into the other, leaving a single sine or cosine term in the integrand. We summarize the general technique in the following Key Idea.

Key Idea 6.3.1   Integrals Involving Powers of Sine and Cosine

Consider $\int \sin^m x \cos^n x \, dx$, where m, n are nonnegative integers.

1. If m is odd, then $m = 2k + 1$ for some integer k. Rewrite

$$\sin^m x = \sin^{2k+1} x = \sin^{2k} x \sin x = (\sin^2 x)^k \sin x = (1 - \cos^2 x)^k \sin x.$$

   Then

$$\int \sin^m x \cos^n x \, dx = \int (1 - \cos^2 x)^k \sin x \cos^n x \, dx = -\int (1 - u^2)^k u^n \, du,$$

   where $u = \cos x$ and $du = -\sin x \, dx$.

2. If n is odd, then using substitutions similar to that outlined above we have

$$\int \sin^m x \cos^n x \, dx = \int u^m (1 - u^2)^k \, du,$$

   where $u = \sin x$ and $du = \cos x \, dx$.

3. If both m and n are even, use the power–reducing identities

$$\cos^2 x = \frac{1 + \cos(2x)}{2} \quad \text{and} \quad \sin^2 x = \frac{1 - \cos(2x)}{2}$$

   to reduce the degree of the integrand. Expand the result and apply the principles of this Key Idea again.

Notes:

We practice applying Key Idea 6.3.1 in the next examples.

Example 6.3.1   **Integrating powers of sine and cosine**

Evaluate $\displaystyle\int \sin^5 x \cos^8 x \, dx.$

SOLUTION   The power of the sine term is odd, so we rewrite $\sin^5 x$ as

$$\sin^5 x = \sin^4 x \sin x = (\sin^2 x)^2 \sin x = (1 - \cos^2 x)^2 \sin x.$$

Our integral is now $\displaystyle\int (1 - \cos^2 x)^2 \cos^8 x \sin x \, dx.$ Let $u = \cos x$, hence $du = -\sin x \, dx$. Making the substitution and expanding the integrand gives

$$\int (1-\cos^2)^2 \cos^8 x \sin x \, dx = -\int (1-u^2)^2 u^8 \, du = -\int \left(1 - 2u^2 + u^4\right)u^8 \, du = -\int \left(u^8 - 2u^{10} + u^{12}\right) du.$$

This final integral is not difficult to evaluate, giving

$$-\int \left(u^8 - 2u^{10} + u^{12}\right) du = -\frac{1}{9}u^9 + \frac{2}{11}u^{11} - \frac{1}{13}u^{13} + C$$

$$= -\frac{1}{9}\cos^9 x + \frac{2}{11}\cos^{11} x - \frac{1}{13}\cos^{13} x + C.$$

Example 6.3.2   **Integrating powers of sine and cosine**

Evaluate $\displaystyle\int \sin^5 x \cos^9 x \, dx.$

SOLUTION   The powers of both the sine and cosine terms are odd, therefore we can apply the techniques of Key Idea 6.3.1 to either power. We choose to work with the power of the cosine term since the previous example used the sine term's power.

We rewrite $\cos^9 x$ as

$$\begin{aligned}
\cos^9 x &= \cos^8 x \cos x \\
&= (\cos^2 x)^4 \cos x \\
&= (1 - \sin^2 x)^4 \cos x \\
&= (1 - 4\sin^2 x + 6\sin^4 x - 4\sin^6 x + \sin^8 x)\cos x.
\end{aligned}$$

We rewrite the integral as

$$\int \sin^5 x \cos^9 x \, dx = \int \sin^5 x \left(1 - 4\sin^2 x + 6\sin^4 x - 4\sin^6 x + \sin^8 x\right)\cos x \, dx.$$

Notes:

Now substitute and integrate, using $u = \sin x$ and $du = \cos x\, dx$.

$$\int \sin^5 x \left(1 - 4\sin^2 x + 6\sin^4 x - 4\sin^6 x + \sin^8 x \right) \cos x\, dx =$$

$$\int u^5 (1 - 4u^2 + 6u^4 - 4u^6 + u^8)\, du = \int \left(u^5 - 4u^7 + 6u^9 - 4u^{11} + u^{13} \right) du$$

$$= \frac{1}{6}u^6 - \frac{1}{2}u^8 + \frac{3}{5}u^{10} - \frac{1}{3}u^{12} + \frac{1}{14}u^{14} + C$$

$$= \frac{1}{6}\sin^6 x - \frac{1}{2}\sin^8 x + \frac{3}{5}\sin^{10} x + \dots$$

$$- \frac{1}{3}\sin^{12} x + \frac{1}{14}\sin^{14} x + C.$$

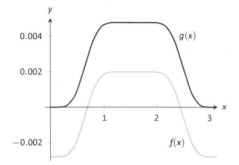

Figure 6.3.1: A plot of $f(x)$ and $g(x)$ from Example 6.3.2 and the Technology Note.

Technology Note: The work we are doing here can be a bit tedious, but the skills developed (problem solving, algebraic manipulation, etc.) are important. Nowadays problems of this sort are often solved using a computer algebra system. The powerful program *Mathematica*® integrates $\int \sin^5 x \cos^9 x\, dx$ as

$$f(x) = -\frac{45\cos(2x)}{16384} - \frac{5\cos(4x)}{8192} + \frac{19\cos(6x)}{49152} + \frac{\cos(8x)}{4096} - \frac{\cos(10x)}{81920} - \frac{\cos(12x)}{24576} - \frac{\cos(14x)}{114688},$$

which clearly has a different form than our answer in Example 6.3.2, which is

$$g(x) = \frac{1}{6}\sin^6 x - \frac{1}{2}\sin^8 x + \frac{3}{5}\sin^{10} x - \frac{1}{3}\sin^{12} x + \frac{1}{14}\sin^{14} x.$$

Figure 6.3.1 shows a graph of f and g; they are clearly not equal, but they differ *only by a constant*. That is $g(x) = f(x) + C$ for some constant C. So we have two different antiderivatives of the same function, meaning both answers are correct.

Example 6.3.3   Integrating powers of sine and cosine

Evaluate $\int \cos^4 x \sin^2 x\, dx$.

SOLUTION   The powers of sine and cosine are both even, so we employ the power–reducing formulas and algebra as follows.

$$\int \cos^4 x \sin^2 x\, dx = \int \left(\frac{1 + \cos(2x)}{2} \right)^2 \left(\frac{1 - \cos(2x)}{2} \right) dx$$

$$= \int \frac{1 + 2\cos(2x) + \cos^2(2x)}{4} \cdot \frac{1 - \cos(2x)}{2}\, dx$$

$$= \int \frac{1}{8} \left(1 + \cos(2x) - \cos^2(2x) - \cos^3(2x) \right) dx$$

The $\cos(2x)$ term is easy to integrate, especially with Key Idea 6.1.1. The $\cos^2(2x)$ term is another trigonometric integral with an even power, requiring the power–reducing formula again. The $\cos^3(2x)$ term is a cosine function with an odd power, requiring a substitution as done before. We integrate each in turn below.

Notes:

$$\int \cos(2x)\,dx = \frac{1}{2}\sin(2x) + C.$$

$$\int \cos^2(2x)\,dx = \int \frac{1 + \cos(4x)}{2}\,dx = \frac{1}{2}\left(x + \frac{1}{4}\sin(4x)\right) + C.$$

Finally, we rewrite $\cos^3(2x)$ as

$$\cos^3(2x) = \cos^2(2x)\cos(2x) = \left(1 - \sin^2(2x)\right)\cos(2x).$$

Letting $u = \sin(2x)$, we have $du = 2\cos(2x)\,dx$, hence

$$\int \cos^3(2x)\,dx = \int \left(1 - \sin^2(2x)\right)\cos(2x)\,dx$$

$$= \int \frac{1}{2}(1 - u^2)\,du$$

$$= \frac{1}{2}\left(u - \frac{1}{3}u^3\right) + C$$

$$= \frac{1}{2}\left(\sin(2x) - \frac{1}{3}\sin^3(2x)\right) + C$$

Putting all the pieces together, we have

$$\int \cos^4 x \sin^2 x\,dx = \int \frac{1}{8}\left(1 + \cos(2x) - \cos^2(2x) - \cos^3(2x)\right)\,dx$$

$$= \frac{1}{8}\left[x + \frac{1}{2}\sin(2x) - \frac{1}{2}\left(x + \frac{1}{4}\sin(4x)\right) - \frac{1}{2}\left(\sin(2x) - \frac{1}{3}\sin^3(2x)\right)\right] + C$$

$$= \frac{1}{8}\left[\frac{1}{2}x - \frac{1}{8}\sin(4x) + \frac{1}{6}\sin^3(2x)\right] + C$$

The process above was a bit long and tedious, but being able to work a problem such as this from start to finish is important.

Integrals of the form $\displaystyle\int \sin(mx)\sin(nx)\,dx,\ \int \cos(mx)\cos(nx)\,dx,$

and $\displaystyle\int \sin(mx)\cos(nx)\,dx.$

Functions that contain products of sines and cosines of differing periods are important in many applications including the analysis of sound waves. Integrals of the form

$$\int \sin(mx)\sin(nx)\,dx, \quad \int \cos(mx)\cos(nx)\,dx \quad \text{and} \quad \int \sin(mx)\cos(nx)\,dx$$

Notes:

are best approached by first applying the Product to Sum Formulas found in the back cover of this text, namely

$$\sin(mx)\sin(nx) = \frac{1}{2}\Big[\cos\big((m-n)x\big) - \cos\big((m+n)x\big)\Big]$$

$$\cos(mx)\cos(nx) = \frac{1}{2}\Big[\cos\big((m-n)x\big) + \cos\big((m+n)x\big)\Big]$$

$$\sin(mx)\cos(nx) = \frac{1}{2}\Big[\sin\big((m-n)x\big) + \sin\big((m+n)x\big)\Big]$$

Example 6.3.4   **Integrating products of** $\sin(mx)$ **and** $\cos(nx)$

Evaluate $\displaystyle\int \sin(5x)\cos(2x)\,dx$.

SOLUTION   The application of the formula and subsequent integration are straightforward:

$$\int \sin(5x)\cos(2x)\,dx = \int \frac{1}{2}\Big[\sin(3x) + \sin(7x)\Big]\,dx$$

$$= -\frac{1}{6}\cos(3x) - \frac{1}{14}\cos(7x) + C$$

Integrals of the form $\displaystyle\int \tan^m x \sec^n x\,dx$.

When evaluating integrals of the form $\int \sin^m x \cos^n x\,dx$, the Pythagorean Theorem allowed us to convert even powers of sine into even powers of cosine, and vise–versa. If, for instance, the power of sine was odd, we pulled out one $\sin x$ and converted the remaining even power of $\sin x$ into a function using powers of $\cos x$, leading to an easy substitution.

The same basic strategy applies to integrals of the form $\int \tan^m x \sec^n x\,dx$, albeit a bit more nuanced. The following three facts will prove useful:

- $\frac{d}{dx}(\tan x) = \sec^2 x$,

- $\frac{d}{dx}(\sec x) = \sec x \tan x$, and

- $1 + \tan^2 x = \sec^2 x$ (the Pythagorean Theorem).

If the integrand can be manipulated to separate a $\sec^2 x$ term with the remaining secant power even, or if a $\sec x \tan x$ term can be separated with the remaining $\tan x$ power even, the Pythagorean Theorem can be employed, leading to a simple substitution. This strategy is outlined in the following Key Idea.

Notes:

Key Idea 6.3.2   **Integrals Involving Powers of Tangent and Secant**

Consider $\displaystyle\int \tan^m x \sec^n x\, dx$, where m, n are nonnegative integers.

1. If n is even, then $n = 2k$ for some integer k. Rewrite $\sec^n x$ as

$$\sec^n x = \sec^{2k} x = \sec^{2k-2} x \sec^2 x = (1 + \tan^2 x)^{k-1} \sec^2 x.$$

   Then

$$\int \tan^m x \sec^n x\, dx = \int \tan^m x (1 + \tan^2 x)^{k-1} \sec^2 x\, dx = \int u^m (1 + u^2)^{k-1}\, du,$$

   where $u = \tan x$ and $du = \sec^2 x\, dx$.

2. If m is odd, then $m = 2k + 1$ for some integer k. Rewrite $\tan^m x \sec^n x$ as

$$\tan^m x \sec^n x = \tan^{2k+1} x \sec^n x = \tan^{2k} x \sec^{n-1} x \sec x \tan x = (\sec^2 x - 1)^k \sec^{n-1} x \sec x \tan x.$$

   Then

$$\int \tan^m x \sec^n x\, dx = \int (\sec^2 x - 1)^k \sec^{n-1} x \sec x \tan x\, dx = \int (u^2 - 1)^k u^{n-1}\, du,$$

   where $u = \sec x$ and $du = \sec x \tan x\, dx$.

3. If n is odd and m is even, then $m = 2k$ for some integer k. Convert $\tan^m x$ to $(\sec^2 x - 1)^k$. Expand the new integrand and use Integration By Parts, with $dv = \sec^2 x\, dx$.

4. If m is even and $n = 0$, rewrite $\tan^m x$ as

$$\tan^m x = \tan^{m-2} x \tan^2 x = \tan^{m-2} x (\sec^2 x - 1) = \tan^{m-2} \sec^2 x - \tan^{m-2} x.$$

   So

$$\int \tan^m x\, dx = \underbrace{\int \tan^{m-2} \sec^2 x\, dx}_{\text{apply rule \#1}} - \underbrace{\int \tan^{m-2} x\, dx}_{\text{apply rule \#4 again}}.$$

The techniques described in items 1 and 2 of Key Idea 6.3.2 are relatively straightforward, but the techniques in items 3 and 4 can be rather tedious. A few examples will help with these methods.

Notes:

Example 6.3.5   **Integrating powers of tangent and secant**

Evaluate $\displaystyle\int \tan^2 x \sec^6 x \, dx$.

Solution   Since the power of secant is even, we use rule #1 from Key Idea 6.3.2 and pull out a $\sec^2 x$ in the integrand. We convert the remaining powers of secant into powers of tangent.

$$\int \tan^2 x \sec^6 x \, dx = \int \tan^2 x \sec^4 x \sec^2 x \, dx$$

$$= \int \tan^2 x \left(1 + \tan^2 x\right)^2 \sec^2 x \, dx$$

Now substitute, with $u = \tan x$, with $du = \sec^2 x \, dx$.

$$= \int u^2 \left(1 + u^2\right)^2 du$$

We leave the integration and subsequent substitution to the reader. The final answer is

$$= \frac{1}{3}\tan^3 x + \frac{2}{5}\tan^5 x + \frac{1}{7}\tan^7 x + C.$$

Example 6.3.6   **Integrating powers of tangent and secant**

Evaluate $\displaystyle\int \sec^3 x \, dx$.

Solution   We apply rule #3 from Key Idea 6.3.2 as the power of secant is odd and the power of tangent is even (0 is an even number). We use Integration by Parts; the rule suggests letting $dv = \sec^2 x \, dx$, meaning that $u = \sec x$.

$$
\begin{array}{ll}
u = \sec x & v = \, ? \\
du = \, ? & dv = \sec^2 x \, dx
\end{array}
\quad \Rightarrow \quad
\begin{array}{ll}
u = \sec x & v = \tan x \\
du = \sec x \tan x \, dx & dv = \sec^2 x \, dx
\end{array}
$$

Figure 6.3.2: Setting up Integration by Parts.

Employing Integration by Parts, we have

$$\int \sec^3 x \, dx = \int \underbrace{\sec x}_{u} \cdot \underbrace{\sec^2 x \, dx}_{dv}$$

$$= \sec x \tan x - \int \sec x \tan^2 x \, dx.$$

Notes:

This new integral also requires applying rule #3 of Key Idea 6.3.2:

$$= \sec x \tan x - \int \sec x \left(\sec^2 x - 1 \right) dx$$

$$= \sec x \tan x - \int \sec^3 x \, dx + \int \sec x \, dx$$

$$= \sec x \tan x - \int \sec^3 x \, dx + \ln|\sec x + \tan x|$$

In previous applications of Integration by Parts, we have seen where the original integral has reappeared in our work. We resolve this by adding $\int \sec^3 x \, dx$ to both sides, giving:

$$2 \int \sec^3 x \, dx = \sec x \tan x + \ln|\sec x + \tan x|$$

$$\int \sec^3 x \, dx = \frac{1}{2} \left(\sec x \tan x + \ln|\sec x + \tan x| \right) + C$$

We give one more example.

Example 6.3.7   **Integrating powers of tangent and secant**

Evaluate $\int \tan^6 x \, dx$.

    SOLUTION   We employ rule #4 of Key Idea 6.3.2.

$$\int \tan^6 x \, dx = \int \tan^4 x \tan^2 x \, dx$$

$$= \int \tan^4 x \left(\sec^2 x - 1 \right) dx$$

$$= \int \tan^4 x \sec^2 x \, dx - \int \tan^4 x \, dx$$

Integrate the first integral with substitution, $u = \tan x$; integrate the second by employing rule #4 again.

$$= \frac{1}{5} \tan^5 x - \int \tan^2 x \tan^2 x \, dx$$

$$= \frac{1}{5} \tan^5 x - \int \tan^2 x \left(\sec^2 x - 1 \right) dx$$

$$= \frac{1}{5} \tan^5 x - \int \tan^2 x \sec^2 x \, dx + \int \tan^2 x \, dx$$

Notes:

Again, use substitution for the first integral and rule #4 for the second.

$$= \frac{1}{5}\tan^5 x - \frac{1}{3}\tan^3 x + \int \left(\sec^2 x - 1 \right) dx$$

$$= \frac{1}{5}\tan^5 x - \frac{1}{3}\tan^3 x + \tan x - x + C.$$

These latter examples were admittedly long, with repeated applications of the same rule. Try to not be overwhelmed by the length of the problem, but rather admire how robust this solution method is. A trigonometric function of a high power can be systematically reduced to trigonometric functions of lower powers until all antiderivatives can be computed.

The next section introduces an integration technique known as Trigonometric Substitution, a clever combination of Substitution and the Pythagorean Theorem.

Notes:

Exercises 6.3

Terms and Concepts

1. T/F: $\int \sin^2 x \cos^2 x \, dx$ cannot be evaluated using the techniques described in this section since both powers of $\sin x$ and $\cos x$ are even.

2. T/F: $\int \sin^3 x \cos^3 x \, dx$ cannot be evaluated using the techniques described in this section since both powers of $\sin x$ and $\cos x$ are odd.

3. T/F: This section addresses how to evaluate indefinite integrals such as $\int \sin^5 x \tan^3 x \, dx$.

4. T/F: Sometimes computer programs evaluate integrals involving trigonometric functions differently than one would using the techniques of this section. When this is the case, the techniques of this section have failed and one should only trust the answer given by the computer.

Problems

In Exercises 5 – 28, evaluate the indefinite integral.

5. $\int \sin x \cos^4 x \, dx$

6. $\int \sin^3 x \cos x \, dx$

7. $\int \sin^3 x \cos^2 x \, dx$

8. $\int \sin^3 x \cos^3 x \, dx$

9. $\int \sin^6 x \cos^5 x \, dx$

10. $\int \sin^2 x \cos^7 x \, dx$

11. $\int \sin^2 x \cos^2 x \, dx$

12. $\int \sin x \cos x \, dx$

13. $\int \sin(5x) \cos(3x) \, dx$

14. $\int \sin(x) \cos(2x) \, dx$

15. $\int \sin(3x) \sin(7x) \, dx$

16. $\int \sin(\pi x) \sin(2\pi x) \, dx$

17. $\int \cos(x) \cos(2x) \, dx$

18. $\int \cos \left(\frac{\pi}{2} x \right) \cos(\pi x) \, dx$

19. $\int \tan^4 x \sec^2 x \, dx$

20. $\int \tan^2 x \sec^4 x \, dx$

21. $\int \tan^3 x \sec^4 x \, dx$

22. $\int \tan^3 x \sec^2 x \, dx$

23. $\int \tan^3 x \sec^3 x \, dx$

24. $\int \tan^5 x \sec^5 x \, dx$

25. $\int \tan^4 x \, dx$

26. $\int \sec^5 x \, dx$

27. $\int \tan^2 x \sec x \, dx$

28. $\int \tan^2 x \sec^3 x \, dx$

In Exercises 29 – 35, evaluate the definite integral. Note: the corresponding indefinite integrals appear in the previous set.

29. $\int_0^\pi \sin x \cos^4 x \, dx$

30. $\int_{-\pi}^\pi \sin^3 x \cos x \, dx$

31. $\int_{-\pi/2}^{\pi/2} \sin^2 x \cos^7 x \, dx$

32. $\int_0^{\pi/2} \sin(5x) \cos(3x) \, dx$

33. $\int_{-\pi/2}^{\pi/2} \cos(x) \cos(2x) \, dx$

34. $\int_0^{\pi/4} \tan^4 x \sec^2 x \, dx$

35. $\int_{-\pi/4}^{\pi/4} \tan^2 x \sec^4 x \, dx$

6.4   Trigonometric Substitution

In Section 5.2 we defined the definite integral as the "signed area under the curve." In that section we had not yet learned the Fundamental Theorem of Calculus, so we only evaluated special definite integrals which described nice, geometric shapes. For instance, we were able to evaluate

$$\int_{-3}^{3} \sqrt{9 - x^2}\, dx = \frac{9\pi}{2} \tag{6.1}$$

as we recognized that $f(x) = \sqrt{9 - x^2}$ described the upper half of a circle with radius 3.

We have since learned a number of integration techniques, including Substitution and Integration by Parts, yet we are still unable to evaluate the above integral without resorting to a geometric interpretation. This section introduces Trigonometric Substitution, a method of integration that fills this gap in our integration skill. This technique works on the same principle as Substitution as found in Section 6.1, though it can feel "backward." In Section 6.1, we set $u = f(x)$, for some function f, and replaced $f(x)$ with u. In this section, we will set $x = f(\theta)$, where f is a trigonometric function, then replace x with $f(\theta)$.

We start by demonstrating this method in evaluating the integral in Equation (6.1). After the example, we will generalize the method and give more examples.

Example 6.4.1   **Using Trigonometric Substitution**

Evaluate $\int_{-3}^{3} \sqrt{9 - x^2}\, dx$.

SOLUTION   We begin by noting that $9\sin^2\theta + 9\cos^2\theta = 9$, and hence $9\cos^2\theta = 9 - 9\sin^2\theta$. If we let $x = 3\sin\theta$, then $9 - x^2 = 9 - 9\sin^2\theta = 9\cos^2\theta$.

Setting $x = 3\sin\theta$ gives $dx = 3\cos\theta\, d\theta$. We are almost ready to substitute. We also wish to change our bounds of integration. The bound $x = -3$ corresponds to $\theta = -\pi/2$ (for when $\theta = -\pi/2$, $x = 3\sin\theta = -3$). Likewise, the bound of $x = 3$ is replaced by the bound $\theta = \pi/2$. Thus

$$\int_{-3}^{3} \sqrt{9 - x^2}\, dx = \int_{-\pi/2}^{\pi/2} \sqrt{9 - 9\sin^2\theta}(3\cos\theta)\, d\theta$$

$$= \int_{-\pi/2}^{\pi/2} 3\sqrt{9\cos^2\theta}\cos\theta\, d\theta$$

$$= \int_{-\pi/2}^{\pi/2} 3|3\cos\theta|\cos\theta\, d\theta.$$

On $[-\pi/2, \pi/2]$, $\cos\theta$ is always positive, so we can drop the absolute value bars, then employ a power–reducing formula:

Notes:

$$= \int_{-\pi/2}^{\pi/2} 9\cos^2\theta \, d\theta$$

$$= \int_{-\pi/2}^{\pi/2} \frac{9}{2}\big(1 + \cos(2\theta)\big) \, d\theta$$

$$= \frac{9}{2}\Big(\theta + \frac{1}{2}\sin(2\theta)\Big)\Big|_{-\pi/2}^{\pi/2} = \frac{9}{2}\pi.$$

This matches our answer from before.

We now describe in detail Trigonometric Substitution. This method excels when dealing with integrands that contain $\sqrt{a^2 - x^2}$, $\sqrt{x^2 - a^2}$ and $\sqrt{x^2 + a^2}$. The following Key Idea outlines the procedure for each case, followed by more examples. Each right triangle acts as a reference to help us understand the relationships between x and θ.

Key Idea 6.4.1   Trigonometric Substitution

(a) For integrands containing $\sqrt{a^2 - x^2}$:

Let $x = a\sin\theta$, $dx = a\cos\theta \, d\theta$

Thus $\theta = \sin^{-1}(x/a)$, for $-\pi/2 \le \theta \le \pi/2$.

On this interval, $\cos\theta \ge 0$, so

$\sqrt{a^2 - x^2} = a\cos\theta$

(b) For integrands containing $\sqrt{x^2 + a^2}$:

Let $x = a\tan\theta$, $dx = a\sec^2\theta \, d\theta$

Thus $\theta = \tan^{-1}(x/a)$, for $-\pi/2 < \theta < \pi/2$.

On this interval, $\sec\theta > 0$, so

$\sqrt{x^2 + a^2} = a\sec\theta$

(c) For integrands containing $\sqrt{x^2 - a^2}$:

Let $x = a\sec\theta$, $dx = a\sec\theta\tan\theta \, d\theta$

Thus $\theta = \sec^{-1}(x/a)$. If $x/a \ge 1$, then $0 \le \theta < \pi/2$; if $x/a \le -1$, then $\pi/2 < \theta \le \pi$.

We restrict our work to where $x \ge a$, so $x/a \ge 1$, and $0 \le \theta < \pi/2$. On this interval, $\tan\theta \ge 0$, so

$\sqrt{x^2 - a^2} = a\tan\theta$

Notes:

Example 6.4.2 **Using Trigonometric Substitution**

Evaluate $\int \dfrac{1}{\sqrt{5+x^2}}\, dx$.

SOLUTION Using Key Idea 6.4.1(b), we recognize $a = \sqrt{5}$ and set $x = \sqrt{5}\tan\theta$. This makes $dx = \sqrt{5}\sec^2\theta\, d\theta$. We will use the fact that $\sqrt{5+x^2} = \sqrt{5+5\tan^2\theta} = \sqrt{5\sec^2\theta} = \sqrt{5}\sec\theta$. Substituting, we have:

$$\int \frac{1}{\sqrt{5+x^2}}\, dx = \int \frac{1}{\sqrt{5+5\tan^2\theta}}\sqrt{5}\sec^2\theta\, d\theta$$

$$= \int \frac{\sqrt{5}\sec^2\theta}{\sqrt{5}\sec\theta}\, d\theta$$

$$= \int \sec\theta\, d\theta$$

$$= \ln\left|\sec\theta + \tan\theta\right| + C.$$

While the integration steps are over, we are not yet done. The original problem was stated in terms of x, whereas our answer is given in terms of θ. We must convert back to x.

The reference triangle given in Key Idea 6.4.1(b) helps. With $x = \sqrt{5}\tan\theta$, we have

$$\tan\theta = \frac{x}{\sqrt{5}} \quad \text{and} \quad \sec\theta = \frac{\sqrt{x^2+5}}{\sqrt{5}}.$$

This gives

$$\int \frac{1}{\sqrt{5+x^2}}\, dx = \ln\left|\sec\theta + \tan\theta\right| + C$$

$$= \ln\left|\frac{\sqrt{x^2+5}}{\sqrt{5}} + \frac{x}{\sqrt{5}}\right| + C.$$

We can leave this answer as is, or we can use a logarithmic identity to simplify it. Note:

$$\ln\left|\frac{\sqrt{x^2+5}}{\sqrt{5}} + \frac{x}{\sqrt{5}}\right| + C = \ln\left|\frac{1}{\sqrt{5}}\left(\sqrt{x^2+5}+x\right)\right| + C$$

$$= \ln\left|\frac{1}{\sqrt{5}}\right| + \ln\left|\sqrt{x^2+5}+x\right| + C$$

$$= \ln\left|\sqrt{x^2+5}+x\right| + C,$$

where the $\ln\left(1/\sqrt{5}\right)$ term is absorbed into the constant C. (In Section 6.6 we will learn another way of approaching this problem.)

Notes:

Example 6.4.3   **Using Trigonometric Substitution**

Evaluate $\displaystyle\int \sqrt{4x^2 - 1}\,dx$.

SOLUTION   We start by rewriting the integrand so that it looks like $\sqrt{x^2 - a^2}$ for some value of a:

$$\sqrt{4x^2 - 1} = \sqrt{4\left(x^2 - \frac{1}{4}\right)}$$

$$= 2\sqrt{x^2 - \left(\frac{1}{2}\right)^2}.$$

So we have $a = 1/2$, and following Key Idea 6.4.1(c), we set $x = \frac{1}{2}\sec\theta$, and hence $dx = \frac{1}{2}\sec\theta\tan\theta\,d\theta$. We now rewrite the integral with these substitutions:

$$\int \sqrt{4x^2 - 1}\,dx = \int 2\sqrt{x^2 - \left(\frac{1}{2}\right)^2}\,dx$$

$$= \int 2\sqrt{\frac{1}{4}\sec^2\theta - \frac{1}{4}}\left(\frac{1}{2}\sec\theta\tan\theta\right)d\theta$$

$$= \int \sqrt{\frac{1}{4}(\sec^2\theta - 1)}\left(\sec\theta\tan\theta\right)d\theta$$

$$= \int \sqrt{\frac{1}{4}\tan^2\theta}\left(\sec\theta\tan\theta\right)d\theta$$

$$= \int \frac{1}{2}\tan^2\theta\sec\theta\,d\theta$$

$$= \frac{1}{2}\int \left(\sec^2\theta - 1\right)\sec\theta\,d\theta$$

$$= \frac{1}{2}\int \left(\sec^3\theta - \sec\theta\right)d\theta.$$

We integrated $\sec^3\theta$ in Example 6.3.6, finding its antiderivatives to be

$$\int \sec^3\theta\,d\theta = \frac{1}{2}\left(\sec\theta\tan\theta + \ln|\sec\theta + \tan\theta|\right) + C.$$

Thus

$$\int \sqrt{4x^2 - 1}\,dx = \frac{1}{2}\int \left(\sec^3\theta - \sec\theta\right)d\theta$$

$$= \frac{1}{2}\left(\frac{1}{2}\left(\sec\theta\tan\theta + \ln|\sec\theta + \tan\theta|\right) - \ln|\sec\theta + \tan\theta|\right) + C$$

$$= \frac{1}{4}\left(\sec\theta\tan\theta - \ln|\sec\theta + \tan\theta|\right) + C.$$

Notes:

We are not yet done. Our original integral is given in terms of x, whereas our final answer, as given, is in terms of θ. We need to rewrite our answer in terms of x. With $a = 1/2$, and $x = \frac{1}{2}\sec\theta$, the reference triangle in Key Idea 6.4.1(c) shows that

$$\tan\theta = \sqrt{x^2 - 1/4}\Big/(1/2) = 2\sqrt{x^2 - 1/4} \quad\text{and}\quad \sec\theta = 2x.$$

Thus

$$\frac{1}{4}\left(\sec\theta\tan\theta - \ln\left|\sec\theta + \tan\theta\right|\right) + C = \frac{1}{4}\left(2x\cdot 2\sqrt{x^2 - 1/4} - \ln\left|2x + 2\sqrt{x^2 - 1/4}\right|\right) + C$$

$$= \frac{1}{4}\left(4x\sqrt{x^2 - 1/4} - \ln\left|2x + 2\sqrt{x^2 - 1/4}\right|\right) + C.$$

The final answer is given in the last line above, repeated here:

$$\int \sqrt{4x^2 - 1}\, dx = \frac{1}{4}\left(4x\sqrt{x^2 - 1/4} - \ln\left|2x + 2\sqrt{x^2 - 1/4}\right|\right) + C.$$

Example 6.4.4   **Using Trigonometric Substitution**

Evaluate $\displaystyle\int \frac{\sqrt{4 - x^2}}{x^2}\, dx.$

SOLUTION   We use Key Idea 6.4.1(a) with $a = 2$, $x = 2\sin\theta$, $dx = 2\cos\theta$ and hence $\sqrt{4 - x^2} = 2\cos\theta$. This gives

$$\int \frac{\sqrt{4 - x^2}}{x^2}\, dx = \int \frac{2\cos\theta}{4\sin^2\theta}(2\cos\theta)\, d\theta$$

$$= \int \cot^2\theta\, d\theta$$

$$= \int (\csc^2\theta - 1)\, d\theta$$

$$= -\cot\theta - \theta + C.$$

We need to rewrite our answer in terms of x. Using the reference triangle found in Key Idea 6.4.1(a), we have $\cot\theta = \sqrt{4 - x^2}/x$ and $\theta = \sin^{-1}(x/2)$. Thus

$$\int \frac{\sqrt{4 - x^2}}{x^2}\, dx = -\frac{\sqrt{4 - x^2}}{x} - \sin^{-1}\left(\frac{x}{2}\right) + C.$$

Trigonometric Substitution can be applied in many situations, even those not of the form $\sqrt{a^2 - x^2}$, $\sqrt{x^2 - a^2}$ or $\sqrt{x^2 + a^2}$. In the following example, we apply it to an integral we already know how to handle.

Notes:

Example 6.4.5 **Using Trigonometric Substitution**

Evaluate $\displaystyle\int \frac{1}{x^2 + 1}\, dx$.

SOLUTION We know the answer already as $\tan^{-1} x + C$. We apply Trigonometric Substitution here to show that we get the same answer without inherently relying on knowledge of the derivative of the arctangent function.

Using Key Idea 6.4.1(b), let $x = \tan\theta$, $dx = \sec^2\theta\, d\theta$ and note that $x^2 + 1 = \tan^2\theta + 1 = \sec^2\theta$. Thus

$$\int \frac{1}{x^2 + 1}\, dx = \int \frac{1}{\sec^2\theta}\sec^2\theta\, d\theta$$

$$= \int 1\, d\theta$$

$$= \theta + C.$$

Since $x = \tan\theta$, $\theta = \tan^{-1} x$, and we conclude that $\displaystyle\int \frac{1}{x^2 + 1}\, dx = \tan^{-1} x + C$.

The next example is similar to the previous one in that it does not involve a square–root. It shows how several techniques and identities can be combined to obtain a solution.

Example 6.4.6 **Using Trigonometric Substitution**

Evaluate $\displaystyle\int \frac{1}{(x^2 + 6x + 10)^2}\, dx$.

SOLUTION We start by completing the square, then make the substitution $u = x + 3$, followed by the trigonometric substitution of $u = \tan\theta$:

$$\int \frac{1}{(x^2 + 6x + 10)^2}\, dx = \int \frac{1}{\left((x+3)^2 + 1\right)^2}\, dx = \int \frac{1}{(u^2 + 1)^2}\, du.$$

Now make the substitution $u = \tan\theta$, $du = \sec^2\theta\, d\theta$:

$$= \int \frac{1}{(\tan^2\theta + 1)^2}\sec^2\theta\, d\theta$$

$$= \int \frac{1}{(\sec^2\theta)^2}\sec^2\theta\, d\theta$$

$$= \int \cos^2\theta\, d\theta.$$

Notes:

Applying a power reducing formula, we have

$$= \int \left(\frac{1}{2} + \frac{1}{2} \cos(2\theta) \right) d\theta$$

$$= \frac{1}{2}\theta + \frac{1}{4}\sin(2\theta) + C. \quad (6.2)$$

We need to return to the variable x. As $u = \tan\theta$, $\theta = \tan^{-1} u$. Using the identity $\sin(2\theta) = 2\sin\theta\cos\theta$ and using the reference triangle found in Key Idea 6.4.1(b), we have

$$\frac{1}{4}\sin(2\theta) = \frac{1}{2}\frac{u}{\sqrt{u^2+1}} \cdot \frac{1}{\sqrt{u^2+1}} = \frac{1}{2}\frac{u}{u^2+1}.$$

Finally, we return to x with the substitution $u = x + 3$. We start with the expression in Equation (6.2):

$$\frac{1}{2}\theta + \frac{1}{4}\sin(2\theta) + C = \frac{1}{2}\tan^{-1} u + \frac{1}{2}\frac{u}{u^2+1} + C$$

$$= \frac{1}{2}\tan^{-1}(x+3) + \frac{x+3}{2(x^2+6x+10)} + C.$$

Stating our final result in one line,

$$\int \frac{1}{(x^2+6x+10)^2}\, dx = \frac{1}{2}\tan^{-1}(x+3) + \frac{x+3}{2(x^2+6x+10)} + C.$$

Our last example returns us to definite integrals, as seen in our first example. Given a definite integral that can be evaluated using Trigonometric Substitution, we could first evaluate the corresponding indefinite integral (by changing from an integral in terms of x to one in terms of θ, then converting back to x) and then evaluate using the original bounds. It is much more straightforward, though, to change the bounds as we substitute.

Example 6.4.7   Definite integration and Trigonometric Substitution

Evaluate $\displaystyle\int_0^5 \frac{x^2}{\sqrt{x^2+25}}\, dx$.

SOLUTION   Using Key Idea 6.4.1(b), we set $x = 5\tan\theta$, $dx = 5\sec^2\theta\, d\theta$, and note that $\sqrt{x^2+25} = 5\sec\theta$. As we substitute, we can also change the bounds of integration.

The lower bound of the original integral is $x = 0$. As $x = 5\tan\theta$, we solve for θ and find $\theta = \tan^{-1}(x/5)$. Thus the new lower bound is $\theta = \tan^{-1}(0) = 0$. The

Notes:

original upper bound is $x = 5$, thus the new upper bound is $\theta = \tan^{-1}(5/5) = \pi/4$.

Thus we have

$$\int_0^5 \frac{x^2}{\sqrt{x^2 + 25}} \, dx = \int_0^{\pi/4} \frac{25 \tan^2 \theta}{5 \sec \theta} 5 \sec^2 \theta \, d\theta$$

$$= 25 \int_0^{\pi/4} \tan^2 \theta \sec \theta \, d\theta.$$

We encountered this indefinite integral in Example 6.4.3 where we found

$$\int \tan^2 \theta \sec \theta \, d\theta = \frac{1}{2} \left(\sec \theta \tan \theta - \ln|\sec \theta + \tan \theta| \right).$$

So

$$25 \int_0^{\pi/4} \tan^2 \theta \sec \theta \, d\theta = \frac{25}{2} \left(\sec \theta \tan \theta - \ln|\sec \theta + \tan \theta| \right) \Big|_0^{\pi/4}$$

$$= \frac{25}{2} \left(\sqrt{2} - \ln(\sqrt{2} + 1) \right)$$

$$\approx 6.661.$$

The following equalities are very useful when evaluating integrals using Trigonometric Substitution.

Key Idea 6.4.2   Useful Equalities with Trigonometric Substitution

1. $\sin(2\theta) = 2 \sin \theta \cos \theta$

2. $\cos(2\theta) = \cos^2 \theta - \sin^2 \theta = 2 \cos^2 \theta - 1 = 1 - 2 \sin^2 \theta$

3. $\displaystyle\int \sec^3 \theta \, d\theta = \frac{1}{2} \left(\sec \theta \tan \theta + \ln|\sec \theta + \tan \theta| \right) + C$

4. $\displaystyle\int \cos^2 \theta \, d\theta = \int \frac{1}{2} (1 + \cos(2\theta)) \, d\theta = \frac{1}{2} (\theta + \sin \theta \cos \theta) + C.$

The next section introduces Partial Fraction Decomposition, which is an algebraic technique that turns "complicated" fractions into sums of "simpler" fractions, making integration easier.

Notes:

Exercises 6.4

Terms and Concepts

1. Trigonometric Substitution works on the same principles as Integration by Substitution, though it can feel "_____".

2. If one uses Trigonometric Substitution on an integrand containing $\sqrt{25 - x^2}$, then one should set $x =$_____.

3. Consider the Pythagorean Identity $\sin^2\theta + \cos^2\theta = 1$.

   (a) What identity is obtained when both sides are divided by $\cos^2\theta$?

   (b) Use the new identity to simplify $9\tan^2\theta + 9$.

4. Why does Key Idea 6.4.1(a) state that $\sqrt{a^2 - x^2} = a\cos\theta$, and not $|a\cos\theta|$?

Problems

In Exercises 5 – 16, apply Trigonometric Substitution to evaluate the indefinite integrals.

5. $\int \sqrt{x^2 + 1}\, dx$

6. $\int \sqrt{x^2 + 4}\, dx$

7. $\int \sqrt{1 - x^2}\, dx$

8. $\int \sqrt{9 - x^2}\, dx$

9. $\int \sqrt{x^2 - 1}\, dx$

10. $\int \sqrt{x^2 - 16}\, dx$

11. $\int \sqrt{4x^2 + 1}\, dx$

12. $\int \sqrt{1 - 9x^2}\, dx$

13. $\int \sqrt{16x^2 - 1}\, dx$

14. $\int \dfrac{8}{\sqrt{x^2 + 2}}\, dx$

15. $\int \dfrac{3}{\sqrt{7 - x^2}}\, dx$

16. $\int \dfrac{5}{\sqrt{x^2 - 8}}\, dx$

In Exercises 17 – 26, evaluate the indefinite integrals. Some may be evaluated without Trigonometric Substitution.

17. $\int \dfrac{\sqrt{x^2 - 11}}{x}\, dx$

18. $\int \dfrac{1}{(x^2 + 1)^2}\, dx$

19. $\int \dfrac{x}{\sqrt{x^2 - 3}}\, dx$

20. $\int x^2\sqrt{1 - x^2}\, dx$

21. $\int \dfrac{x}{(x^2 + 9)^{3/2}}\, dx$

22. $\int \dfrac{5x^2}{\sqrt{x^2 - 10}}\, dx$

23. $\int \dfrac{1}{(x^2 + 4x + 13)^2}\, dx$

24. $\int x^2(1 - x^2)^{-3/2}\, dx$

25. $\int \dfrac{\sqrt{5 - x^2}}{7x^2}\, dx$

26. $\int \dfrac{x^2}{\sqrt{x^2 + 3}}\, dx$

In Exercises 27 – 32, evaluate the definite integrals by making the proper trigonometric substitution *and* changing the bounds of integration. (Note: each of the corresponding indefinite integrals has appeared previously in this Exercise set.)

27. $\displaystyle\int_{-1}^{1} \sqrt{1 - x^2}\, dx$

28. $\displaystyle\int_{4}^{8} \sqrt{x^2 - 16}\, dx$

29. $\displaystyle\int_{0}^{2} \sqrt{x^2 + 4}\, dx$

30. $\displaystyle\int_{-1}^{1} \dfrac{1}{(x^2 + 1)^2}\, dx$

31. $\displaystyle\int_{-1}^{1} \sqrt{9 - x^2}\, dx$

32. $\displaystyle\int_{-1}^{1} x^2\sqrt{1 - x^2}\, dx$

6.5 Partial Fraction Decomposition

In this section we investigate the antiderivatives of rational functions. Recall that rational functions are functions of the form $f(x) = \frac{p(x)}{q(x)}$, where $p(x)$ and $q(x)$ are polynomials and $q(x) \neq 0$. Such functions arise in many contexts, one of which is the solving of certain fundamental differential equations.

We begin with an example that demonstrates the motivation behind this section. Consider the integral $\displaystyle\int \frac{1}{x^2 - 1}\, dx$. We do not have a simple formula for this (if the denominator were $x^2 + 1$, we would recognize the antiderivative as being the arctangent function). It can be solved using Trigonometric Substitution, but note how the integral is easy to evaluate once we realize:

$$\frac{1}{x^2 - 1} = \frac{1/2}{x - 1} - \frac{1/2}{x + 1}.$$

Thus

$$\int \frac{1}{x^2 - 1}\, dx = \int \frac{1/2}{x - 1}\, dx - \int \frac{1/2}{x + 1}\, dx$$
$$= \frac{1}{2}\ln|x - 1| - \frac{1}{2}\ln|x + 1| + C.$$

This section teaches how to *decompose*

$$\frac{1}{x^2 - 1} \quad \text{into} \quad \frac{1/2}{x - 1} - \frac{1/2}{x + 1}.$$

We start with a rational function $f(x) = \frac{p(x)}{q(x)}$, where p and q do not have any common factors and the degree of p is less than the degree of q. It can be shown that any polynomial, and hence q, can be factored into a product of linear and irreducible quadratic terms. The following Key Idea states how to decompose a rational function into a sum of rational functions whose denominators are all of lower degree than q.

Notes:

Key Idea 6.5.1   Partial Fraction Decomposition

Let $\dfrac{p(x)}{q(x)}$ be a rational function, where the degree of p is less than the degree of q.

1. **Linear Terms:** Let $(x-a)$ divide $q(x)$, where $(x-a)^n$ is the highest power of $(x-a)$ that divides $q(x)$. Then the decomposition of $\frac{p(x)}{q(x)}$ will contain the sum

$$\frac{A_1}{(x-a)} + \frac{A_2}{(x-a)^2} + \cdots + \frac{A_n}{(x-a)^n}.$$

2. **Quadratic Terms:** Let $x^2 + bx + c$ divide $q(x)$, where $(x^2 + bx + c)^n$ is the highest power of $x^2 + bx + c$ that divides $q(x)$. Then the decomposition of $\frac{p(x)}{q(x)}$ will contain the sum

$$\frac{B_1 x + C_1}{x^2 + bx + c} + \frac{B_2 x + C_2}{(x^2 + bx + c)^2} + \cdots + \frac{B_n x + C_n}{(x^2 + bx + c)^n}.$$

To find the coefficients A_i, B_i and C_i:

1. Multiply all fractions by $q(x)$, clearing the denominators. Collect like terms.

2. Equate the resulting coefficients of the powers of x and solve the resulting system of linear equations.

The following examples will demonstrate how to put this Key Idea into practice. Example 6.5.1 stresses the decomposition aspect of the Key Idea.

Example 6.5.1   Decomposing into partial fractions

Decompose $f(x) = \dfrac{1}{(x+5)(x-2)^3(x^2+x+2)(x^2+x+7)^2}$ without solving for the resulting coefficients.

SOLUTION   The denominator is already factored, as both $x^2 + x + 2$ and $x^2 + x + 7$ cannot be factored further. We need to decompose $f(x)$ properly. Since $(x+5)$ is a linear term that divides the denominator, there will be a

$$\frac{A}{x+5}$$

Notes:

term in the decomposition.

As $(x-2)^3$ divides the denominator, we will have the following terms in the decomposition:

$$\frac{B}{x-2}, \quad \frac{C}{(x-2)^2} \quad \text{and} \quad \frac{D}{(x-2)^3}.$$

The $x^2 + x + 2$ term in the denominator results in a $\dfrac{Ex+F}{x^2+x+2}$ term.

Finally, the $(x^2 + x + 7)^2$ term results in the terms

$$\frac{Gx+H}{x^2+x+7} \quad \text{and} \quad \frac{Ix+J}{(x^2+x+7)^2}.$$

All together, we have

$$\frac{1}{(x+5)(x-2)^3(x^2+x+2)(x^2+x+7)^2} = \frac{A}{x+5} + \frac{B}{x-2} + \frac{C}{(x-2)^2} + \frac{D}{(x-2)^3} +$$
$$\frac{Ex+F}{x^2+x+2} + \frac{Gx+H}{x^2+x+7} + \frac{Ix+J}{(x^2+x+7)^2}$$

Solving for the coefficients $A, B \ldots J$ would be a bit tedious but not "hard."

Example 6.5.2   **Decomposing into partial fractions**

Perform the partial fraction decomposition of $\dfrac{1}{x^2-1}$.

SOLUTION   The denominator factors into two linear terms: $x^2 - 1 = (x-1)(x+1)$. Thus

$$\frac{1}{x^2-1} = \frac{A}{x-1} + \frac{B}{x+1}.$$

To solve for A and B, first multiply through by $x^2 - 1 = (x-1)(x+1)$:

$$1 = \frac{A(x-1)(x+1)}{x-1} + \frac{B(x-1)(x+1)}{x+1}$$
$$= A(x+1) + B(x-1)$$
$$= Ax + A + Bx - B$$

Now collect like terms.

$$= (A+B)x + (A-B).$$

The next step is key. Note the equality we have:

$$1 = (A+B)x + (A-B).$$

Notes:

For clarity's sake, rewrite the left hand side as

$$0x + 1 = (A + B)x + (A - B).$$

On the left, the coefficient of the x term is 0; on the right, it is $(A + B)$. Since both sides are equal, we must have that $0 = A + B$.

Likewise, on the left, we have a constant term of 1; on the right, the constant term is $(A - B)$. Therefore we have $1 = A - B$.

We have two linear equations with two unknowns. This one is easy to solve by hand, leading to

$$\begin{matrix} A + B = 0 \\ A - B = 1 \end{matrix} \Rightarrow \begin{matrix} A = 1/2 \\ B = -1/2 \end{matrix}.$$

Thus

$$\frac{1}{x^2 - 1} = \frac{1/2}{x - 1} - \frac{1/2}{x + 1}.$$

Example 6.5.3   Integrating using partial fractions

Use partial fraction decomposition to integrate $\displaystyle\int \frac{1}{(x - 1)(x + 2)^2}\, dx$.

Note: Equation 6.3 offers a direct route to finding the values of A, B and C. Since the equation holds for all values of x, it holds in particular when $x = 1$. However, when $x = 1$, the right hand side simplifies to $A(1 + 2)^2 = 9A$. Since the left hand side is still 1, we have $1 = 9A$. Hence $A = 1/9$. Likewise, the equality holds when $x = -2$; this leads to the equation $1 = -3C$. Thus $C = -1/3$.

Knowing A and C, we can find the value of B by choosing yet another value of x, such as $x = 0$, and solving for B.

SOLUTION   We decompose the integrand as follows, as described by Key Idea 6.5.1:

$$\frac{1}{(x - 1)(x + 2)^2} = \frac{A}{x - 1} + \frac{B}{x + 2} + \frac{C}{(x + 2)^2}.$$

To solve for A, B and C, we multiply both sides by $(x - 1)(x + 2)^2$ and collect like terms:

$$1 = A(x + 2)^2 + B(x - 1)(x + 2) + C(x - 1) \tag{6.3}$$
$$= Ax^2 + 4Ax + 4A + Bx^2 + Bx - 2B + Cx - C$$
$$= (A + B)x^2 + (4A + B + C)x + (4A - 2B - C)$$

We have

$$0x^2 + 0x + 1 = (A + B)x^2 + (4A + B + C)x + (4A - 2B - C)$$

leading to the equations

$$A + B = 0, \quad 4A + B + C = 0 \quad \text{and} \quad 4A - 2B - C = 1.$$

These three equations of three unknowns lead to a unique solution:

$$A = 1/9, \quad B = -1/9 \quad \text{and} \quad C = -1/3.$$

Notes:

Thus

$$\int \frac{1}{(x-1)(x+2)^2}\, dx = \int \frac{1/9}{x-1}\, dx + \int \frac{-1/9}{x+2}\, dx + \int \frac{-1/3}{(x+2)^2}\, dx.$$

Each can be integrated with a simple substitution with $u = x - 1$ or $u = x + 2$ (or by directly applying Key Idea 6.1.1 as the denominators are linear functions). The end result is

$$\int \frac{1}{(x-1)(x+2)^2}\, dx = \frac{1}{9}\ln|x-1| - \frac{1}{9}\ln|x+2| + \frac{1}{3(x+2)} + C.$$

Example 6.5.4   Integrating using partial fractions

Use partial fraction decomposition to integrate $\displaystyle\int \frac{x^3}{(x-5)(x+3)}\, dx.$

SOLUTION   Key Idea 6.5.1 presumes that the degree of the numerator is less than the degree of the denominator. Since this is not the case here, we begin by using polynomial division to reduce the degree of the numerator. We omit the steps, but encourage the reader to verify that

$$\frac{x^3}{(x-5)(x+3)} = x + 2 + \frac{19x+30}{(x-5)(x+3)}.$$

Using Key Idea 6.5.1, we can rewrite the new rational function as:

$$\frac{19x+30}{(x-5)(x+3)} - \frac{A}{x-5} + \frac{B}{x+3}$$

for appropriate values of A and B. Clearing denominators, we have

$$19x + 30 = A(x+3) + B(x-5)$$
$$= (A+B)x + (3A - 5B).$$

This implies that:

$$19 = A + B$$
$$30 = 3A - 5B.$$

Solving this system of linear equations gives

$$125/8 = A$$
$$27/8 = B.$$

Note: The values of A and B can be quickly found using the technique described in the margin of Example 6.5.3.

Notes:

We can now integrate.

$$\int \frac{x^3}{(x-5)(x+3)}\, dx = \int \left(x + 2 + \frac{125/8}{x-5} + \frac{27/8}{x+3} \right) dx$$

$$= \frac{x^2}{2} + 2x + \frac{125}{8} \ln|x-5| + \frac{27}{8} \ln|x+3| + C.$$

Example 6.5.5   Integrating using partial fractions

Use partial fraction decomposition to evaluate $\displaystyle\int \frac{7x^2 + 31x + 54}{(x+1)(x^2+6x+11)}\, dx$.

SOLUTION   The degree of the numerator is less than the degree of the denominator so we begin by applying Key Idea 6.5.1. We have:

$$\frac{7x^2 + 31x + 54}{(x+1)(x^2+6x+11)} = \frac{A}{x+1} + \frac{Bx + C}{x^2 + 6x + 11}.$$

Now clear the denominators.

$$7x^2 + 31x + 54 = A(x^2 + 6x + 11) + (Bx + C)(x+1)$$

$$= (A+B)x^2 + (6A + B + C)x + (11A + C).$$

This implies that:

$$7 = A + B$$
$$31 = 6A + B + C$$
$$54 = 11A + C.$$

Solving this system of linear equations gives the nice result of $A = 5$, $B = 2$ and $C = -1$. Thus

$$\int \frac{7x^2 + 31x + 54}{(x+1)(x^2+6x+11)}\, dx = \int \left(\frac{5}{x+1} + \frac{2x - 1}{x^2 + 6x + 11} \right) dx.$$

The first term of this new integrand is easy to evaluate; it leads to a $5 \ln|x+1|$ term. The second term is not hard, but takes several steps and uses substitution techniques.

The integrand $\dfrac{2x - 1}{x^2 + 6x + 11}$ has a quadratic in the denominator and a linear term in the numerator. This leads us to try substitution. Let $u = x^2 + 6x + 11$, so $du = (2x + 6)\, dx$. The numerator is $2x - 1$, not $2x + 6$, but we can get a $2x + 6$

Notes:

term in the numerator by adding 0 in the form of "$7 - 7$."

$$\frac{2x - 1}{x^2 + 6x + 11} = \frac{2x - 1 + 7 - 7}{x^2 + 6x + 11}$$
$$= \frac{2x + 6}{x^2 + 6x + 11} - \frac{7}{x^2 + 6x + 11}.$$

We can now integrate the first term with substitution, leading to a $\ln|x^2+6x+11|$ term. The final term can be integrated using arctangent. First, complete the square in the denominator:

$$\frac{7}{x^2 + 6x + 11} = \frac{7}{(x + 3)^2 + 2}.$$

An antiderivative of the latter term can be found using Theorem 6.1.3 and substitution:

$$\int \frac{7}{x^2 + 6x + 11}\, dx = \frac{7}{\sqrt{2}} \tan^{-1}\left(\frac{x + 3}{\sqrt{2}}\right) + C.$$

Let's start at the beginning and put all of the steps together.

$$\int \frac{7x^2 + 31x + 54}{(x + 1)(x^2 + 6x + 11)}\, dx = \int \left(\frac{5}{x + 1} + \frac{2x - 1}{x^2 + 6x + 11}\right) dx$$
$$= \int \frac{5}{x + 1}\, dx + \int \frac{2x + 6}{x^2 + 6x + 11}\, dx - \int \frac{7}{x^2 + 6x + 11}\, dx$$
$$= 5\ln|x + 1| + \ln|x^2 + 6x + 11| - \frac{7}{\sqrt{2}} \tan^{-1}\left(\frac{x + 3}{\sqrt{2}}\right) + C.$$

As with many other problems in calculus, it is important to remember that one is not expected to "see" the final answer immediately after seeing the problem. Rather, given the initial problem, we break it down into smaller problems that are easier to solve. The final answer is a combination of the answers of the smaller problems.

Partial Fraction Decomposition is an important tool when dealing with rational functions. Note that at its heart, it is a technique of algebra, not calculus, as we are rewriting a fraction in a new form. Regardless, it is very useful in the realm of calculus as it lets us evaluate a certain set of "complicated" integrals.

The next section introduces new functions, called the Hyperbolic Functions. They will allow us to make substitutions similar to those found when studying Trigonometric Substitution, allowing us to approach even more integration problems.

Notes:

Exercises 6.5

Terms and Concepts

1. Fill in the blank: Partial Fraction Decomposition is a method of rewriting _____ functions.

2. T/F: It is sometimes necessary to use polynomial division before using Partial Fraction Decomposition.

3. Decompose $\dfrac{1}{x^2 - 3x}$ without solving for the coefficients, as done in Example 6.5.1.

4. Decompose $\dfrac{7 - x}{x^2 - 9}$ without solving for the coefficients, as done in Example 6.5.1.

5. Decompose $\dfrac{x - 3}{x^2 - 7}$ without solving for the coefficients, as done in Example 6.5.1.

6. Decompose $\dfrac{2x + 5}{x^3 + 7x}$ without solving for the coefficients, as done in Example 6.5.1.

Problems

In Exercises 7 – 26, evaluate the indefinite integral.

7. $\displaystyle\int \dfrac{7x + 7}{x^2 + 3x - 10}\, dx$

8. $\displaystyle\int \dfrac{7x - 2}{x^2 + x}\, dx$

9. $\displaystyle\int \dfrac{-4}{3x^2 - 12}\, dx$

10. $\displaystyle\int \dfrac{6x + 4}{3x^2 + 4x + 1}\, dx$

11. $\displaystyle\int \dfrac{x + 7}{(x + 5)^2}\, dx$

12. $\displaystyle\int \dfrac{-3x - 20}{(x + 8)^2}\, dx$

13. $\displaystyle\int \dfrac{9x^2 + 11x + 7}{x(x + 1)^2}\, dx$

14. $\displaystyle\int \dfrac{-12x^2 - x + 33}{(x - 1)(x + 3)(3 - 2x)}\, dx$

15. $\displaystyle\int \dfrac{94x^2 - 10x}{(7x + 3)(5x - 1)(3x - 1)}\, dx$

16. $\displaystyle\int \dfrac{x^2 + x + 1}{x^2 + x - 2}\, dx$

17. $\displaystyle\int \dfrac{x^3}{x^2 - x - 20}\, dx$

18. $\displaystyle\int \dfrac{2x^2 - 4x + 6}{x^2 - 2x + 3}\, dx$

19. $\displaystyle\int \dfrac{1}{x^3 + 2x^2 + 3x}\, dx$

20. $\displaystyle\int \dfrac{x^2 + x + 5}{x^2 + 4x + 10}\, dx$

21. $\displaystyle\int \dfrac{12x^2 + 21x + 3}{(x + 1)(3x^2 + 5x - 1)}\, dx$

22. $\displaystyle\int \dfrac{6x^2 + 8x - 4}{(x - 3)(x^2 + 6x + 10)}\, dx$

23. $\displaystyle\int \dfrac{2x^2 + x + 1}{(x + 1)(x^2 + 9)}\, dx$

24. $\displaystyle\int \dfrac{x^2 - 20x - 69}{(x - 7)(x^2 + 2x + 17)}\, dx$

25. $\displaystyle\int \dfrac{9x^2 - 60x + 33}{(x - 9)(x^2 - 2x + 11)}\, dx$

26. $\displaystyle\int \dfrac{6x^2 + 45x + 121}{(x + 2)(x^2 + 10x + 27)}\, dx$

In Exercises 27 – 30, evaluate the definite integral.

27. $\displaystyle\int_1^2 \dfrac{8x + 21}{(x + 2)(x + 3)}\, dx$

28. $\displaystyle\int_0^5 \dfrac{14x + 6}{(3x + 2)(x + 4)}\, dx$

29. $\displaystyle\int_{-1}^1 \dfrac{x^2 + 5x - 5}{(x - 10)(x^2 + 4x + 5)}\, dx$

30. $\displaystyle\int_0^1 \dfrac{x}{(x + 1)(x^2 + 2x + 1)}\, dx$

6.6 Hyperbolic Functions

The **hyperbolic functions** are a set of functions that have many applications to mathematics, physics, and engineering. Among many other applications, they are used to describe the formation of satellite rings around planets, to describe the shape of a rope hanging from two points, and have application to the theory of special relativity. This section defines the hyperbolic functions and describes many of their properties, especially their usefulness to calculus.

These functions are sometimes referred to as the "hyperbolic trigonometric functions" as there are many, many connections between them and the standard trigonometric functions. Figure 6.6.1 demonstrates one such connection. Just as cosine and sine are used to define points on the circle defined by $x^2 + y^2 = 1$, the functions **hyperbolic cosine** and **hyperbolic sine** are used to define points on the hyperbola $x^2 - y^2 = 1$.

We begin with their definition.

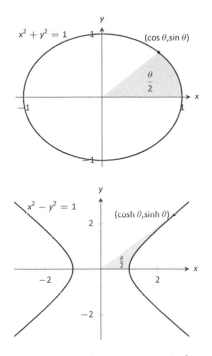

Definition 6.6.1   **Hyperbolic Functions**

1. $\cosh x = \dfrac{e^x + e^{-x}}{2}$

2. $\sinh x = \dfrac{e^x - e^{-x}}{2}$

3. $\tanh x = \dfrac{\sinh x}{\cosh x}$

4. $\operatorname{sech} x = \dfrac{1}{\cosh x}$

5. $\operatorname{csch} x = \dfrac{1}{\sinh x}$

6. $\coth x = \dfrac{\cosh x}{\sinh x}$

Figure 6.6.1: Using trigonometric functions to define points on a circle and hyperbolic functions to define points on a hyperbola. The area of the shaded regions are included in them.

These hyperbolic functions are graphed in Figure 6.6.2. In the graphs of $\cosh x$ and $\sinh x$, graphs of $e^x/2$ and $e^{-x}/2$ are included with dashed lines. As x gets "large," $\cosh x$ and $\sinh x$ each act like $e^x/2$; when x is a large negative number, $\cosh x$ acts like $e^{-x}/2$ whereas $\sinh x$ acts like $-e^{-x}/2$.

Notice the domains of $\tanh x$ and $\operatorname{sech} x$ are $(-\infty, \infty)$, whereas both $\coth x$ and $\operatorname{csch} x$ have vertical asymptotes at $x = 0$. Also note the ranges of these functions, especially $\tanh x$: as $x \to \infty$, both $\sinh x$ and $\cosh x$ approach $e^x/2$, hence $\tanh x$ approaches 1.

The following example explores some of the properties of these functions that bear remarkable resemblance to the properties of their trigonometric counterparts.

Pronunciation Note:
"cosh" rhymes with "gosh,"
"sinh" rhymes with "pinch," and
"tanh" rhymes with "ranch."

Notes:

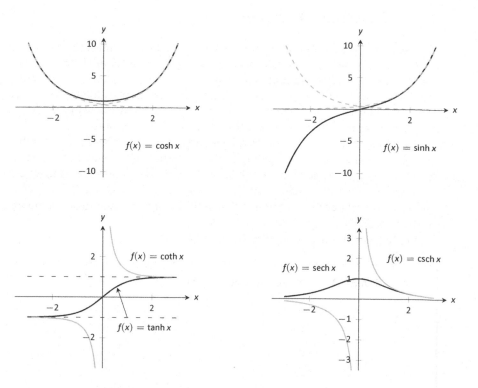

Figure 6.6.2: Graphs of the hyperbolic functions.

Example 6.6.1   Exploring properties of hyperbolic functions
Use Definition 6.6.1 to rewrite the following expressions.

1. $\cosh^2 x - \sinh^2 x$

2. $\tanh^2 x + \text{sech}^2 x$

3. $2 \cosh x \sinh x$

4. $\frac{d}{dx}\left(\cosh x \right)$

5. $\frac{d}{dx}\left(\sinh x \right)$

6. $\frac{d}{dx}\left(\tanh x \right)$

SOLUTION

1.
$$\cosh^2 x - \sinh^2 x = \left(\frac{e^x + e^{-x}}{2} \right)^2 - \left(\frac{e^x - e^{-x}}{2} \right)^2$$
$$= \frac{e^{2x} + 2e^x e^{-x} + e^{-2x}}{4} - \frac{e^{2x} - 2e^x e^{-x} + e^{-2x}}{4}$$
$$= \frac{4}{4} = 1.$$

So $\cosh^2 x - \sinh^2 x = 1$.

Notes:

2.
$$\tanh^2 x + \text{sech}^2 x = \frac{\sinh^2 x}{\cosh^2 x} + \frac{1}{\cosh^2 x}$$
$$= \frac{\sinh^2 x + 1}{\cosh^2 x} \qquad \text{Now use identity from \#1.}$$
$$= \frac{\cosh^2 x}{\cosh^2 x} = 1.$$

So $\tanh^2 x + \text{sech}^2 x = 1$.

3.
$$2\cosh x \sinh x = 2\left(\frac{e^x + e^{-x}}{2}\right)\left(\frac{e^x - e^{-x}}{2}\right)$$
$$= 2 \cdot \frac{e^{2x} - e^{-2x}}{4}$$
$$= \frac{e^{2x} - e^{-2x}}{2} = \sinh(2x).$$

Thus $2\cosh x \sinh x = \sinh(2x)$.

4.
$$\frac{d}{dx}\left(\cosh x\right) = \frac{d}{dx}\left(\frac{e^x + e^{-x}}{2}\right)$$
$$= \frac{e^x - e^{-x}}{2}$$
$$= \sinh x.$$

So $\frac{d}{dx}\left(\cosh x\right) = \sinh x$.

5.
$$\frac{d}{dx}\left(\sinh x\right) = \frac{d}{dx}\left(\frac{e^x - e^{-x}}{2}\right)$$
$$= \frac{e^x + e^{-x}}{2}$$
$$= \cosh x.$$

So $\frac{d}{dx}\left(\sinh x\right) = \cosh x$.

6.
$$\frac{d}{dx}\left(\tanh x\right) = \frac{d}{dx}\left(\frac{\sinh x}{\cosh x}\right)$$
$$= \frac{\cosh x \cosh x - \sinh x \sinh x}{\cosh^2 x}$$
$$= \frac{1}{\cosh^2 x}$$
$$= \text{sech}^2 x.$$

So $\frac{d}{dx}\left(\tanh x\right) = \text{sech}^2 x$.

Notes:

The following Key Idea summarizes many of the important identities relating to hyperbolic functions. Each can be verified by referring back to Definition 6.6.1.

Key Idea 6.6.1   **Useful Hyperbolic Function Properties**

Basic Identities

1. $\cosh^2 x - \sinh^2 x = 1$

2. $\tanh^2 x + \operatorname{sech}^2 x = 1$

3. $\coth^2 x - \operatorname{csch}^2 x = 1$

4. $\cosh 2x = \cosh^2 x + \sinh^2 x$

5. $\sinh 2x = 2\sinh x \cosh x$

6. $\cosh^2 x = \dfrac{\cosh 2x + 1}{2}$

7. $\sinh^2 x = \dfrac{\cosh 2x - 1}{2}$

Derivatives

1. $\dfrac{d}{dx}\left(\cosh x\right) = \sinh x$

2. $\dfrac{d}{dx}\left(\sinh x\right) = \cosh x$

3. $\dfrac{d}{dx}\left(\tanh x\right) = \operatorname{sech}^2 x$

4. $\dfrac{d}{dx}\left(\operatorname{sech} x\right) = -\operatorname{sech} x \tanh x$

5. $\dfrac{d}{dx}\left(\operatorname{csch} x\right) = -\operatorname{csch} x \coth x$

6. $\dfrac{d}{dx}\left(\coth x\right) = -\operatorname{csch}^2 x$

Integrals

1. $\displaystyle\int \cosh x \, dx = \sinh x + C$

2. $\displaystyle\int \sinh x \, dx = \cosh x + C$

3. $\displaystyle\int \tanh x \, dx = \ln(\cosh x) + C$

4. $\displaystyle\int \coth x \, dx = \ln|\sinh x| + C$

We practice using Key Idea 6.6.1.

Example 6.6.2   **Derivatives and integrals of hyperbolic functions**
Evaluate the following derivatives and integrals.

1. $\dfrac{d}{dx}\left(\cosh 2x\right)$

2. $\displaystyle\int \operatorname{sech}^2(7t - 3) \, dt$

3. $\displaystyle\int_0^{\ln 2} \cosh x \, dx$

SOLUTION

1. Using the Chain Rule directly, we have $\dfrac{d}{dx}\left(\cosh 2x\right) = 2\sinh 2x$.

   Just to demonstrate that it works, let's also use the Basic Identity found in Key Idea 6.6.1: $\cosh 2x = \cosh^2 x + \sinh^2 x$.

   $$\frac{d}{dx}\left(\cosh 2x\right) = \frac{d}{dx}\left(\cosh^2 x + \sinh^2 x\right) = 2\cosh x \sinh x + 2\sinh x \cosh x$$

   $$= 4\cosh x \sinh x.$$

Notes:

Using another Basic Identity, we can see that $4\cosh x \sinh x = 2\sinh 2x$. We get the same answer either way.

2. We employ substitution, with $u = 7t - 3$ and $du = 7dt$. Applying Key Ideas 6.1.1 and 6.6.1 we have:

$$\int \operatorname{sech}^2(7t - 3)\, dt = \frac{1}{7}\tanh(7t - 3) + C.$$

3.

$$\int_0^{\ln 2} \cosh x\, dx = \sinh x \Big|_0^{\ln 2} = \sinh(\ln 2) - \sinh 0 = \sinh(\ln 2).$$

We can simplify this last expression as $\sinh x$ is based on exponentials:

$$\sinh(\ln 2) = \frac{e^{\ln 2} - e^{-\ln 2}}{2} = \frac{2 - 1/2}{2} = \frac{3}{4}.$$

Inverse Hyperbolic Functions

Just as the inverse trigonometric functions are useful in certain integrations, the inverse hyperbolic functions are useful with others. Figure 6.6.3 shows the restrictions on the domains to make each function one-to-one and the resulting domains and ranges of their inverse functions. Their graphs are shown in Figure 6.6.4.

Because the hyperbolic functions are defined in terms of exponential functions, their inverses can be expressed in terms of logarithms as shown in Key Idea 6.6.2. It is often more convenient to refer to $\sinh^{-1} x$ than to $\ln\left(x + \sqrt{x^2 + 1}\right)$, especially when one is working on theory and does not need to compute actual values. On the other hand, when computations are needed, technology is often helpful but many hand-held calculators lack a *convenient* $\sinh^{-1} x$ button. (Often it can be accessed under a menu system, but not conveniently.) In such a situation, the logarithmic representation is useful. The reader is not encouraged to memorize these, but rather know they exist and know how to use them when needed.

Notes:

Function	Domain	Range
$\cosh x$	$[0, \infty)$	$[1, \infty)$
$\sinh x$	$(-\infty, \infty)$	$(-\infty, \infty)$
$\tanh x$	$(-\infty, \infty)$	$(-1, 1)$
$\operatorname{sech} x$	$[0, \infty)$	$(0, 1]$
$\operatorname{csch} x$	$(-\infty, 0) \cup (0, \infty)$	$(-\infty, 0) \cup (0, \infty)$
$\coth x$	$(-\infty, 0) \cup (0, \infty)$	$(-\infty, -1) \cup (1, \infty)$

Function	Domain	Range
$\cosh^{-1} x$	$[1, \infty)$	$[0, \infty)$
$\sinh^{-1} x$	$(-\infty, \infty)$	$(-\infty, \infty)$
$\tanh^{-1} x$	$(-1, 1)$	$(-\infty, \infty)$
$\operatorname{sech}^{-1} x$	$(0, 1]$	$[0, \infty)$
$\operatorname{csch}^{-1} x$	$(-\infty, 0) \cup (0, \infty)$	$(-\infty, 0) \cup (0, \infty)$
$\coth^{-1} x$	$(-\infty, -1) \cup (1, \infty)$	$(-\infty, 0) \cup (0, \infty)$

Figure 6.6.3: Domains and ranges of the hyperbolic and inverse hyperbolic functions.

Figure 6.6.4: Graphs of the hyperbolic functions and their inverses.

Key Idea 6.6.2   Logarithmic definitions of Inverse Hyperbolic Functions

1. $\cosh^{-1} x = \ln\left(x + \sqrt{x^2 - 1}\right)$; $x \geq 1$

2. $\tanh^{-1} x = \dfrac{1}{2} \ln\left(\dfrac{1 + x}{1 - x}\right)$; $|x| < 1$

3. $\operatorname{sech}^{-1} x = \ln\left(\dfrac{1 + \sqrt{1 - x^2}}{x}\right)$; $0 < x \leq 1$

4. $\sinh^{-1} x = \ln\left(x + \sqrt{x^2 + 1}\right)$

5. $\coth^{-1} x = \dfrac{1}{2} \ln\left(\dfrac{x + 1}{x - 1}\right)$; $|x| > 1$

6. $\operatorname{csch}^{-1} x = \ln\left(\dfrac{1}{x} + \dfrac{\sqrt{1 + x^2}}{|x|}\right)$; $x \neq 0$

Notes:

The following Key Ideas give the derivatives and integrals relating to the inverse hyperbolic functions. In Key Idea 6.6.4, both the inverse hyperbolic and logarithmic function representations of the antiderivative are given, based on Key Idea 6.6.2. Again, these latter functions are often more useful than the former. Note how inverse hyperbolic functions can be used to solve integrals we used Trigonometric Substitution to solve in Section 6.4.

Key Idea 6.6.3   Derivatives Involving Inverse Hyperbolic Functions

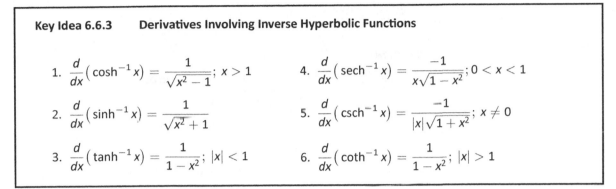

1. $\dfrac{d}{dx}\left(\cosh^{-1}x\right) = \dfrac{1}{\sqrt{x^2-1}}; \; x > 1$

2. $\dfrac{d}{dx}\left(\sinh^{-1}x\right) = \dfrac{1}{\sqrt{x^2+1}}$

3. $\dfrac{d}{dx}\left(\tanh^{-1}x\right) = \dfrac{1}{1-x^2}; \; |x| < 1$

4. $\dfrac{d}{dx}\left(\text{sech}^{-1}x\right) = \dfrac{-1}{x\sqrt{1-x^2}}; \; 0 < x < 1$

5. $\dfrac{d}{dx}\left(\text{csch}^{-1}x\right) = \dfrac{-1}{|x|\sqrt{1+x^2}}; \; x \neq 0$

6. $\dfrac{d}{dx}\left(\coth^{-1}x\right) = \dfrac{1}{1-x^2}; \; |x| > 1$

Key Idea 6.6.4   Integrals Involving Inverse Hyperbolic Functions

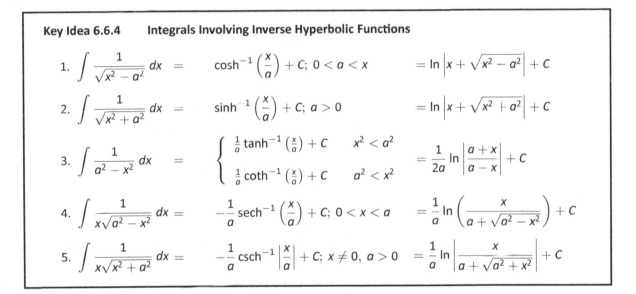

1. $\displaystyle\int \dfrac{1}{\sqrt{x^2-a^2}}\,dx = \cosh^{-1}\left(\dfrac{x}{a}\right) + C; \; 0 < a < x \qquad = \ln\left|x + \sqrt{x^2-a^2}\right| + C$

2. $\displaystyle\int \dfrac{1}{\sqrt{x^2+a^2}}\,dx = \sinh^{-1}\left(\dfrac{x}{a}\right) + C; \; a > 0 \qquad = \ln\left|x + \sqrt{x^2+a^2}\right| + C$

3. $\displaystyle\int \dfrac{1}{a^2-x^2}\,dx = \begin{cases} \frac{1}{a}\tanh^{-1}\left(\frac{x}{a}\right) + C & x^2 < a^2 \\ \frac{1}{a}\coth^{-1}\left(\frac{x}{a}\right) + C & a^2 < x^2 \end{cases} = \dfrac{1}{2a}\ln\left|\dfrac{a+x}{a-x}\right| + C$

4. $\displaystyle\int \dfrac{1}{x\sqrt{a^2-x^2}}\,dx = -\dfrac{1}{a}\,\text{sech}^{-1}\left(\dfrac{x}{a}\right) + C; \; 0 < x < a \qquad = \dfrac{1}{a}\ln\left(\dfrac{x}{a+\sqrt{a^2-x^2}}\right) + C$

5. $\displaystyle\int \dfrac{1}{x\sqrt{x^2+a^2}}\,dx = -\dfrac{1}{a}\,\text{csch}^{-1}\left|\dfrac{x}{a}\right| + C; \; x \neq 0, \, a > 0 \qquad = \dfrac{1}{a}\ln\left|\dfrac{x}{a+\sqrt{a^2+x^2}}\right| + C$

We practice using the derivative and integral formulas in the following example.

Notes:

Example 6.6.3 **Derivatives and integrals involving inverse hyperbolic functions**

Evaluate the following.

1. $\dfrac{d}{dx}\left[\cosh^{-1}\left(\dfrac{3x-2}{5}\right)\right]$

2. $\displaystyle\int \dfrac{1}{x^2-1}\,dx$

3. $\displaystyle\int \dfrac{1}{\sqrt{9x^2+10}}\,dx$

SOLUTION

1. Applying Key Idea 6.6.3 with the Chain Rule gives:

$$\frac{d}{dx}\left[\cosh^{-1}\left(\frac{3x-2}{5}\right)\right] = \frac{1}{\sqrt{\left(\frac{3x-2}{5}\right)^2-1}}\cdot\frac{3}{5}.$$

2. Multiplying the numerator and denominator by (-1) gives: $\displaystyle\int \dfrac{1}{x^2-1}\,dx =$ $\displaystyle\int \dfrac{-1}{1-x^2}\,dx$. The second integral can be solved with a direct application of item #3 from Key Idea 6.6.4, with $a=1$. Thus

$$\int \frac{1}{x^2-1}\,dx = -\int \frac{1}{1-x^2}\,dx$$

$$= \begin{cases} -\tanh^{-1}(x)+C & x^2<1 \\[2mm] -\coth^{-1}(x)+C & 1<x^2 \end{cases}$$

$$= -\frac{1}{2}\ln\left|\frac{x+1}{x-1}\right|+C$$

$$= \frac{1}{2}\ln\left|\frac{x-1}{x+1}\right|+C. \tag{6.4}$$

We should note that this exact problem was solved at the beginning of Section 6.5. In that example the answer was given as $\frac{1}{2}\ln|x-1|-\frac{1}{2}\ln|x+1|+C$. Note that this is equivalent to the answer given in Equation 6.4, as $\ln(a/b) = \ln a - \ln b$.

3. This requires a substitution, then item #2 of Key Idea 6.6.4 can be applied. Let $u=3x$, hence $du=3dx$. We have

$$\int \frac{1}{\sqrt{9x^2+10}}\,dx = \frac{1}{3}\int \frac{1}{\sqrt{u^2+10}}\,du.$$

Notes:

Note $a^2 = 10$, hence $a = \sqrt{10}$. Now apply the integral rule.

$$= \frac{1}{3} \sinh^{-1}\left(\frac{3x}{\sqrt{10}}\right) + C$$
$$= \frac{1}{3} \ln\left|3x + \sqrt{9x^2 + 10}\right| + C.$$

This section covers a lot of ground. New functions were introduced, along with some of their fundamental identities, their derivatives and antiderivatives, their inverses, and the derivatives and antiderivatives of these inverses. Four Key Ideas were presented, each including quite a bit of information.

Do not view this section as containing a source of information to be memorized, but rather as a reference for future problem solving. Key Idea 6.6.4 contains perhaps the most useful information. Know the integration forms it helps evaluate and understand how to use the inverse hyperbolic answer and the logarithmic answer.

The next section takes a brief break from demonstrating new integration techniques. It instead demonstrates a technique of evaluating limits that return indeterminate forms. This technique will be useful in Section 6.8, where limits will arise in the evaluation of certain definite integrals.

Notes:

Exercises 6.6

Terms and Concepts

1. In Key Idea 6.6.1, the equation $\int \tanh x \, dx = \ln(\cosh x) + C$ is given. Why is "$\ln | \cosh x |$" not used – i.e., why are absolute values not necessary?

2. The hyperbolic functions are used to define points on the right hand portion of the hyperbola $x^2 - y^2 = 1$, as shown in Figure 6.6.1. How can we use the hyperbolic functions to define points on the left hand portion of the hyperbola?

Problems

In Exercises 3 – 10, verify the given identity using Definition 6.6.1, as done in Example 6.6.1.

3. $\coth^2 x - \text{csch}^2 x = 1$

4. $\cosh 2x = \cosh^2 x + \sinh^2 x$

5. $\cosh^2 x = \dfrac{\cosh 2x + 1}{2}$

6. $\sinh^2 x = \dfrac{\cosh 2x - 1}{2}$

7. $\dfrac{d}{dx} [\text{sech} \, x] = -\text{sech} \, x \tanh x$

8. $\dfrac{d}{dx} [\coth x] = -\text{csch}^2 x$

9. $\int \tanh x \, dx = \ln(\cosh x) + C$

10. $\int \coth x \, dx = \ln | \sinh x | + C$

In Exercises 11 – 22, find the derivative of the given function.

11. $f(x) = \sinh 2x$

12. $f(x) = \cosh^2 x$

13. $f(x) = \tanh(x^2)$

14. $f(x) = \ln(\sinh x)$

15. $f(x) = \sinh x \cosh x$

16. $f(x) = x \sinh x - \cosh x$

17. $f(x) = \text{sech}^{-1}(x^2)$

18. $f(x) = \sinh^{-1}(3x)$

19. $f(x) = \cosh^{-1}(2x^2)$

20. $f(x) = \tanh^{-1}(x + 5)$

21. $f(x) = \tanh^{-1}(\cos x)$

22. $f(x) = \cosh^{-1}(\sec x)$

In Exercises 23 – 28, find the equation of the line tangent to the function at the given x-value.

23. $f(x) = \sinh x$ at $x = 0$

24. $f(x) = \cosh x$ at $x = \ln 2$

25. $f(x) = \tanh x$ at $x = -\ln 3$

26. $f(x) = \text{sech}^2 x$ at $x = \ln 3$

27. $f(x) = \sinh^{-1} x$ at $x = 0$

28. $f(x) = \cosh^{-1} x$ at $x = \sqrt{2}$

In Exercises 29 – 44, evaluate the given indefinite integral.

29. $\int \tanh(2x) \, dx$

30. $\int \cosh(3x - 7) \, dx$

31. $\int \sinh x \cosh x \, dx$

32. $\int x \cosh x \, dx$

33. $\int x \sinh x \, dx$

34. $\int \dfrac{1}{\sqrt{x^2 + 1}} \, dx$

35. $\int \dfrac{1}{\sqrt{x^2 - 9}} \, dx$

36. $\int \dfrac{1}{9 - x^2} \, dx$

37. $\int \dfrac{2x}{\sqrt{x^4 - 4}} \, dx$

38. $\int \dfrac{\sqrt{x}}{\sqrt{1 + x^3}} \, dx$

39. $\int \dfrac{1}{x^4 - 16} \, dx$

40. $\int \dfrac{1}{x^2 + x} \, dx$

41. $\displaystyle\int \frac{e^x}{e^{2x}+1}\, dx$

42. $\displaystyle\int \sinh^{-1} x\, dx$

43. $\displaystyle\int \tanh^{-1} x\, dx$

44. $\displaystyle\int \operatorname{sech} x\, dx$ (Hint: multiply by $\frac{\cosh x}{\cosh x}$; set $u = \sinh x$.)

In Exercises 45 – 48, evaluate the given definite integral.

45. $\displaystyle\int_{-1}^{1} \sinh x\, dx$

46. $\displaystyle\int_{-\ln 2}^{\ln 2} \cosh x\, dx$

47. $\displaystyle\int_{0}^{1} \tanh^{-1} x\, dx$

48. $\displaystyle\int_{0}^{2} \frac{1}{\sqrt{x^2+1}}\, dx$

6.7   L'Hôpital's Rule

While this chapter is devoted to learning techniques of integration, this section is not about integration. Rather, it is concerned with a technique of evaluating certain limits that will be useful in the following section, where integration is once more discussed.

Our treatment of limits exposed us to the notion of "0/0", an indeterminate form. If $\lim_{x \to c} f(x) = 0$ and $\lim_{x \to c} g(x) = 0$, we do not conclude that $\lim_{x \to c} f(x)/g(x)$ is 0/0; rather, we use 0/0 as notation to describe the fact that both the numerator and denominator approach 0. The expression 0/0 has no numeric value; other work must be done to evaluate the limit.

Other indeterminate forms exist; they are: ∞/∞, $0 \cdot \infty$, $\infty - \infty$, 0^0, 1^∞ and ∞^0. Just as "0/0" does not mean "divide 0 by 0," the expression "∞/∞" does not mean "divide infinity by infinity." Instead, it means "a quantity is growing without bound and is being divided by another quantity that is growing without bound." We cannot determine from such a statement what value, if any, results in the limit. Likewise, "$0 \cdot \infty$" does not mean "multiply zero by infinity." Instead, it means "one quantity is shrinking to zero, and is being multiplied by a quantity that is growing without bound." We cannot determine from such a description what the result of such a limit will be.

This section introduces l'Hôpital's Rule, a method of resolving limits that produce the indeterminate forms 0/0 and ∞/∞. We'll also show how algebraic manipulation can be used to convert other indeterminate expressions into one of these two forms so that our new rule can be applied.

Theorem 6.7.1   L'Hôpital's Rule, Part 1

Let $\lim_{x \to c} f(x) = 0$ and $\lim_{x \to c} g(x) = 0$, where f and g are differentiable functions on an open interval I containing c, and $g'(x) \neq 0$ on I except possibly at c. Then

$$\lim_{x \to c} \frac{f(x)}{g(x)} = \lim_{x \to c} \frac{f'(x)}{g'(x)}.$$

We demonstrate the use of l'Hôpital's Rule in the following examples; we will often use "LHR" as an abbreviation of "l'Hôpital's Rule."

Notes:

Example 6.7.1   Using l'Hôpital's Rule

Evaluate the following limits, using l'Hôpital's Rule as needed.

1. $\lim_{x\to 0} \dfrac{\sin x}{x}$

2. $\lim_{x\to 1} \dfrac{\sqrt{x+3}-2}{1-x}$

3. $\lim_{x\to 0} \dfrac{x^2}{1-\cos x}$

4. $\lim_{x\to 2} \dfrac{x^2+x-6}{x^2-3x+2}$

Solution

1. We proved this limit is 1 in Example 1.3.4 using the Squeeze Theorem. Here we use l'Hôpital's Rule to show its power.

$$\lim_{x\to 0} \frac{\sin x}{x} \overset{\text{by LHR}}{=} \lim_{x\to 0} \frac{\cos x}{1} = 1.$$

2.
$$\lim_{x\to 1} \frac{\sqrt{x+3}-2}{1-x} \overset{\text{by LHR}}{=} \lim_{x\to 1} \frac{\frac{1}{2}(x+3)^{-1/2}}{-1} = -\frac{1}{4}.$$

3.
$$\lim_{x\to 0} \frac{x^2}{1-\cos x} \overset{\text{by LHR}}{=} \lim_{x\to 0} \frac{2x}{\sin x}.$$

This latter limit also evaluates to the 0/0 indeterminate form. To evaluate it, we apply l'Hôpital's Rule again.

$$\lim_{x\to 0} \frac{2x}{\sin x} \overset{\text{by LHR}}{=} \frac{2}{\cos x} = 2.$$

Thus $\lim_{x\to 0} \dfrac{x^2}{1-\cos x} = 2.$

4. We already know how to evaluate this limit; first factor the numerator and denominator. We then have:

$$\lim_{x\to 2} \frac{x^2+x-6}{x^2-3x+2} = \lim_{x\to 2} \frac{(x-2)(x+3)}{(x-2)(x-1)} = \lim_{x\to 2} \frac{x+3}{x-1} = 5.$$

We now show how to solve this using l'Hôpital's Rule.

$$\lim_{x\to 2} \frac{x^2+x-6}{x^2-3x+2} \overset{\text{by LHR}}{=} \lim_{x\to 2} \frac{2x+1}{2x-3} = 5.$$

Note that at each step where l'Hôpital's Rule was applied, it was *needed*: the initial limit returned the indeterminate form of "0/0." If the initial limit returns, for example, 1/2, then l'Hôpital's Rule does not apply.

Notes:

The following theorem extends our initial version of l'Hôpital's Rule in two ways. It allows the technique to be applied to the indeterminate form ∞/∞ and to limits where x approaches $\pm\infty$.

Theorem 6.7.2   L'Hôpital's Rule, Part 2

1. Let $\lim\limits_{x\to a} f(x) = \pm\infty$ and $\lim\limits_{x\to a} g(x) = \pm\infty$, where f and g are differentiable on an open interval I containing a. Then

$$\lim_{x\to a} \frac{f(x)}{g(x)} = \lim_{x\to a} \frac{f'(x)}{g'(x)}.$$

2. Let f and g be differentiable functions on the open interval (a, ∞) for some value a, where $g'(x) \neq 0$ on (a, ∞) and $\lim\limits_{x\to\infty} f(x)/g(x)$ returns either "0/0" or "∞/∞". Then

$$\lim_{x\to\infty} \frac{f(x)}{g(x)} = \lim_{x\to\infty} \frac{f'(x)}{g'(x)}.$$

A similar statement can be made for limits where x approaches $-\infty$.

Example 6.7.2   Using l'Hôpital's Rule with limits involving ∞

Evaluate the following limits.

1. $\lim\limits_{x\to\infty} \dfrac{3x^2 - 100x + 2}{4x^2 + 5x - 1000}$

2. $\lim\limits_{x\to\infty} \dfrac{e^x}{x^3}$.

SOLUTION

1. We can evaluate this limit already using Theorem 1.6.1; the answer is 3/4. We apply l'Hôpital's Rule to demonstrate its applicability.

$$\lim_{x\to\infty} \frac{3x^2 - 100x + 2}{4x^2 + 5x - 1000} \overset{\text{by LHR}}{=} \lim_{x\to\infty} \frac{6x - 100}{8x + 5} \overset{\text{by LHR}}{=} \lim_{x\to\infty} \frac{6}{8} = \frac{3}{4}.$$

2. $\lim\limits_{x\to\infty} \dfrac{e^x}{x^3} \overset{\text{by LHR}}{=} \lim\limits_{x\to\infty} \dfrac{e^x}{3x^2} \overset{\text{by LHR}}{=} \lim\limits_{x\to\infty} \dfrac{e^x}{6x} \overset{\text{by LHR}}{=} \lim\limits_{x\to\infty} \dfrac{e^x}{6} = \infty.$

Recall that this means that the limit does not exist; as x approaches ∞, the expression e^x/x^3 grows without bound. We can infer from this that e^x grows "faster" than x^3; as x gets large, e^x is far larger than x^3. (This

---

Notes:

has important implications in computing when considering efficiency of algorithms.)

Indeterminate Forms $0 \cdot \infty$ and $\infty - \infty$

L'Hôpital's Rule can only be applied to ratios of functions. When faced with an indeterminate form such as $0 \cdot \infty$ or $\infty - \infty$, we can sometimes apply algebra to rewrite the limit so that l'Hôpital's Rule can be applied. We demonstrate the general idea in the next example.

Example 6.7.3   **Applying l'Hôpital's Rule to other indeterminate forms**
Evaluate the following limits.

1. $\displaystyle\lim_{x \to 0^+} x \cdot e^{1/x}$

2. $\displaystyle\lim_{x \to 0^-} x \cdot e^{1/x}$

3. $\displaystyle\lim_{x \to \infty} \ln(x + 1) - \ln x$

4. $\displaystyle\lim_{x \to \infty} x^2 - e^x$

SOLUTION

1. As $x \to 0^+$, $x \to 0$ and $e^{1/x} \to \infty$. Thus we have the indeterminate form $0 \cdot \infty$. We rewrite the expression $x \cdot e^{1/x}$ as $\dfrac{e^{1/x}}{1/x}$; now, as $x \to 0^+$, we get the indeterminate form ∞/∞ to which l'Hôpital's Rule can be applied.

$$\lim_{x \to 0^+} x \cdot e^{1/x} = \lim_{x \to 0^+} \frac{e^{1/x}}{1/x} \stackrel{\text{by LHR}}{=} \lim_{x \to 0^+} \frac{(-1/x^2)e^{1/x}}{-1/x^2} = \lim_{x \to 0^+} e^{1/x} = \infty.$$

Interpretation: $e^{1/x}$ grows "faster" than x shrinks to zero, meaning their product grows without bound.

2. As $x \to 0^-$, $x \to 0$ and $e^{1/x} \to e^{-\infty} \to 0$. The the limit evaluates to $0 \cdot 0$ which is not an indeterminate form. We conclude then that

$$\lim_{x \to 0^-} x \cdot e^{1/x} = 0.$$

3. This limit initially evaluates to the indeterminate form $\infty - \infty$. By applying a logarithmic rule, we can rewrite the limit as

$$\lim_{x \to \infty} \ln(x + 1) - \ln x = \lim_{x \to \infty} \ln\left(\frac{x + 1}{x}\right).$$

As $x \to \infty$, the argument of the ln term approaches ∞/∞, to which we can apply l'Hôpital's Rule.

$$\lim_{x \to \infty} \frac{x + 1}{x} \stackrel{\text{by LHR}}{=} \frac{1}{1} = 1.$$

Notes:

Since $x \to \infty$ implies $\dfrac{x+1}{x} \to 1$, it follows that

$$x \to \infty \quad \text{implies} \quad \ln\left(\frac{x+1}{x}\right) \to \ln 1 = 0.$$

Thus

$$\lim_{x\to\infty} \ln(x+1) - \ln x = \lim_{x\to\infty} \ln\left(\frac{x+1}{x}\right) = 0.$$

Interpretation: since this limit evaluates to 0, it means that for large x, there is essentially no difference between $\ln(x+1)$ and $\ln x$; their difference is essentially 0.

4. The limit $\lim\limits_{x\to\infty} x^2 - e^x$ initially returns the indeterminate form $\infty - \infty$. We can rewrite the expression by factoring out x^2; $x^2 - e^x = x^2\left(1 - \dfrac{e^x}{x^2}\right)$.

   We need to evaluate how e^x/x^2 behaves as $x \to \infty$:

   $$\lim_{x\to\infty} \frac{e^x}{x^2} \overset{\text{by LHR}}{=} \lim_{x\to\infty} \frac{e^x}{2x} \overset{\text{by LHR}}{=} \lim_{x\to\infty} \frac{e^x}{2} = \infty.$$

   Thus $\lim_{x\to\infty} x^2(1 - e^x/x^2)$ evaluates to $\infty \cdot (-\infty)$, which is not an indeterminate form; rather, $\infty \cdot (-\infty)$ evaluates to $-\infty$. We conclude that $\lim\limits_{x\to\infty} x^2 - e^x = -\infty$.

   Interpretation: as x gets large, the difference between x^2 and e^x grows very large.

Indeterminate Forms 0^0, 1^∞ and ∞^0

When faced with an indeterminate form that involves a power, it often helps to employ the natural logarithmic function. The following Key Idea expresses the concept, which is followed by an example that demonstrates its use.

Key Idea 6.7.1   **Evaluating Limits Involving Indeterminate Forms 0^0, 1^∞ and ∞^0**

If $\lim\limits_{x\to c} \ln\big(f(x)\big) = L$, then $\lim\limits_{x\to c} f(x) = \lim\limits_{x\to c} e^{\ln(f(x))} = e^L$.

Notes:

Example 6.7.4   **Using l'Hôpital's Rule with indeterminate forms involving exponents**

Evaluate the following limits.

1. $\displaystyle\lim_{x\to\infty}\left(1+\frac{1}{x}\right)^x$ 2. $\displaystyle\lim_{x\to 0^+} x^x$.

SOLUTION

1. This is equivalent to a special limit given in Theorem 1.3.5; these limits have important applications within mathematics and finance. Note that the exponent approaches ∞ while the base approaches 1, leading to the indeterminate form 1^∞. Let $f(x) = (1 + 1/x)^x$; the problem asks to evaluate $\displaystyle\lim_{x\to\infty} f(x)$. Let's first evaluate $\displaystyle\lim_{x\to\infty} \ln\left(f(x)\right)$.

$$\lim_{x\to\infty} \ln\left(f(x)\right) = \lim_{x\to\infty} \ln\left(1+\frac{1}{x}\right)^x$$
$$= \lim_{x\to\infty} x\ln\left(1+\frac{1}{x}\right)$$
$$= \lim_{x\to\infty} \frac{\ln\left(1+\frac{1}{x}\right)}{1/x}$$

This produces the indeterminate form 0/0, so we apply l'Hôpital's Rule.

$$= \lim_{x\to\infty} \frac{\frac{1}{1+1/x}\cdot(-1/x^2)}{(-1/x^2)}$$
$$= \lim_{x\to\infty} \frac{1}{1+1/x}$$
$$= 1.$$

Thus $\displaystyle\lim_{x\to\infty} \ln\left(f(x)\right) = 1$. We return to the original limit and apply Key Idea 6.7.1.

$$\lim_{x\to\infty}\left(1+\frac{1}{x}\right)^x = \lim_{x\to\infty} f(x) = \lim_{x\to\infty} e^{\ln(f(x))} = e^1 = e.$$

2. This limit leads to the indeterminate form 0^0. Let $f(x) = x^x$ and consider

---

Notes:

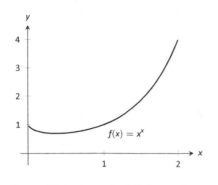

Figure 6.7.1: A graph of $f(x) = x^x$ supporting the fact that as $x \to 0^+$, $f(x) \to 1$.

first $\displaystyle\lim_{x \to 0^+} \ln\left(f(x)\right)$.

$$\lim_{x \to 0^+} \ln\left(f(x)\right) = \lim_{x \to 0^+} \ln\left(x^x\right)$$
$$= \lim_{x \to 0^+} x \ln x$$
$$= \lim_{x \to 0^+} \frac{\ln x}{1/x}.$$

This produces the indeterminate form $-\infty/\infty$ so we apply l'Hôpital's Rule.

$$= \lim_{x \to 0^+} \frac{1/x}{-1/x^2}$$
$$= \lim_{x \to 0^+} -x$$
$$= 0.$$

Thus $\displaystyle\lim_{x \to 0^+} \ln\left(f(x)\right) = 0$. We return to the original limit and apply Key Idea 6.7.1.

$$\lim_{x \to 0^+} x^x = \lim_{x \to 0^+} f(x) = \lim_{x \to 0^+} e^{\ln(f(x))} = e^0 = 1.$$

This result is supported by the graph of $f(x) = x^x$ given in Figure 6.7.1.

Our brief revisit of limits will be rewarded in the next section where we consider *improper integration*. So far, we have only considered definite integrals where the bounds are finite numbers, such as $\int_0^1 f(x)\, dx$. Improper integration considers integrals where one, or both, of the bounds are "infinity." Such integrals have many uses and applications, in addition to generating ideas that are enlightening.

Notes:

Exercises 6.7

Terms and Concepts

1. List the different indeterminate forms described in this section.

2. T/F: l'Hôpital's Rule provides a faster method of computing derivatives.

3. T/F: l'Hôpital's Rule states that $\dfrac{d}{dx}\left[\dfrac{f(x)}{g(x)}\right] = \dfrac{f'(x)}{g'(x)}$.

4. Explain what the indeterminate form "1^{∞}" means.

5. Fill in the blanks:

   The Quotient Rule is applied to $\dfrac{f(x)}{g(x)}$ when taking _____;

   l'Hôpital's Rule is applied to $\dfrac{f(x)}{g(x)}$ when taking certain _____.

6. Create (but do not evaluate!) a limit that returns "∞^{0}".

7. Create a function $f(x)$ such that $\lim\limits_{x\to 1} f(x)$ returns "0^{0}".

8. Create a function $f(x)$ such that $\lim\limits_{x\to\infty} f(x)$ returns "$0 \cdot \infty$".

Problems

In Exercises 9 – 54, evaluate the given limit.

9. $\lim\limits_{x\to 1} \dfrac{x^2 + x - 2}{x - 1}$

10. $\lim\limits_{x\to 2} \dfrac{x^2 + x - 6}{x^2 - 7x + 10}$

11. $\lim\limits_{x\to\pi} \dfrac{\sin x}{x - \pi}$

12. $\lim\limits_{x\to\pi/4} \dfrac{\sin x - \cos x}{\cos(2x)}$

13. $\lim\limits_{x\to 0} \dfrac{\sin(5x)}{x}$

14. $\lim\limits_{x\to 0} \dfrac{\sin(2x)}{x + 2}$

15. $\lim\limits_{x\to 0} \dfrac{\sin(2x)}{\sin(3x)}$

16. $\lim\limits_{x\to 0} \dfrac{\sin(ax)}{\sin(bx)}$

17. $\lim\limits_{x\to 0^+} \dfrac{e^x - 1}{x^2}$

18. $\lim\limits_{x\to 0^+} \dfrac{e^x - x - 1}{x^2}$

19. $\lim\limits_{x\to 0^+} \dfrac{x - \sin x}{x^3 - x^2}$

20. $\lim\limits_{x\to\infty} \dfrac{x^4}{e^x}$

21. $\lim\limits_{x\to\infty} \dfrac{\sqrt{x}}{e^x}$

22. $\lim\limits_{x\to\infty} \dfrac{1}{x^2} e^x$

23. $\lim\limits_{x\to\infty} \dfrac{e^x}{\sqrt{x}}$

24. $\lim\limits_{x\to\infty} \dfrac{e^x}{2^x}$

25. $\lim\limits_{x\to\infty} \dfrac{e^x}{3^x}$

26. $\lim\limits_{x\to 3} \dfrac{x^3 - 5x^2 + 3x + 9}{x^3 - 7x^2 + 15x - 9}$

27. $\lim\limits_{x\to -2} \dfrac{x^3 + 4x^2 + 4x}{x^3 + 7x^2 + 16x + 12}$

28. $\lim\limits_{x\to\infty} \dfrac{\ln x}{x}$

29. $\lim\limits_{x\to\infty} \dfrac{\ln(x^2)}{x}$

30. $\lim\limits_{x\to\infty} \dfrac{\left(\ln x\right)^2}{x}$

31. $\lim\limits_{x\to 0^+} x \cdot \ln x$

32. $\lim\limits_{x\to 0^+} \sqrt{x} \cdot \ln x$

33. $\lim\limits_{x\to 0^+} xe^{1/x}$

34. $\lim\limits_{x\to\infty} x^3 - x^2$

35. $\lim\limits_{x\to\infty} \sqrt{x} - \ln x$

36. $\lim\limits_{x\to -\infty} xe^x$

37. $\lim\limits_{x\to 0^+} \dfrac{1}{x^2} e^{-1/x}$

38. $\lim\limits_{x\to 0^+} (1 + x)^{1/x}$

39. $\lim\limits_{x \to 0^+} (2x)^x$

40. $\lim\limits_{x \to 0^+} (2/x)^x$

41. $\lim\limits_{x \to 0^+} (\sin x)^x$ Hint: use the Squeeze Theorem.

42. $\lim\limits_{x \to 1^+} (1-x)^{1-x}$

43. $\lim\limits_{x \to \infty} (x)^{1/x}$

44. $\lim\limits_{x \to \infty} (1/x)^x$

45. $\lim\limits_{x \to 1^+} (\ln x)^{1-x}$

46. $\lim\limits_{x \to \infty} (1+x)^{1/x}$

47. $\lim\limits_{x \to \infty} (1+x^2)^{1/x}$

48. $\lim\limits_{x \to \pi/2} \tan x \cos x$

49. $\lim\limits_{x \to \pi/2} \tan x \sin(2x)$

50. $\lim\limits_{x \to 1^+} \dfrac{1}{\ln x} - \dfrac{1}{x-1}$

51. $\lim\limits_{x \to 3^+} \dfrac{5}{x^2-9} - \dfrac{x}{x-3}$

52. $\lim\limits_{x \to \infty} x \tan(1/x)$

53. $\lim\limits_{x \to \infty} \dfrac{(\ln x)^3}{x}$

54. $\lim\limits_{x \to 1} \dfrac{x^2+x-2}{\ln x}$

6.8 Improper Integration

We begin this section by considering the following definite integrals:

- $\displaystyle\int_0^{100} \frac{1}{1+x^2}\, dx \approx 1.5608,$

- $\displaystyle\int_0^{1000} \frac{1}{1+x^2}\, dx \approx 1.5698,$

- $\displaystyle\int_0^{10,000} \frac{1}{1+x^2}\, dx \approx 1.5707.$

Notice how the integrand is $1/(1+x^2)$ in each integral (which is sketched in Figure 6.8.1). As the upper bound gets larger, one would expect the "area under the curve" would also grow. While the definite integrals do increase in value as the upper bound grows, they are not increasing by much. In fact, consider:

$$\int_0^b \frac{1}{1+x^2}\, dx = \tan^{-1} x\Big|_0^b = \tan^{-1} b - \tan^{-1} 0 = \tan^{-1} b.$$

As $b \rightarrow \infty$, $\tan^{-1} b \rightarrow \pi/2$. Therefore it seems that as the upper bound b grows, the value of the definite integral $\displaystyle\int_0^b \frac{1}{1+x^2}\, dx$ approaches $\pi/2 \approx 1.5708$. This should strike the reader as being a bit amazing: even though the curve extends "to infinity," it has a finite amount of area underneath it.

When we defined the definite integral $\displaystyle\int_a^b f(x)\, dx$, we made two stipulations:

1. The interval over which we integrated, $[a, b]$, was a finite interval, and

2. The function $f(x)$ was continuous on $[a, b]$ (ensuring that the range of f was finite).

In this section we consider integrals where one or both of the above conditions do not hold. Such integrals are called **improper integrals.**

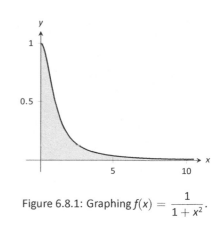

Figure 6.8.1: Graphing $f(x) = \dfrac{1}{1+x^2}$.

Notes:

Improper Integrals with Infinite Bounds

Definition 6.8.1   **Improper Integrals with Infinite Bounds; Converge, Diverge**

1. Let f be a continuous function on $[a, \infty)$. Define

$$\int_a^\infty f(x)\, dx \quad \text{to be} \quad \lim_{b \to \infty} \int_a^b f(x)\, dx.$$

2. Let f be a continuous function on $(-\infty, b]$. Define

$$\int_{-\infty}^b f(x)\, dx \quad \text{to be} \quad \lim_{a \to -\infty} \int_a^b f(x)\, dx.$$

3. Let f be a continuous function on $(-\infty, \infty)$. Let c be any real number; define

$$\int_{-\infty}^\infty f(x)\, dx \quad \text{to be} \quad \lim_{a \to -\infty} \int_a^c f(x)\, dx + \lim_{b \to \infty} \int_c^b f(x)\, dx.$$

An improper integral is said to **converge** if its corresponding limit exists; otherwise, it **diverges**. The improper integral in part 3 converges if and only if both of its limits exist.

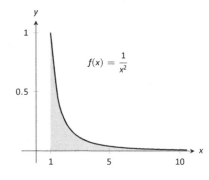

Figure 6.8.2: A graph of $f(x) = \frac{1}{x^2}$ in Example 6.8.1.

Example 6.8.1   **Evaluating improper integrals**
Evaluate the following improper integrals.

1. $\int_1^\infty \dfrac{1}{x^2}\, dx$

2. $\int_1^\infty \dfrac{1}{x}\, dx$

3. $\int_{-\infty}^0 e^x\, dx$

4. $\int_{-\infty}^\infty \dfrac{1}{1+x^2}\, dx$

SOLUTION

1.
$$\int_1^\infty \frac{1}{x^2}\, dx = \lim_{b \to \infty} \int_1^b \frac{1}{x^2}\, dx = \lim_{b \to \infty} \left. \frac{-1}{x} \right|_1^b$$
$$= \lim_{b \to \infty} \frac{-1}{b} + 1$$
$$= 1.$$

A graph of the area defined by this integral is given in Figure 6.8.2.

Notes:

2.

$$\int_1^\infty \frac{1}{x}\,dx = \lim_{b\to\infty} \int_1^b \frac{1}{x}\,dx$$

$$= \lim_{b\to\infty} \ln|x|\Big|_1^b$$

$$= \lim_{b\to\infty} \ln(b)$$

$$= \infty.$$

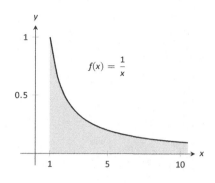

Figure 6.8.3: A graph of $f(x) = \frac{1}{x}$ in Example 6.8.1.

The limit does not exist, hence the improper integral $\int_1^\infty \frac{1}{x}\,dx$ diverges. Compare the graphs in Figures 6.8.2 and 6.8.3; notice how the graph of $f(x) = 1/x$ is noticeably larger. This difference is enough to cause the improper integral to diverge.

3.

$$\int_{-\infty}^0 e^x\,dx = \lim_{a\to-\infty} \int_a^0 e^x\,dx$$

$$= \lim_{a\to-\infty} e^x\Big|_a^0$$

$$= \lim_{a\to-\infty} e^0 - e^a$$

$$= 1.$$

A graph of the area defined by this integral is given in Figure 6.8.4.

4. We will need to break this into two improper integrals and choose a value of c as in part 3 of Definition 6.8.1. Any value of c is fine; we choose $c = 0$.

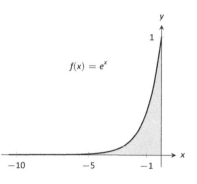

Figure 6.8.4: A graph of $f(x) = e^x$ in Example 6.8.1.

$$\int_{-\infty}^\infty \frac{1}{1+x^2}\,dx = \lim_{a\to-\infty} \int_a^0 \frac{1}{1+x^2}\,dx + \lim_{b\to\infty} \int_0^b \frac{1}{1+x^2}\,dx$$

$$= \lim_{a\to-\infty} \tan^{-1}x\Big|_a^0 + \lim_{b\to\infty} \tan^{-1}x\Big|_0^b$$

$$= \lim_{a\to-\infty} \left(\tan^{-1}0 - \tan^{-1}a\right) + \lim_{b\to\infty} \left(\tan^{-1}b - \tan^{-1}0\right)$$

$$= \left(0 - \frac{-\pi}{2}\right) + \left(\frac{\pi}{2} - 0\right).$$

Each limit exists, hence the original integral converges and has value:

$$= \pi.$$

A graph of the area defined by this integral is given in Figure 6.8.5.

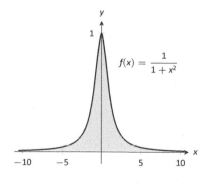

Figure 6.8.5: A graph of $f(x) = \frac{1}{1+x^2}$ in Example 6.8.1.

Notes:

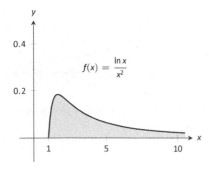

Figure 6.8.6: A graph of $f(x) = \frac{\ln x}{x^2}$ in Example 6.8.2.

The previous section introduced l'Hôpital's Rule, a method of evaluating limits that return indeterminate forms. It is not uncommon for the limits resulting from improper integrals to need this rule as demonstrated next.

Example 6.8.2   Improper integration and l'Hôpital's Rule

Evaluate the improper integral $\int_1^\infty \frac{\ln x}{x^2}\, dx$.

SOLUTION   This integral will require the use of Integration by Parts. Let $u = \ln x$ and $dv = 1/x^2\, dx$. Then

$$\int_1^\infty \frac{\ln x}{x^2}\, dx = \lim_{b\to\infty} \int_1^b \frac{\ln x}{x^2}\, dx$$

$$= \lim_{b\to\infty} \left(-\frac{\ln x}{x}\Big|_1^b + \int_1^b \frac{1}{x^2}\, dx \right)$$

$$= \lim_{b\to\infty} \left(-\frac{\ln x}{x} - \frac{1}{x} \right)\Big|_1^b$$

$$= \lim_{b\to\infty} \left(-\frac{\ln b}{b} - \frac{1}{b} - (-\ln 1 - 1) \right).$$

The $1/b$ and $\ln 1$ terms go to 0, leaving $\displaystyle\lim_{b\to\infty} -\frac{\ln b}{b} + 1$. We need to evaluate $\displaystyle\lim_{b\to\infty} \frac{\ln b}{b}$ with l'Hôpital's Rule. We have:

$$\lim_{b\to\infty} \frac{\ln b}{b} \overset{\text{by LHR}}{=} \lim_{b\to\infty} \frac{1/b}{1}$$

$$= 0.$$

Thus the improper integral evaluates as:

$$\int_1^\infty \frac{\ln x}{x^2}\, dx = 1.$$

Improper Integrals with Infinite Range

We have just considered definite integrals where the interval of integration was infinite. We now consider another type of improper integration, where the range of the integrand is infinite.

Notes:

Let $f(x)$ be a continuous function on $[a, b]$ except at c, $a \leq c \leq b$, where $x = c$ is a vertical asymptote of f. Define

$$\int_a^b f(x)\, dx = \lim_{t \to c^-} \int_a^t f(x)\, dx + \lim_{t \to c^+} \int_t^b f(x)\, dx.$$

Note: In Definition 6.8.2, c can be one of the endpoints (a or b). In that case, there is only one limit to consider as part of the definition.

Example 6.8.3   Improper integration of functions with infinite range

Evaluate the following improper integrals:

1. $\displaystyle\int_0^1 \frac{1}{\sqrt{x}}\, dx$ 2. $\displaystyle\int_{-1}^1 \frac{1}{x^2}\, dx$

SOLUTION

1. A graph of $f(x) = 1/\sqrt{x}$ is given in Figure 6.8.7. Notice that f has a vertical asymptote at $x = 0$; in some sense, we are trying to compute the area of a region that has no "top." Could this have a finite value?

$$\int_0^1 \frac{1}{\sqrt{x}}\, dx = \lim_{a \to 0^+} \int_a^1 \frac{1}{\sqrt{x}}\, dx$$
$$= \lim_{a \to 0^+} 2\sqrt{x}\Big|_a^1$$
$$= \lim_{a \to 0^+} 2\left(\sqrt{1} - \sqrt{a}\right)$$
$$= 2.$$

It turns out that the region does have a finite area even though it has no upper bound (strange things can occur in mathematics when considering the infinite).

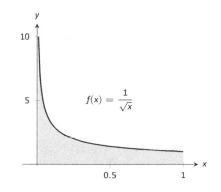

Figure 6.8.7: A graph of $f(x) = \frac{1}{\sqrt{x}}$ in Example 6.8.3.

2. The function $f(x) = 1/x^2$ has a vertical asymptote at $x = 0$, as shown in Figure 6.8.8, so this integral is an improper integral. Let's eschew using limits for a moment and proceed without recognizing the improper nature of the integral. This leads to:

$$\int_{-1}^1 \frac{1}{x^2}\, dx = -\frac{1}{x}\Big|_{-1}^1$$
$$= -1 - (1)$$
$$= -2. \ (!)$$

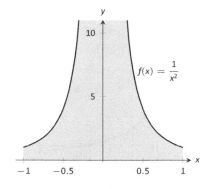

Figure 6.8.8: A graph of $f(x) = \frac{1}{x^2}$ in Example 6.8.3.

Notes:

Clearly the area in question is above the x-axis, yet the area is supposedly negative! Why does our answer not match our intuition? To answer this, evaluate the integral using Definition 6.8.2.

$$\int_{-1}^{1} \frac{1}{x^2}\, dx = \lim_{t\to 0^-} \int_{-1}^{t} \frac{1}{x^2}\, dx + \lim_{t\to 0^+} \int_{t}^{1} \frac{1}{x^2}\, dx$$

$$= \lim_{t\to 0^-} \left. -\frac{1}{x}\right|_{-1}^{t} + \lim_{t\to 0^+} \left. -\frac{1}{x}\right|_{t}^{1}$$

$$= \lim_{t\to 0^-} -\frac{1}{t} - 1 + \lim_{t\to 0^+} -1 + \frac{1}{t}$$

$$\Rightarrow \Big(\infty - 1\Big) + \Big(-1 + \infty\Big).$$

Neither limit converges hence the original improper integral diverges. The nonsensical answer we obtained by ignoring the improper nature of the integral is just that: nonsensical.

Understanding Convergence and Divergence

Oftentimes we are interested in knowing simply whether or not an improper integral converges, and not necessarily the value of a convergent integral. We provide here several tools that help determine the convergence or divergence of improper integrals without integrating.

Our first tool is to understand the behavior of functions of the form $\dfrac{1}{x^p}$.

Example 6.8.4   Improper integration of $1/x^p$

Determine the values of p for which $\displaystyle\int_{1}^{\infty} \frac{1}{x^p}\, dx$ converges.

SOLUTION   We begin by integrating and then evaluating the limit.

$$\int_{1}^{\infty} \frac{1}{x^p}\, dx = \lim_{b\to\infty} \int_{1}^{b} \frac{1}{x^p}\, dx$$

$$= \lim_{b\to\infty} \int_{1}^{b} x^{-p}\, dx \qquad \text{(assume } p \neq 1\text{)}$$

$$= \lim_{b\to\infty} \frac{1}{-p+1} x^{-p+1} \Big|_{1}^{b}$$

$$= \lim_{b\to\infty} \frac{1}{1-p}\left(b^{1-p} - 1^{1-p}\right).$$

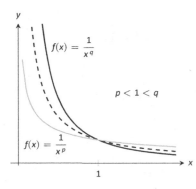

Figure 6.8.9: Plotting functions of the form $1/x^p$ in Example 6.8.4.

When does this limit converge – i.e., when is this limit *not* ∞? This limit converges precisely when the power of b is less than 0: when $1 - p < 0 \Rightarrow 1 < p$.

Notes:

Our analysis shows that if $p > 1$, then $\int_1^\infty \dfrac{1}{x^p}\,dx$ converges. When $p < 1$ the improper integral diverges; we showed in Example 6.8.1 that when $p = 1$ the integral also diverges.

Figure 6.8.9 graphs $y = 1/x$ with a dashed line, along with graphs of $y = 1/x^p$, $p < 1$, and $y = 1/x^q$, $q > 1$. Somehow the dashed line forms a dividing line between convergence and divergence.

The result of Example 6.8.4 provides an important tool in determining the convergence of other integrals. A similar result is proved in the exercises about improper integrals of the form $\int_0^1 \dfrac{1}{x^p}\,dx$. These results are summarized in the following Key Idea.

Key Idea 6.8.1   **Convergence of Improper Integrals** $\int_1^\infty \dfrac{1}{x^p}\,dx$ **and** $\int_0^1 \dfrac{1}{x^p}\,dx.$

1. The improper integral $\int_1^\infty \dfrac{1}{x^p}\,dx$ converges when $p > 1$ and diverges when $p \leq 1$.

2. The improper integral $\int_0^1 \dfrac{1}{x^p}\,dx$ converges when $p < 1$ and diverges when $p \geq 1$.

A basic technique in determining convergence of improper integrals is to compare an integrand whose convergence is unknown to an integrand whose convergence is known. We often use integrands of the form $1/x^p$ to compare to as their convergence on certain intervals is known. This is described in the following theorem.

Note: We used the upper and lower bound of "1" in Key Idea 6.8.1 for convenience. It can be replaced by any a where $a > 0$.

Theorem 6.8.1   **Direct Comparison Test for Improper Integrals**

Let f and g be continuous on $[a, \infty)$ where $0 \leq f(x) \leq g(x)$ for all x in $[a, \infty)$.

1. If $\int_a^\infty g(x)\,dx$ converges, then $\int_a^\infty f(x)\,dx$ converges.

2. If $\int_a^\infty f(x)\,dx$ diverges, then $\int_a^\infty g(x)\,dx$ diverges.

Notes:

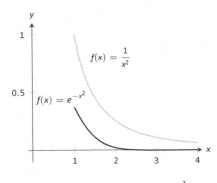

Figure 6.8.10: Graphs of $f(x) = e^{-x^2}$ and $f(x) = 1/x^2$ in Example 6.8.5.

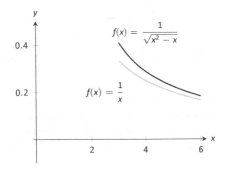

Figure 6.8.11:   Graphs of $f(x) = 1/\sqrt{x^2 - x}$ and $f(x) = 1/x$ in Example 6.8.5.

Example 6.8.5   **Determining convergence of improper integrals**

Determine the convergence of the following improper integrals.

1. $\displaystyle\int_1^\infty e^{-x^2}\, dx$ 2. $\displaystyle\int_3^\infty \frac{1}{\sqrt{x^2 - x}}\, dx$

SOLUTION

1. The function $f(x) = e^{-x^2}$ does not have an antiderivative expressible in terms of elementary functions, so we cannot integrate directly. It is comparable to $g(x) = 1/x^2$, and as demonstrated in Figure 6.8.10, $e^{-x^2} < 1/x^2$ on $[1, \infty)$. We know from Key Idea 6.8.1 that $\displaystyle\int_1^\infty \frac{1}{x^2}\, dx$ converges, hence $\displaystyle\int_1^\infty e^{-x^2}\, dx$ also converges.

2. Note that for large values of x, $\dfrac{1}{\sqrt{x^2 - x}} \approx \dfrac{1}{\sqrt{x^2}} = \dfrac{1}{x}$. We know from Key Idea 6.8.1 and the subsequent note that $\displaystyle\int_3^\infty \frac{1}{x}\, dx$ diverges, so we seek to compare the original integrand to $1/x$.

It is easy to see that when $x > 0$, we have $x = \sqrt{x^2} > \sqrt{x^2 - x}$. Taking reciprocals reverses the inequality, giving

$$\frac{1}{x} < \frac{1}{\sqrt{x^2 - x}}.$$

Using Theorem 6.8.1, we conclude that since $\displaystyle\int_3^\infty \frac{1}{x}\, dx$ diverges, $\displaystyle\int_3^\infty \frac{1}{\sqrt{x^2 - x}}\, dx$ diverges as well. Figure 6.8.11 illustrates this.

Being able to compare "unknown" integrals to "known" integrals is very useful in determining convergence. However, some of our examples were a little "too nice." For instance, it was convenient that $\dfrac{1}{x} < \dfrac{1}{\sqrt{x^2 - x}}$, but what if the "$-x$" were replaced with a "$+2x + 5$"? That is, what can we say about the convergence of $\displaystyle\int_3^\infty \frac{1}{\sqrt{x^2 + 2x + 5}}\, dx$? We have $\dfrac{1}{x} > \dfrac{1}{\sqrt{x^2 + 2x + 5}}$, so we cannot use Theorem 6.8.1.

In cases like this (and many more) it is useful to employ the following theorem.

Notes:

Theorem 6.8.2 **Limit Comparison Test for Improper Integrals**

Let f and g be continuous functions on $[a, \infty)$ where $f(x) > 0$ and $g(x) > 0$ for all x. If

$$\lim_{x \to \infty} \frac{f(x)}{g(x)} = L, \qquad 0 < L < \infty,$$

then

$$\int_a^\infty f(x)\, dx \quad \text{and} \quad \int_a^\infty g(x)\, dx$$

either both converge or both diverge.

Example 6.8.6 **Determining convergence of improper integrals**

Determine the convergence of $\displaystyle\int_3^\infty \frac{1}{\sqrt{x^2 + 2x + 5}}\, dx$.

SOLUTION   As x gets large, the denominator of the integrand will begin to behave much like $y = x$. So we compare $\dfrac{1}{\sqrt{x^2 + 2x + 5}}$ to $\dfrac{1}{x}$ with the Limit Comparison Test:

$$\lim_{x \to \infty} \frac{1/\sqrt{x^2 + 2x + 5}}{1/x} = \lim_{x \to \infty} \frac{x}{\sqrt{x^2 + 2x + 5}}.$$

The immediate evaluation of this limit returns ∞/∞, an indeterminate form. Using l'Hôpital's Rule seems appropriate, but in this situation, it does not lead to useful results. (We encourage the reader to employ l'Hôpital's Rule at least once to verify this.)

The trouble is the square root function. To get rid of it, we employ the following fact: If $\lim_{x \to c} f(x) = L$, then $\lim_{x \to c} f(x)^2 = L^2$. (This is true when either c or L is ∞.) So we consider now the limit

$$\lim_{x \to \infty} \frac{x^2}{x^2 + 2x + 5}.$$

This converges to 1, meaning the original limit also converged to 1. As x gets very large, the function $\dfrac{1}{\sqrt{x^2 + 2x + 5}}$ looks very much like $\dfrac{1}{x}$. Since we know that $\displaystyle\int_3^\infty \frac{1}{x}\, dx$ diverges, by the Limit Comparison Test we know that $\displaystyle\int_3^\infty \frac{1}{\sqrt{x^2 + 2x + 5}}\, dx$ also diverges. Figure 6.8.12 graphs $f(x) = 1/\sqrt{x^2 + 2x + 5}$ and $f(x) = 1/x$, illustrating that as x gets large, the functions become indistinguishable.

Figure 6.8.12: Graphing $f(x) = \dfrac{1}{\sqrt{x^2 + 2x + 5}}$ and $f(x) = \frac{1}{x}$ in Example 6.8.6.

Notes:

Both the Direct and Limit Comparison Tests were given in terms of integrals over an infinite interval. There are versions that apply to improper integrals with an infinite range, but as they are a bit wordy and a little more difficult to employ, they are omitted from this text.

This chapter has explored many integration techniques. We learned Substitution, which "undoes" the Chain Rule of differentiation, as well as Integration by Parts, which "undoes" the Product Rule. We learned specialized techniques for handling trigonometric functions and introduced the hyperbolic functions, which are closely related to the trigonometric functions. All techniques effectively have this goal in common: rewrite the integrand in a new way so that the integration step is easier to see and implement.

As stated before, integration is, in general, hard. It is easy to write a function whose antiderivative is impossible to write in terms of elementary functions, and even when a function does have an antiderivative expressible by elementary functions, it may be really hard to discover what it is. The powerful computer algebra system *Mathematica*® has approximately 1,000 pages of code dedicated to integration.

Do not let this difficulty discourage you. There is great value in learning integration techniques, as they allow one to manipulate an integral in ways that can illuminate a concept for greater understanding. There is also great value in understanding the need for good numerical techniques: the Trapezoidal and Simpson's Rules are just the beginning of powerful techniques for approximating the value of integration.

The next chapter stresses the uses of integration. We generally do not find antiderivatives for antiderivative's sake, but rather because they provide the solution to some type of problem. The following chapter introduces us to a number of different problems whose solution is provided by integration.

Notes:

Exercises 6.8

Terms and Concepts

1. The definite integral was defined with what two stipulations?

2. If $\lim\limits_{b \to \infty} \int_0^b f(x)\, dx$ exists, then the integral $\int_0^\infty f(x)\, dx$ is said to _____.

3. If $\int_1^\infty f(x)\, dx = 10$, and $0 \le g(x) \le f(x)$ for all x, then we know that $\int_1^\infty g(x)\, dx$ _____.

4. For what values of p will $\int_1^\infty \frac{1}{x^p}\, dx$ converge?

5. For what values of p will $\int_{10}^\infty \frac{1}{x^p}\, dx$ converge?

6. For what values of p will $\int_0^1 \frac{1}{x^p}\, dx$ converge?

Problems

In Exercises 7 – 34, evaluate the given improper integral.

7. $\int_0^\infty e^{5-2x}\, dx$

8. $\int_1^\infty \frac{1}{x^3}\, dx$

9. $\int_1^\infty x^{-4}\, dx$

10. $\int_{-\infty}^\infty \frac{1}{x^2 + 9}\, dx$

11. $\int_{-\infty}^0 2^x\, dx$

12. $\int_{-\infty}^0 \left(\frac{1}{2}\right)^x\, dx$

13. $\int_{-\infty}^\infty \frac{x}{x^2 + 1}\, dx$

14. $\int_3^\infty \frac{1}{x^2 - 4}\, dx$

15. $\int_2^\infty \frac{1}{(x-1)^2}\, dx$

16. $\int_1^2 \frac{1}{(x-1)^2}\, dx$

17. $\int_2^\infty \frac{1}{x-1}\, dx$

18. $\int_1^2 \frac{1}{x-1}\, dx$

19. $\int_{-1}^1 \frac{1}{x}\, dx$

20. $\int_1^3 \frac{1}{x-2}\, dx$

21. $\int_0^\pi \sec^2 x\, dx$

22. $\int_{-2}^1 \frac{1}{\sqrt{|x|}}\, dx$

23. $\int_0^\infty xe^{-x}\, dx$

24. $\int_0^\infty xe^{-x^2}\, dx$

25. $\int_{-\infty}^\infty xe^{-x^2}\, dx$

26. $\int_{-\infty}^\infty \frac{1}{e^x + e^{-x}}\, dx$

27. $\int_0^1 x \ln x\, dx$

28. $\int_0^1 x^2 \ln x\, dx$

29. $\int_1^\infty \frac{\ln x}{x}\, dx$

30. $\int_0^1 \ln x\, dx$

31. $\int_1^\infty \frac{\ln x}{x^2}\, dx$

32. $\int_1^\infty \frac{\ln x}{\sqrt{x}}\, dx$

33. $\int_0^\infty e^{-x} \sin x\, dx$

34. $\int_0^\infty e^{-x} \cos x\, dx$

In Exercises 35 – 44, use the Direct Comparison Test or the Limit Comparison Test to determine whether the given definite integral converges or diverges. Clearly state what test is being used and what function the integrand is being compared to.

35. $\displaystyle\int_{10}^{\infty} \frac{3}{\sqrt{3x^2 + 2x - 5}}\,dx$

36. $\displaystyle\int_{2}^{\infty} \frac{4}{\sqrt{7x^3 - x}}\,dx$

37. $\displaystyle\int_{0}^{\infty} \frac{\sqrt{x + 3}}{\sqrt{x^3 - x^2 + x + 1}}\,dx$

38. $\displaystyle\int_{1}^{\infty} e^{-x} \ln x \,dx$

39. $\displaystyle\int_{5}^{\infty} e^{-x^2 + 3x + 1}\,dx$

40. $\displaystyle\int_{0}^{\infty} \frac{\sqrt{x}}{e^x}\,dx$

41. $\displaystyle\int_{2}^{\infty} \frac{1}{x^2 + \sin x}\,dx$

42. $\displaystyle\int_{0}^{\infty} \frac{x}{x^2 + \cos x}\,dx$

43. $\displaystyle\int_{0}^{\infty} \frac{1}{x + e^x}\,dx$

44. $\displaystyle\int_{0}^{\infty} \frac{1}{e^x - x}\,dx$

7: APPLICATIONS OF INTEGRATION

We begin this chapter with a reminder of a few key concepts from Chapter 5. Let f be a continuous function on $[a, b]$ which is partitioned into n equally spaced subintervals as

$$a = x_1 < x_2 < \cdots < x_n < x_{n+1} = b.$$

Let $\Delta x = (b - a)/n$ denote the length of the subintervals, and let c_i be any x-value in the ith subinterval. Definition 5.3.2 states that the sum

$$\sum_{i=1}^{n} f(c_i) \Delta x$$

is a *Riemann Sum*. Riemann Sums are often used to approximate some quantity (area, volume, work, pressure, etc.). The *approximation* becomes *exact* by taking the limit

$$\lim_{n \to \infty} \sum_{i=1}^{n} f(c_i) \Delta x.$$

Theorem 5.3.2 connects limits of Riemann Sums to definite integrals:

$$\lim_{n \to \infty} \sum_{i=1}^{n} f(c_i) \Delta x = \int_{a}^{b} f(x)\, dx.$$

Finally, the Fundamental Theorem of Calculus states how definite integrals can be evaluated using antiderivatives.

This chapter employs the following technique to a variety of applications. Suppose the value Q of a quantity is to be calculated. We first approximate the value of Q using a Riemann Sum, then find the exact value via a definite integral. We spell out this technique in the following Key Idea.

Key Idea 7.0.1   Application of Definite Integrals Strategy

Let a quantity be given whose value Q is to be computed.

1. Divide the quantity into n smaller "subquantities" of value Q_i.

2. Identify a variable x and function $f(x)$ such that each subquantity can be approximated with the product $f(c_i) \Delta x$, where Δx represents a small change in x. Thus $Q_i \approx f(c_i) \Delta x$. A sample approximation $f(c_i) \Delta x$ of Q_i is called a *differential element*.

3. Recognize that $Q = \sum_{i=1}^{n} Q_i \approx \sum_{i=1}^{n} f(c_i) \Delta x$, which is a Riemann Sum.

4. Taking the appropriate limit gives $Q = \int_{a}^{b} f(x)\, dx$

This Key Idea will make more sense after we have had a chance to use it several times. We begin with Area Between Curves, which we addressed briefly in Section 5.4.

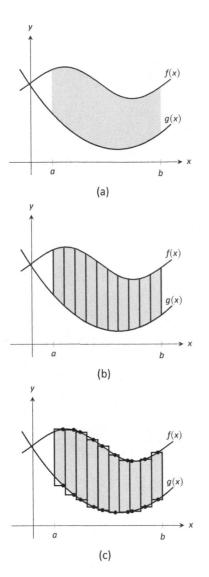

(a)

(b)

(c)

Figure 7.1.1: Subdividing a region into vertical slices and approximating the areas with rectangles.

7.1   Area Between Curves

We are often interested in knowing the area of a region. Forget momentarily that we addressed this already in Section 5.4 and approach it instead using the technique described in Key Idea 7.0.1.

Let Q be the area of a region bounded by continuous functions f and g. If we break the region into many subregions, we have an obvious equation:

Total Area = sum of the areas of the subregions.

The issue to address next is how to systematically break a region into subregions. A graph will help. Consider Figure 7.1.1 (a) where a region between two curves is shaded. While there are many ways to break this into subregions, one particularly efficient way is to "slice" it vertically, as shown in Figure 7.1.1 (b), into n equally spaced slices.

We now approximate the area of a slice. Again, we have many options, but using a rectangle seems simplest. Picking any x-value c_i in the ith slice, we set the height of the rectangle to be $f(c_i) - g(c_i)$, the difference of the corresponding y-values. The width of the rectangle is a small difference in x-values, which we represent with Δx. Figure 7.1.1 (c) shows sample points c_i chosen in each subinterval and appropriate rectangles drawn. (Each of these rectangles represents a differential element.) Each slice has an area approximately equal to $\big(f(c_i) - g(c_i)\big)\Delta x$; hence, the total area is approximately the Riemann Sum

$$Q = \sum_{i=1}^{n} \big(f(c_i) - g(c_i)\big)\Delta x.$$

Taking the limit as $n \to \infty$ gives the exact area as $\int_a^b \big(f(x) - g(x)\big)\,dx$.

Theorem 7.1.1   Area Between Curves (restatement of Theorem 5.4.3)

Let $f(x)$ and $g(x)$ be continuous functions defined on $[a, b]$ where $f(x) \geq g(x)$ for all x in $[a, b]$. The area of the region bounded by the curves $y = f(x)$, $y = g(x)$ and the lines $x = a$ and $x = b$ is

$$\int_a^b \big(f(x) - g(x)\big)\,dx.$$

Example 7.1.1   Finding area enclosed by curves

Find the area of the region bounded by $f(x) = \sin x + 2$, $g(x) = \frac{1}{2}\cos(2x) - 1$, $x = 0$ and $x = 4\pi$, as shown in Figure 7.1.2.

Notes:

SOLUTION The graph verifies that the upper boundary of the region is given by f and the lower bound is given by g. Therefore the area of the region is the value of the integral

$$\int_0^{4\pi} \big(f(x) - g(x)\big)\, dx = \int_0^{4\pi} \left(\sin x + 2 - \left(\frac{1}{2}\cos(2x) - 1\right) \right) dx$$

$$= -\cos x - \frac{1}{4}\sin(2x) + 3x \Big|_0^{4\pi}$$

$$= 12\pi \approx 37.7 \text{ units}^2.$$

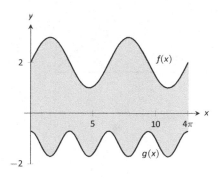

Figure 7.1.2: Graphing an enclosed region in Example 7.1.1.

Example 7.1.2   Finding total area enclosed by curves

Find the total area of the region enclosed by the functions $f(x) = -2x + 5$ and $g(x) = x^3 - 7x^2 + 12x - 3$ as shown in Figure 7.1.3.

SOLUTION A quick calculation shows that $f = g$ at $x = 1, 2$ and 4. One can proceed thoughtlessly by computing $\int_1^4 \big(f(x) - g(x)\big)\, dx$, but this ignores the fact that on $[1, 2]$, $g(x) > f(x)$. (In fact, the thoughtless integration returns $-9/4$, hardly the expected value of an *area*.) Thus we compute the total area by breaking the interval $[1, 4]$ into two subintervals, $[1, 2]$ and $[2, 4]$ and using the proper integrand in each.

$$\text{Total Area} = \int_1^2 \big(g(x) - f(x)\big)\, dx + \int_2^4 \big(f(x) - g(x)\big)\, dx$$

$$= \int_1^2 \big(x^3 - 7x^2 + 14x - 8\big)\, dx + \int_2^4 \big(-x^3 + 7x^2 - 14x + 8\big)\, dx$$

$$= 5/12 + 8/3$$

$$= 37/12 = 3.083 \text{ units}^2.$$

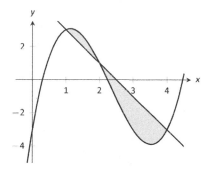

Figure 7.1.3: Graphing a region enclosed by two functions in Example 7.1.2.

The previous example makes note that we are expecting area to be *positive*. When first learning about the definite integral, we interpreted it as "signed area under the curve," allowing for "negative area." That doesn't apply here; area is to be positive.

The previous example also demonstrates that we often have to break a given region into subregions before applying Theorem 7.1.1. The following example shows another situation where this is applicable, along with an alternate view of applying the Theorem.

Example 7.1.3   Finding area: integrating with respect to y

Find the area of the region enclosed by the functions $y = \sqrt{x} + 2$, $y = -(x - 1)^2 + 3$ and $y = 2$, as shown in Figure 7.1.4.

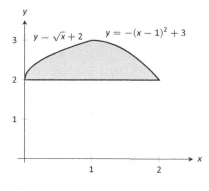

Figure 7.1.4: Graphing a region for Example 7.1.3.

Notes:

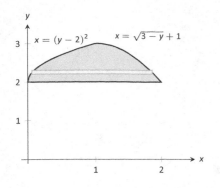

Figure 7.1.5: The region used in Example 7.1.3 with boundaries relabeled as functions of y.

SOLUTION   We give two approaches to this problem. In the first approach, we notice that the region's "top" is defined by two different curves. On $[0, 1]$, the top function is $y = \sqrt{x} + 2$; on $[1, 2]$, the top function is $y = -(x-1)^2 + 3$. Thus we compute the area as the sum of two integrals:

$$\text{Total Area} = \int_0^1 \left((\sqrt{x} + 2) - 2 \right) dx + \int_1^2 \left((-(x-1)^2 + 3) - 2 \right) dx$$
$$= 2/3 + 2/3$$
$$= 4/3.$$

The second approach is clever and very useful in certain situations. We are used to viewing curves as functions of x; we input an x-value and a y-value is returned. Some curves can also be described as functions of y: input a y-value and an x-value is returned. We can rewrite the equations describing the boundary by solving for x:

$$y = \sqrt{x} + 2 \quad \Rightarrow \quad x = (y-2)^2$$
$$y = -(x-1)^2 + 3 \quad \Rightarrow \quad x = \sqrt{3-y} + 1.$$

Figure 7.1.5 shows the region with the boundaries relabeled. A differential element, a horizontal rectangle, is also pictured. The width of the rectangle is a small change in y: Δy. The height of the rectangle is a difference in x-values. The "top" x-value is the largest value, i.e., the rightmost. The "bottom" x-value is the smaller, i.e., the leftmost. Therefore the height of the rectangle is

$$\left(\sqrt{3-y} + 1 \right) - (y-2)^2.$$

The area is found by integrating the above function with respect to y with the appropriate bounds. We determine these by considering the y-values the region occupies. It is bounded below by $y = 2$, and bounded above by $y = 3$. That is, both the "top" and "bottom" functions exist on the y interval $[2, 3]$. Thus

$$\text{Total Area} = \int_2^3 \left(\sqrt{3-y} + 1 - (y-2)^2 \right) dy$$
$$= \left(-\frac{2}{3}(3-y)^{3/2} + y - \frac{1}{3}(y-2)^3 \right) \Big|_2^3$$
$$= 4/3.$$

This calculus–based technique of finding area can be useful even with shapes that we normally think of as "easy." Example 7.1.4 computes the area of a triangle. While the formula "$\frac{1}{2} \times$ base $\times$ height" is well known, in arbitrary triangles it can be nontrivial to compute the height. Calculus makes the problem simple.

Notes:

Example 7.1.4　　**Finding the area of a triangle**

Compute the area of the regions bounded by the lines
$y = x + 1$, $y = -2x + 7$ and $y = -\frac{1}{2}x + \frac{5}{2}$, as shown in Figure 7.1.6.

SOLUTION　　Recognize that there are two "top" functions to this region, causing us to use two definite integrals.

$$\text{Total Area} = \int_1^2 \left((x+1) - (-\frac{1}{2}x + \frac{5}{2})\right) dx + \int_2^3 \left((-2x+7) - (-\frac{1}{2}x + \frac{5}{2})\right) dx$$
$$= 3/4 + 3/4$$
$$= 3/2.$$

We can also approach this by converting each function into a function of y. This also requires 2 integrals, so there isn't really any advantage to doing so. We do it here for demonstration purposes.

The "top" function is always $x = \frac{7-y}{2}$ while there are two "bottom" functions. Being mindful of the proper integration bounds, we have

$$\text{Total Area} = \int_1^2 \left(\frac{7-y}{2} - (5 - 2y)\right) dy + \int_2^3 \left(\frac{7-y}{2} - (y-1)\right) dy$$
$$= 3/4 + 3/4$$
$$= 3/2.$$

Of course, the final answer is the same. (It is interesting to note that the area of all 4 subregions used is 3/4. This is coincidental.)

While we have focused on producing exact answers, we are also able to make approximations using the principle of Theorem 7.1.1. The integrand in the theorem is a distance ("top minus bottom"); integrating this distance function gives an area. By taking discrete measurements of distance, we can approximate an area using numerical integration techniques developed in Section 5.5. The following example demonstrates this.

Example 7.1.5　　**Numerically approximating area**

To approximate the area of a lake, shown in Figure 7.1.7 (a), the "length" of the lake is measured at 200-foot increments as shown in Figure 7.1.7 (b), where the lengths are given in hundreds of feet. Approximate the area of the lake.

SOLUTION　　The measurements of length can be viewed as measuring "top minus bottom" of two functions. The exact answer is found by integrating $\int_0^{12} \left(f(x) - g(x)\right) dx$, but of course we don't know the functions f and g. Our discrete measurements instead allow us to approximate.

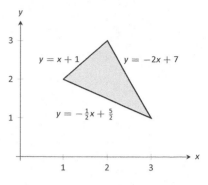

Figure 7.1.6: Graphing a triangular region in Example 7.1.4.

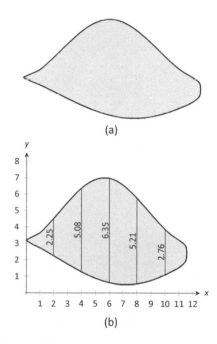

(a)

(b)

Figure 7.1.7: (a) A sketch of a lake, and (b) the lake with length measurements.

Notes:

We have the following data points:

$$(0,0),\ (2,2.25),\ (4,5.08),\ (6,6.35),\ (8,5.21),\ (10,2.76),\ (12,0).$$

We also have that $\Delta x = \frac{b-a}{n} = 2$, so Simpson's Rule gives

$$\text{Area} \approx \frac{2}{3}\Big(1 \cdot 0 + 4 \cdot 2.25 + 2 \cdot 5.08 + 4 \cdot 6.35 + 2 \cdot 5.21 + 4 \cdot 2.76 + 1 \cdot 0\Big)$$
$$= 44.01\overline{3}\ \text{units}^2.$$

Since the measurements are in hundreds of feet, $\text{units}^2 = (100\ \text{ft})^2 = 10,000\ \text{ft}^2$, giving a total area of $440,133\ \text{ft}^2$. (Since we are approximating, we'd likely say the area was about $440,000\ \text{ft}^2$, which is a little more than 10 acres.)

In the next section we apply our applications–of–integration techniques to finding the volumes of certain solids.

Notes:

Exercises 7.1

Terms and Concepts

1. T/F: The area between curves is always positive.

2. T/F: Calculus can be used to find the area of basic geometric shapes.

3. In your own words, describe how to find the total area enclosed by $y = f(x)$ and $y = g(x)$.

4. Describe a situation where it is advantageous to find an area enclosed by curves through integration with respect to y instead of x.

Problems

In Exercises 5 – 12, find the area of the shaded region in the given graph.

5.

6.

7.

8.

9.

10.

11.

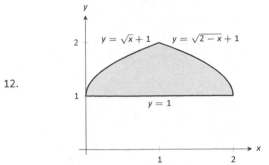

12.

In Exercises 13 – 20, find the total area enclosed by the functions f and g.

13. $f(x) = 2x^2 + 5x - 3$, $g(x) = x^2 + 4x - 1$

14. $f(x) = x^2 - 3x + 2$, $g(x) = -3x + 3$

15. $f(x) = \sin x$, $g(x) = 2x/\pi$

16. $f(x) = x^3 - 4x^2 + x - 1$, $g(x) = -x^2 + 2x - 4$

17. $f(x) = x$, $g(x) = \sqrt{x}$

18. $f(x) = -x^3 + 5x^2 + 2x + 1$, $g(x) = 3x^2 + x + 3$

19. The functions $f(x) = \cos(x)$ and $g(x) = \sin x$ intersect infinitely many times, forming an infinite number of repeated, enclosed regions. Find the areas of these regions.

20. The functions $f(x) = \cos(2x)$ and $g(x) = \sin x$ intersect infinitely many times, forming an infinite number of repeated, enclosed regions. Find the areas of these regions.

In Exercises 21 – 26, find the area of the enclosed region in two ways:

1. by treating the boundaries as functions of x, and

2. by treating the boundaries as functions of y.

21.

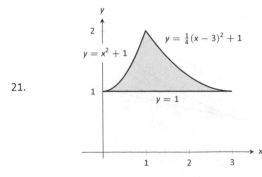

22.

23.

24.

25.

26.

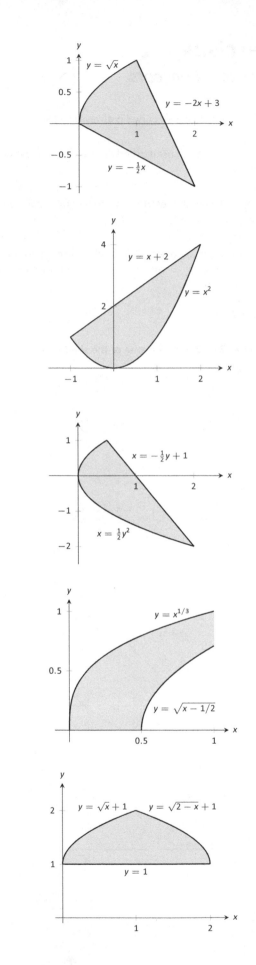

In Exercises 27 – 30, find the area triangle formed by the given three points.

27. $(1, 1)$, $(2, 3)$,   and $(3, 3)$

28. $(-1, 1)$, $(1, 3)$,   and $(2, -1)$

29. $(1, 1)$, $(3, 3)$,   and $(3, 3)$

30. $(0, 0)$, $(2, 5)$,   and $(5, 2)$

31. Use the Trapezoidal Rule to approximate the area of the pictured lake whose lengths, in hundreds of feet, are measured in 100-foot increments.

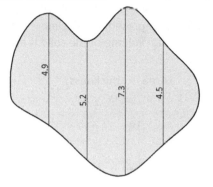

32. Use Simpson's Rule to approximate the area of the pictured lake whose lengths, in hundreds of feet, are measured in 200-foot increments.

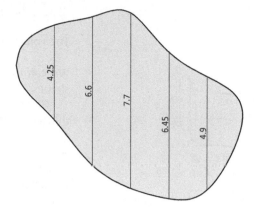

7.2   Volume by Cross-Sectional Area; Disk and Washer Methods

The volume of a general right cylinder, as shown in Figure 7.2.1, is

Area of the base × height.

We can use this fact as the building block in finding volumes of a variety of shapes.

Given an arbitrary solid, we can *approximate* its volume by cutting it into n thin slices. When the slices are thin, each slice can be approximated well by a general right cylinder. Thus the volume of each slice is approximately its cross-sectional area × thickness. (These slices are the differential elements.)

By orienting a solid along the x-axis, we can let $A(x_i)$ represent the cross-sectional area of the ith slice, and let Δx_i represent the thickness of this slice (the thickness is a small change in x). The total volume of the solid is approximately:

$$\text{Volume} \approx \sum_{i=1}^{n} \left[\text{Area} \times \text{thickness} \right]$$

$$= \sum_{i=1}^{n} A(x_i) \Delta x_i.$$

Recognize that this is a Riemann Sum. By taking a limit (as the thickness of the slices goes to 0) we can find the volume exactly.

base area = A
Volume = $A \cdot h$

Figure 7.2.1: The volume of a general right cylinder

Theorem 7.2.1   Volume By Cross-Sectional Area

The volume V of a solid, oriented along the x-axis with cross-sectional area $A(x)$ from $x = a$ to $x = b$, is

$$V = \int_a^b A(x)\, dx.$$

Example 7.2.1   Finding the volume of a solid

Find the volume of a pyramid with a square base of side length 10 in and a height of 5 in.

Solution   There are many ways to "orient" the pyramid along the x-axis; Figure 7.2.2 gives one such way, with the pointed top of the pyramid at the origin and the x-axis going through the center of the base.

Each cross section of the pyramid is a square; this is a sample differential element. To determine its area $A(x)$, we need to determine the side lengths of

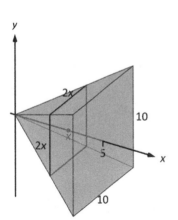

Figure 7.2.2: Orienting a pyramid along the x-axis in Example 7.2.1.

Notes:

the square.

When $x = 5$, the square has side length 10; when $x = 0$, the square has side length 0. Since the edges of the pyramid are lines, it is easy to figure that each cross-sectional square has side length $2x$, giving $A(x) = (2x)^2 = 4x^2$.

If one were to cut a slice out of the pyramid at $x = 3$, as shown in Figure 7.2.3, one would have a shape with square bottom and top with sloped sides. If the slice were thin, both the bottom and top squares would have sides lengths of about 6, and thus the cross–sectional area of the bottom and top would be about 36in^2. Letting Δx_i represent the thickness of the slice, the volume of this slice would then be about $36\Delta x_i$in^3.

Cutting the pyramid into n slices divides the total volume into n equally–spaced smaller pieces, each with volume $(2x_i)^2 \Delta x$, where x_i is the approximate location of the slice along the x-axis and Δx represents the thickness of each slice. One can approximate total volume of the pyramid by summing up the volumes of these slices:

$$\text{Approximate volume } = \sum_{i=1}^{n}(2x_i)^2 \Delta x.$$

Taking the limit as $n \to \infty$ gives the actual volume of the pyramid; recoginizing this sum as a Riemann Sum allows us to find the exact answer using a definite integral, matching the definite integral given by Theorem 7.2.1.

We have

$$V = \lim_{n\to\infty} \sum_{i=1}^{n}(2x_i)^2 \Delta x$$

$$= \int_0^5 4x^2 \, dx$$

$$= \frac{4}{3}x^3 \Big|_0^5$$

$$= \frac{500}{3} \text{ in}^3 \approx 166.67 \text{ in}^3.$$

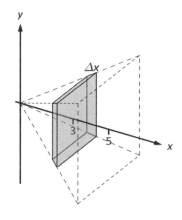

Figure 7.2.3: Cutting a slice in the pyramid in Example 7.2.1 at $x = 3$.

We can check our work by consulting the general equation for the volume of a pyramid (see the back cover under "Volume of A General Cone"):
$$\tfrac{1}{3} \times \text{area of base} \times \text{height}.$$
Certainly, using this formula from geometry is faster than our new method, but the calculus–based method can be applied to much more than just cones.

An important special case of Theorem 7.2.1 is when the solid is a **solid of revolution**, that is, when the solid is formed by rotating a shape around an axis.

Start with a function $y = f(x)$ from $x = a$ to $x = b$. Revolving this curve about a horizontal axis creates a three-dimensional solid whose cross sections

Notes:

are disks (thin circles). Let $R(x)$ represent the radius of the cross-sectional disk at x; the area of this disk is $\pi R(x)^2$. Applying Theorem 7.2.1 gives the Disk Method.

Key Idea 7.2.1   The Disk Method

Let a solid be formed by revolving the curve $y = f(x)$ from $x = a$ to $x = b$ around a horizontal axis, and let $R(x)$ be the radius of the cross-sectional disk at x. The volume of the solid is

$$V = \pi \int_a^b R(x)^2 \, dx.$$

Example 7.2.2   Finding volume using the Disk Method
Find the volume of the solid formed by revolving the curve $y = 1/x$, from $x = 1$ to $x = 2$, around the x-axis.

SOLUTION   A sketch can help us understand this problem. In Figure 7.2.4(a) the curve $y = 1/x$ is sketched along with the differential element – a disk – at x with radius $R(x) = 1/x$. In Figure 7.2.4 (b) the whole solid is pictured, along with the differential element.

The volume of the differential element shown in part (a) of the figure is approximately $\pi R(x_i)^2 \Delta x$, where $R(x_i)$ is the radius of the disk shown and Δx is the thickness of that slice. The radius $R(x_i)$ is the distance from the x-axis to the curve, hence $R(x_i) = 1/x_i$.

Slicing the solid into n equally–spaced slices, we can approximate the total volume by adding up the approximate volume of each slice:

$$\text{Approximate volume} = \sum_{i=1}^{n} \pi \left(\frac{1}{x_i}\right)^2 \Delta x.$$

Taking the limit of the above sum as $n \to \infty$ gives the actual volume; recognizing this sum as a Riemann sum allows us to evaluate the limit with a definite integral, which matches the formula given in Key Idea 7.2.1:

$$V = \lim_{n \to \infty} \sum_{i=1}^{n} \pi \left(\frac{1}{x_i}\right)^2 \Delta x$$

$$= \pi \int_1^2 \left(\frac{1}{x}\right)^2 \, dx$$

$$= \pi \int_1^2 \frac{1}{x^2} \, dx$$

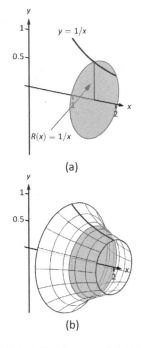

(a)

(b)

Figure 7.2.4: Sketching a solid in Example 7.2.2.

Notes:

$$= \pi \left[-\frac{1}{x} \right] \Big|_1^2$$

$$= \pi \left[-\frac{1}{2} - (-1) \right]$$

$$= \frac{\pi}{2} \text{ units}^3.$$

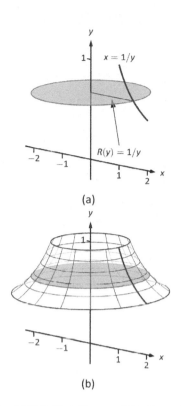

(a)

While Key Idea 7.2.1 is given in terms of functions of x, the principle involved can be applied to functions of y when the axis of rotation is vertical, not horizontal. We demonstrate this in the next example.

Example 7.2.3 Finding volume using the Disk Method

Find the volume of the solid formed by revolving the curve $y = 1/x$, from $x = 1$ to $x = 2$, about the y-axis.

SOLUTION   Since the axis of rotation is vertical, we need to convert the function into a function of y and convert the x-bounds to y-bounds. Since $y = 1/x$ defines the curve, we rewrite it as $x = 1/y$. The bound $x = 1$ corresponds to the y-bound $y = 1$, and the bound $x = 2$ corresponds to the y-bound $y = 1/2$.

Thus we are rotating the curve $x = 1/y$, from $y = 1/2$ to $y = 1$ about the y-axis to form a solid. The curve and sample differential element are sketched in Figure 7.2.5 (a), with a full sketch of the solid in Figure 7.2.5 (b). We integrate to find the volume:

(b)

Figure 7.2.5: Sketching a solid in Example 7.2.3.

$$V = \pi \int_{1/2}^{1} \frac{1}{y^2} \, dy$$

$$= -\frac{\pi}{y} \Big|_{1/2}^1$$

$$= \pi \text{ units}^3.$$

We can also compute the volume of solids of revolution that have a hole in the center. The general principle is simple: compute the volume of the solid irrespective of the hole, then subtract the volume of the hole. If the outside radius of the solid is $R(x)$ and the inside radius (defining the hole) is $r(x)$, then the volume is

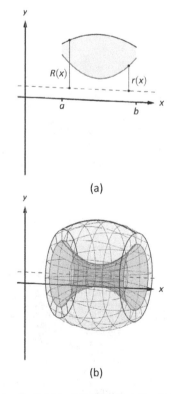

$$V = \pi \int_a^b R(x)^2 \, dx - \pi \int_a^b r(x)^2 \, dx = \pi \int_a^b \left(R(x)^2 - r(x)^2 \right) \, dx.$$

(a)

One can generate a solid of revolution with a hole in the middle by revolving a region about an axis. Consider Figure 7.2.6(a), where a region is sketched along

(b)

Figure 7.2.6: Establishing the Washer Method; see also Figure 7.2.7.

Notes:

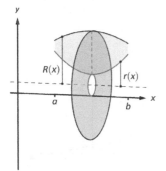

Figure 7.2.7: Establishing the Washer Method; see also Figure 7.2.6.

with a dashed, horizontal axis of rotation. By rotating the region about the axis, a solid is formed as sketched in Figure 7.2.6(b). The outside of the solid has radius $R(x)$, whereas the inside has radius $r(x)$. Each cross section of this solid will be a washer (a disk with a hole in the center) as sketched in Figure 7.2.7. This leads us to the Washer Method.

Key Idea 7.2.2   The Washer Method

Let a region bounded by $y = f(x)$, $y = g(x)$, $x = a$ and $x = b$ be rotated about a horizontal axis that does not intersect the region, forming a solid. Each cross section at x will be a washer with outside radius $R(x)$ and inside radius $r(x)$. The volume of the solid is

$$V = \pi \int_a^b \left(R(x)^2 - r(x)^2 \right) dx.$$

Even though we introduced it first, the Disk Method is just a special case of the Washer Method with an inside radius of $r(x) = 0$.

Example 7.2.4   Finding volume with the Washer Method
Find the volume of the solid formed by rotating the region bounded by $y = x^2 - 2x + 2$ and $y = 2x - 1$ about the x-axis.

SOLUTION   A sketch of the region will help, as given in Figure 7.2.8(a). Rotating about the x-axis will produce cross sections in the shape of washers, as shown in Figure 7.2.8(b); the complete solid is shown in part (c). The outside radius of this washer is $R(x) = 2x + 1$; the inside radius is $r(x) = x^2 - 2x + 2$. As the region is bounded from $x = 1$ to $x = 3$, we integrate as follows to compute the volume.

$$\begin{aligned}
V &= \pi \int_1^3 \left((2x - 1)^2 - (x^2 - 2x + 2)^2 \right) dx \\
&= \pi \int_1^3 \left(-x^4 + 4x^3 - 4x^2 + 4x - 3 \right) dx \\
&= \pi \left[-\frac{1}{5}x^5 + x^4 - \frac{4}{3}x^3 + 2x^2 - 3x \right]\Big|_1^3 \\
&= \frac{104}{15}\pi \approx 21.78 \text{ units}^3.
\end{aligned}$$

When rotating about a vertical axis, the outside and inside radius functions must be functions of y.

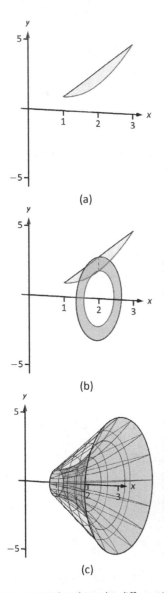

(a)

(b)

(c)

Figure 7.2.8: Sketching the differential element and solid in Example 7.2.4.

Notes:

Example 7.2.5   Finding volume with the Washer Method

Find the volume of the solid formed by rotating the triangular region with vertices at $(1, 1)$, $(2, 1)$ and $(2, 3)$ about the y-axis.

SOLUTION   The triangular region is sketched in Figure 7.2.9(a); the differential element is sketched in (b) and the full solid is drawn in (c). They help us establish the outside and inside radii. Since the axis of rotation is vertical, each radius is a function of y.

The outside radius $R(y)$ is formed by the line connecting $(2, 1)$ and $(2, 3)$; it is a constant function, as regardless of the y-value the distance from the line to the axis of rotation is 2. Thus $R(y) = 2$.

The inside radius is formed by the line connecting $(1, 1)$ and $(2, 3)$. The equation of this line is $y = 2x - 1$, but we need to refer to it as a function of y. Solving for x gives $r(y) = \frac{1}{2}(y + 1)$.

We integrate over the y-bounds of $y = 1$ to $y = 3$. Thus the volume is

$$
\begin{aligned}
V &= \pi \int_1^3 \left(2^2 - \left(\tfrac{1}{2}(y + 1) \right)^2 \right) dy \\
&= \pi \int_1^3 \left(-\tfrac{1}{4}y^2 - \tfrac{1}{2}y + \tfrac{15}{4} \right) dy \\
&= \pi \left[-\tfrac{1}{12}y^3 - \tfrac{1}{4}y^2 + \tfrac{15}{4}y \right]\Big|_1^3 \\
&= \frac{10}{3}\pi \approx 10.47 \text{ units}^3.
\end{aligned}
$$

This section introduced a new application of the definite integral. Our default view of the definite integral is that it gives "the area under the curve." However, we can establish definite integrals that represent other quantities; in this section, we computed volume.

The ultimate goal of this section is not to compute volumes of solids. That can be useful, but what is more useful is the understanding of this basic principle of integral calculus, outlined in Key Idea 7.0.1: to find the exact value of some quantity,

- we start with an approximation (in this section, slice the solid and approximate the volume of each slice),

- then make the approximation better by refining our original approximation (i.e., use more slices),

- then use limits to establish a definite integral which gives the exact value.

We practice this principle in the next section where we find volumes by slicing solids in a different way.

(a)

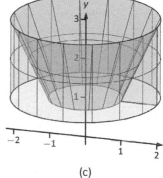

(b)

(c)

Figure 7.2.9: Sketching the solid in Example 7.2.5.

Notes:

Exercises 7.2

Terms and Concepts

1. T/F: A solid of revolution is formed by revolving a shape around an axis.

2. In your own words, explain how the Disk and Washer Methods are related.

3. Explain the how the units of volume are found in the integral of Theorem 7.2.1: if $A(x)$ has units of in^2, how does $\int A(x)\, dx$ have units of in^3?

4. A fundamental principle of this section is "_____ can be found by integrating an area function."

Problems

In Exercises 5 – 8, a region of the Cartesian plane is shaded. Use the Disk/Washer Method to find the volume of the solid of revolution formed by revolving the region about the x-axis.

5.

6.

7.

8.

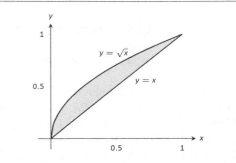

In Exercises 9 – 12, a region of the Cartesian plane is shaded. Use the Disk/Washer Method to find the volume of the solid of revolution formed by revolving the region about the y-axis.

9.

10.

11.

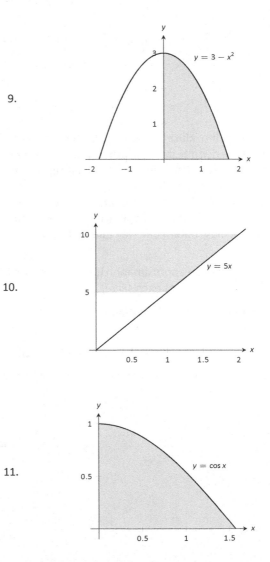

(Hint: Integration By Parts will be necessary, twice. First let $u = \arccos^2 x$, then let $u = \arccos x$.)

12.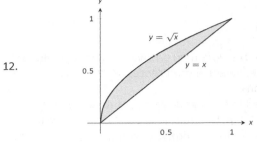

In Exercises 13 – 18, a region of the Cartesian plane is described. Use the Disk/Washer Method to find the volume of the solid of revolution formed by rotating the region about each of the given axes.

13. Region bounded by: $y = \sqrt{x}$, $y = 0$ and $x = 1$.
    Rotate about:

    (a) the x-axis   (c) the y-axis

    (b) $y = 1$ (d) $x = 1$

14. Region bounded by: $y = 4 - x^2$ and $y = 0$.
    Rotate about:

    (a) the x-axis   (c) $y = -1$

    (b) $y = 4$ (d) $x = 2$

15. The triangle with vertices $(1, 1)$, $(1, 2)$ and $(2, 1)$.
    Rotate about:

    (a) the x-axis   (c) the y-axis

    (b) $y = 2$ (d) $x = 1$

16. Region bounded by $y = x^2 - 2x + 2$ and $y = 2x - 1$.
    Rotate about:

    (a) the x-axis   (c) $y = 5$

    (b) $y = 1$

17. Region bounded by $y = 1/\sqrt{x^2 + 1}$, $x = -1$, $x = 1$ and the x-axis.
    Rotate about:

    (a) the x-axis   (c) $y = -1$

    (b) $y = 1$

18. Region bounded by $y = 2x$, $y = x$ and $x = 2$.
    Rotate about:

    (a) the x-axis   (c) the y-axis

    (b) $y = 4$ (d) $x = 2$

In Exercises 19 – 22, a solid is described. Orient the solid along the x-axis such that a cross-sectional area function $A(x)$ can be obtained, then apply Theorem 7.2.1 to find the volume of the solid.

19. A right circular cone with height of 10 and base radius of 5.

20. A skew right circular cone with height of 10 and base radius of 5. (Hint: all cross-sections are circles.)

21. A right triangular cone with height of 10 and whose base is a right, isosceles triangle with side length 4.

22. A solid with length 10 with a rectangular base and triangular top, wherein one end is a square with side length 5 and the other end is a triangle with base and height of 5.

(a)

(b)

(c)

Figure 7.3.1:   Introducing the Shell Method.

7.3   The Shell Method

Often a given problem can be solved in more than one way. A particular method may be chosen out of convenience, personal preference, or perhaps necessity. Ultimately, it is good to have options.

The previous section introduced the Disk and Washer Methods, which computed the volume of solids of revolution by integrating the cross–sectional area of the solid. This section develops another method of computing volume, the **Shell Method.** Instead of slicing the solid perpendicular to the axis of rotation creating cross-sections, we now slice it parallel to the axis of rotation, creating "shells."

Consider Figure 7.3.1, where the region shown in (a) is rotated around the y-axis forming the solid shown in (b). A small slice of the region is drawn in (a), parallel to the axis of rotation. When the region is rotated, this thin slice forms a **cylindrical shell**, as pictured in part (c) of the figure. The previous section approximated a solid with lots of thin disks (or washers); we now approximate a solid with many thin cylindrical shells.

To compute the volume of one shell, first consider the paper label on a soup can with radius r and height h. What is the area of this label? A simple way of determining this is to cut the label and lay it out flat, forming a rectangle with height h and length $2\pi r$. Thus the area is $A = 2\pi rh$; see Figure 7.3.2(a).

Do a similar process with a cylindrical shell, with height h, thickness Δx, and approximate radius r. Cutting the shell and laying it flat forms a rectangular solid with length $2\pi r$, height h and depth Δx. Thus the volume is $V \approx 2\pi rh\Delta x$; see Figure 7.3.2(b). (We say "approximately" since our radius was an approximation.)

By breaking the solid into n cylindrical shells, we can approximate the volume of the solid as

$$V \approx \sum_{i=1}^{n} 2\pi r_i h_i \Delta x_i,$$

where r_i, h_i and Δx_i are the radius, height and thickness of the ith shell, respectively.

This is a Riemann Sum. Taking a limit as the thickness of the shells approaches 0 leads to a definite integral.

Notes:

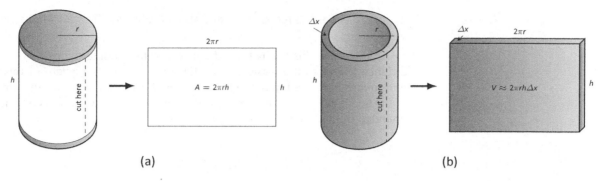

Figure 7.3.2: Determining the volume of a thin cylindrical shell.

Key Idea 7.3.1   The Shell Method

Let a solid be formed by revolving a region R, bounded by $x = a$ and $x = b$, around a vertical axis. Let $r(x)$ represent the distance from the axis of rotation to x (i.e., the radius of a sample shell) and let $h(x)$ represent the height of the solid at x (i.e., the height of the shell). The volume of the solid is

$$V = 2\pi \int_a^b r(x)h(x)\, dx.$$

Special Cases:

1. When the region R is bounded above by $y = f(x)$ and below by $y = g(x)$, then $h(x) = f(x) - g(x)$.

2. When the axis of rotation is the y-axis (i.e., $x = 0$) then $r(x) = x$.

   Let's practice using the Shell Method.

Example 7.3.1   Finding volume using the Shell Method
Find the volume of the solid formed by rotating the region bounded by $y = 0$, $y = 1/(1 + x^2)$, $x = 0$ and $x = 1$ about the y-axis.

    SOLUTION   This is the region used to introduce the Shell Method in Figure 7.3.1, but is sketched again in Figure 7.3.3 for closer reference. A line is drawn in the region parallel to the axis of rotation representing a shell that will

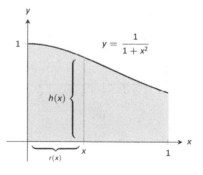

Figure 7.3.3: Graphing a region in Example 7.3.1.

Notes:

be carved out as the region is rotated about the *y*-axis. (This is the differential element.)

The distance this line is from the axis of rotation determines $r(x)$; as the distance from *x* to the *y*-axis is *x*, we have $r(x) = x$. The height of this line determines $h(x)$; the top of the line is at $y = 1/(1+x^2)$, whereas the bottom of the line is at $y = 0$. Thus $h(x) = 1/(1+x^2) - 0 = 1/(1+x^2)$. The region is bounded from $x = 0$ to $x = 1$, so the volume is

$$V = 2\pi \int_0^1 \frac{x}{1+x^2}\, dx.$$

This requires substitution. Let $u = 1 + x^2$, so $du = 2x\, dx$. We also change the bounds: $u(0) = 1$ and $u(1) = 2$. Thus we have:

$$= \pi \int_1^2 \frac{1}{u}\, du$$

$$= \pi \ln u \Big|_1^2$$

$$= \pi \ln 2 \approx 2.178 \text{ units}^3.$$

Note: in order to find this volume using the Disk Method, two integrals would be needed to account for the regions above and below $y = 1/2$.

With the Shell Method, nothing special needs to be accounted for to compute the volume of a solid that has a hole in the middle, as demonstrated next.

Example 7.3.2   Finding volume using the Shell Method
Find the volume of the solid formed by rotating the triangular region determined by the points $(0, 1)$, $(1, 1)$ and $(1, 3)$ about the line $x = 3$.

SOLUTION   The region is sketched in Figure 7.3.4(a) along with the differential element, a line within the region parallel to the axis of rotation. In part (b) of the figure, we see the shell traced out by the differential element, and in part (c) the whole solid is shown.

The height of the differential element is the distance from $y = 1$ to $y = 2x + 1$, the line that connects the points $(0, 1)$ and $(1, 3)$. Thus $h(x) = 2x+1-1 = 2x$. The radius of the shell formed by the differential element is the distance from *x* to $x = 3$; that is, it is $r(x) = 3 - x$. The *x*-bounds of the region are $x = 0$ to

Notes:

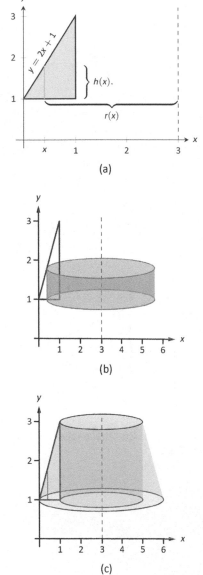

Figure 7.3.4: Graphing a region in Example 7.3.2.

$x = 1$, giving

$$V = 2\pi \int_0^1 (3-x)(2x)\, dx$$

$$= 2\pi \int_0^1 (6x - 2x^2)\, dx$$

$$= 2\pi \left(3x^2 - \frac{2}{3}x^3 \right) \Big|_0^1$$

$$= \frac{14}{3}\pi \approx 14.66 \text{ units}^3.$$

When revolving a region around a horizontal axis, we must consider the radius and height functions in terms of y, not x.

Example 7.3.3   Finding volume using the Shell Method
Find the volume of the solid formed by rotating the region given in Example 7.3.2 about the x-axis.

SOLUTION   The region is sketched in Figure 7.3.5(a) with a sample differential element. In part (b) of the figure the shell formed by the differential element is drawn, and the solid is sketched in (c). (Note that the triangular region looks "short and wide" here, whereas in the previous example the same region looked "tall and narrow." This is because the bounds on the graphs are different.)

The height of the differential element is an x-distance, between $x = \frac{1}{2}y - \frac{1}{2}$ and $x = 1$. Thus $h(y) = 1 - (\frac{1}{2}y - \frac{1}{2}) = -\frac{1}{2}y + \frac{3}{2}$. The radius is the distance from y to the x-axis, so $r(y) = y$. The y bounds of the region are $y = 1$ and $y = 3$, leading to the integral

$$V = 2\pi \int_1^3 \left[y\left(-\frac{1}{2}y + \frac{3}{2}\right) \right] dy$$

$$= 2\pi \int_1^3 \left[-\frac{1}{2}y^2 + \frac{3}{2}y \right] dy$$

$$= 2\pi \left[-\frac{1}{6}y^3 + \frac{3}{4}y^2 \right] \Big|_1^3$$

$$= 2\pi \left[\frac{9}{4} - \frac{7}{12} \right]$$

$$= \frac{10}{3}\pi \approx 10.472 \text{ units}^3.$$

Notes:

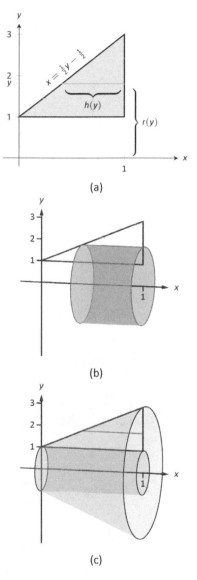

(a)

(b)

(c)

Figure 7.3.5: Graphing a region in Example 7.3.3.

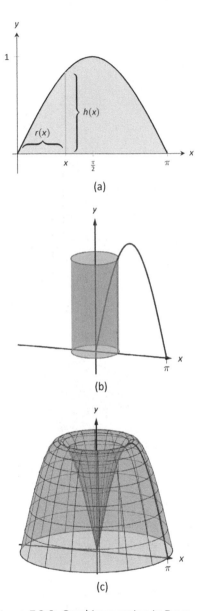

(a)

(b)

(c)

Figure 7.3.6: Graphing a region in Example 7.3.4.

At the beginning of this section it was stated that "it is good to have options." The next example finds the volume of a solid rather easily with the Shell Method, but using the Washer Method would be quite a chore.

Example 7.3.4   Finding volume using the Shell Method

Find the volume of the solid formed by revolving the region bounded by $y = \sin x$ and the x-axis from $x = 0$ to $x = \pi$ about the y-axis.

SOLUTION   The region and a differential element, the shell formed by this differential element, and the resulting solid are given in Figure 7.3.6.   The radius of a sample shell is $r(x) = x$; the height of a sample shell is $h(x) = \sin x$, each from $x = 0$ to $x = \pi$. Thus the volume of the solid is

$$V = 2\pi \int_0^\pi x \sin x \, dx.$$

This requires Integration By Parts. Set $u = x$ and $dv = \sin x \, dx$; we leave it to the reader to fill in the rest. We have:

$$= 2\pi \left[-x \cos x \Big|_0^\pi + \int_0^\pi \cos x \, dx \right]$$
$$= 2\pi \left[\pi + \sin x \Big|_0^\pi \right]$$
$$= 2\pi \left[\pi + 0 \right]$$
$$= 2\pi^2 \approx 19.74 \text{ units}^3.$$

Note that in order to use the Washer Method, we would need to solve $y = \sin x$ for x, requiring the use of the arcsine function. We leave it to the reader to verify that the outside radius function is $R(y) = \pi - \arcsin y$ and the inside radius function is $r(y) = \arcsin y$. Thus the volume can be computed as

$$\pi \int_0^1 \left[(\pi - \arcsin y)^2 - (\arcsin y)^2 \right] dy.$$

This integral isn't terrible given that the $\arcsin^2 y$ terms cancel, but it is more onerous than the integral created by the Shell Method.

We end this section with a table summarizing the usage of the Washer and Shell Methods.

Notes:

Key Idea 7.3.2   **Summary of the Washer and Shell Methods**

Let a region R be given with x-bounds $x = a$ and $x = b$ and y-bounds $y = c$ and $y = d$.

	Washer Method	Shell Method
Horizontal Axis	$\pi \displaystyle\int_a^b \left(R(x)^2 - r(x)^2 \right) dx$	$2\pi \displaystyle\int_c^d r(y)h(y)\, dy$
Vertical Axis	$\pi \displaystyle\int_c^d \left(R(y)^2 - r(y)^2 \right) dy$	$2\pi \displaystyle\int_a^b r(x)h(x)\, dx$

As in the previous section, the real goal of this section is not to be able to compute volumes of certain solids. Rather, it is to be able to solve a problem by first approximating, then using limits to refine the approximation to give the exact value. In this section, we approximate the volume of a solid by cutting it into thin cylindrical shells. By summing up the volumes of each shell, we get an approximation of the volume. By taking a limit as the number of equally spaced shells goes to infinity, our summation can be evaluated as a definite integral, giving the exact value.

We use this same principle again in the next section, where we find the length of curves in the plane.

Notes:

Exercises 7.3

Terms and Concepts

1. T/F: A solid of revolution is formed by revolving a shape around an axis.

2. T/F: The Shell Method can only be used when the Washer Method fails.

3. T/F: The Shell Method works by integrating cross–sectional areas of a solid.

4. T/F: When finding the volume of a solid of revolution that was revolved around a vertical axis, the Shell Method integrates with respect to x.

7.

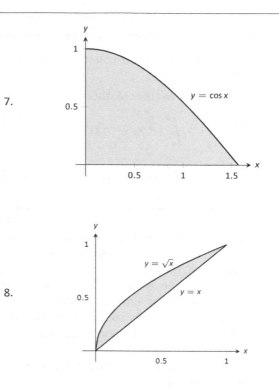

8.

Problems

In Exercises 5 – 8, a region of the Cartesian plane is shaded. Use the Shell Method to find the volume of the solid of revolution formed by revolving the region about the y-axis.

In Exercises 9 – 12, a region of the Cartesian plane is shaded. Use the Shell Method to find the volume of the solid of revolution formed by revolving the region about the x-axis.

5.

9.

6.

10.

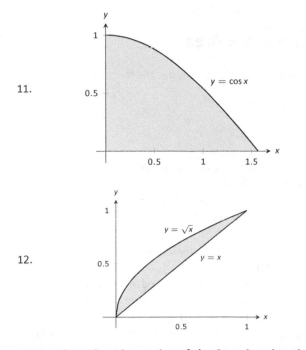

11.

12.

In Exercises 13 – 18, a region of the Cartesian plane is described. Use the Shell Method to find the volume of the solid of revolution formed by rotating the region about each of the given axes.

13. Region bounded by: $y = \sqrt{x}$, $y = 0$ and $x = 1$.
    Rotate about:

    (a) the y-axis   (c) the x-axis

    (b) $x = 1$ (d) $y = 1$

14. Region bounded by: $y = 4 - x^2$ and $y = 0$.
    Rotate about:

    (a) $x = 2$ (c) the x-axis

    (b) $x = -2$ (d) $y = 4$

15. The triangle with vertices $(1, 1)$, $(1, 2)$ and $(2, 1)$.
    Rotate about:

    (a) the y-axis   (c) the x-axis

    (b) $x = 1$ (d) $y = 2$

16. Region bounded by $y = x^2 - 2x + 2$ and $y = 2x - 1$.
    Rotate about:

    (a) the y-axis   (c) $x = -1$

    (b) $x = 1$

17. Region bounded by $y = 1/\sqrt{x^2 + 1}$, $x = 1$ and the x and y-axes.
    Rotate about:

    (a) the y-axis   (b) $x = 1$

18. Region bounded by $y = 2x$, $y = x$ and $x = 2$.
    Rotate about:

    (a) the y-axis   (c) the x-axis

    (b) $x = 2$ (d) $y = 4$

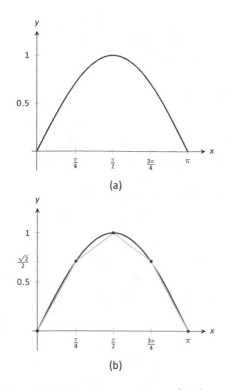

(a)

(b)

Figure 7.4.1: Graphing $y = \sin x$ on $[0, \pi]$ and approximating the curve with line segments.

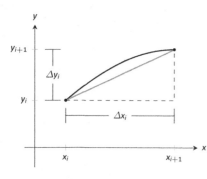

Figure 7.4.2: Zooming in on the ith subinterval $[x_i, x_{i+1}]$ of a partition of $[a, b]$.

7.4 Arc Length and Surface Area

In previous sections we have used integration to answer the following questions:

1. Given a region, what is its area?

2. Given a solid, what is its volume?

In this section, we address a related question: Given a curve, what is its length? This is often referred to as **arc length**.

Consider the graph of $y = \sin x$ on $[0, \pi]$ given in Figure 7.4.1(a). How long is this curve? That is, if we were to use a piece of string to exactly match the shape of this curve, how long would the string be?

As we have done in the past, we start by approximating; later, we will refine our answer using limits to get an exact solution.

The length of straight–line segments is easy to compute using the Distance Formula. We can approximate the length of the given curve by approximating the curve with straight lines and measuring their lengths.

In Figure 7.4.1(b), the curve $y = \sin x$ has been approximated with 4 line segments (the interval $[0, \pi]$ has been divided into 4 equally–lengthed subintervals). It is clear that these four line segments approximate $y = \sin x$ very well on the first and last subinterval, though not so well in the middle. Regardless, the sum of the lengths of the line segments is 3.79, so we approximate the arc length of $y = \sin x$ on $[0, \pi]$ to be 3.79.

In general, we can approximate the arc length of $y = f(x)$ on $[a, b]$ in the following manner. Let $a = x_1 < x_2 < \ldots < x_n < x_{n+1} = b$ be a partition of $[a, b]$ into n subintervals. Let Δx_i represent the length of the ith subinterval $[x_i, x_{i+1}]$.

Figure 7.4.2 zooms in on the ith subinterval where $y = f(x)$ is approximated by a straight line segment. The dashed lines show that we can view this line segment as the hypotenuse of a right triangle whose sides have length Δx_i and Δy_i. Using the Pythagorean Theorem, the length of this line segment is $\sqrt{\Delta x_i^2 + \Delta y_i^2}$. Summing over all subintervals gives an arc length approximation

$$L \approx \sum_{i=1}^{n} \sqrt{\Delta x_i^2 + \Delta y_i^2}.$$

As shown here, this is *not* a Riemann Sum. While we could conclude that taking a limit as the subinterval length goes to zero gives the exact arc length, we would not be able to compute the answer with a definite integral. We need first to do a little algebra.

Notes:

In the above expression factor out a Δx_i^2 term:

$$\sum_{i=1}^{n} \sqrt{\Delta x_i^2 + \Delta y_i^2} = \sum_{i=1}^{n} \sqrt{\Delta x_i^2 \left(1 + \frac{\Delta y_i^2}{\Delta x_i^2}\right)}.$$

Now pull the Δx_i^2 term out of the square root:

$$= \sum_{i=1}^{n} \sqrt{1 + \frac{\Delta y_i^2}{\Delta x_i^2}} \, \Delta x_i.$$

This is nearly a Riemann Sum. Consider the $\Delta y_i^2 / \Delta x_i^2$ term. The expression $\Delta y_i / \Delta x_i$ measures the "change in y/change in x," that is, the "rise over run" of f on the i th subinterval. The Mean Value Theorem of Differentiation (Theorem 3.2.1) states that there is a c_i in the i th subinterval where $f'(c_i) = \Delta y_i / \Delta x_i$. Thus we can rewrite our above expression as:

$$= \sum_{i=1}^{n} \sqrt{1 + f'(c_i)^2} \, \Delta x_i.$$

This *is* a Riemann Sum. As long as f' is continuous, we can invoke Theorem 5.3.2 and conclude

$$= \int_{a}^{b} \sqrt{1 + f'(x)^2} \, dx.$$

Theorem 7.4.1   Arc Length

Let f be differentiable on $[a, b]$, where f' is also continuous on $[a, b]$. Then the arc length of f from $x = a$ to $x = b$ is

$$L = \int_{a}^{b} \sqrt{1 + f'(x)^2} \, dx.$$

Note: This is our first use of differentiability on a closed interval since Section 2.1.

The theorem also requires that f' be continuous on $[a, b]$; while examples are arcane, it is possible for f to be differentiable yet f' is not continuous.

As the integrand contains a square root, it is often difficult to use the formula in Theorem 7.4.1 to find the length exactly. When exact answers are difficult to come by, we resort to using numerical methods of approximating definite integrals. The following examples will demonstrate this.

Notes:

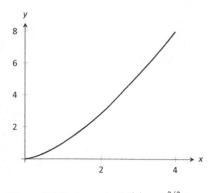

Figure 7.4.3: A graph of $f(x) = x^{3/2}$ from Example 7.4.1.

Example 7.4.1   **Finding arc length**
Find the arc length of $f(x) = x^{3/2}$ from $x = 0$ to $x = 4$.

SOLUTION   We find $f'(x) = \frac{3}{2}x^{1/2}$; note that on $[0, 4]$, f is differentiable and f' is also continuous. Using the formula, we find the arc length L as

$$L = \int_0^4 \sqrt{1 + \left(\frac{3}{2}x^{1/2}\right)^2} \, dx$$

$$= \int_0^4 \sqrt{1 + \frac{9}{4}x} \, dx$$

$$= \int_0^4 \left(1 + \frac{9}{4}x\right)^{1/2} \, dx$$

$$= \frac{2}{3} \cdot \frac{4}{9} \cdot \left(1 + \frac{9}{4}x\right)^{3/2} \Big|_0^4$$

$$= \frac{8}{27}\left(10^{3/2} - 1\right) \approx 9.07 \text{units}.$$

A graph of f is given in Figure 7.4.3.

Example 7.4.2   **Finding arc length**
Find the arc length of $f(x) = \frac{1}{8}x^2 - \ln x$ from $x = 1$ to $x = 2$.

SOLUTION   This function was chosen specifically because the resulting integral can be evaluated exactly. We begin by finding $f'(x) = x/4 - 1/x$. The arc length is

$$L = \int_1^2 \sqrt{1 + \left(\frac{x}{4} - \frac{1}{x}\right)^2} \, dx$$

$$= \int_1^2 \sqrt{1 + \frac{x^2}{16} - \frac{1}{2} + \frac{1}{x^2}} \, dx$$

$$= \int_1^2 \sqrt{\frac{x^2}{16} + \frac{1}{2} + \frac{1}{x^2}} \, dx$$

$$= \int_1^2 \sqrt{\left(\frac{x}{4} + \frac{1}{x}\right)^2} \, dx$$

Notes:

$$= \int_1^2 \left(\frac{x}{4} + \frac{1}{x} \right) dx$$

$$= \left. \left(\frac{x^2}{8} + \ln x \right) \right|_1^2$$

$$= \frac{3}{8} + \ln 2 \approx 1.07 \text{ units.}$$

A graph of f is given in Figure 7.4.4; the portion of the curve measured in this problem is in bold.

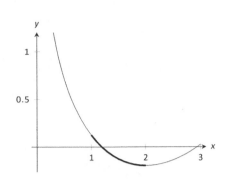

Figure 7.4.4: A graph of $f(x) = \frac{1}{8}x^2 - \ln x$ from Example 7.4.2.

The previous examples found the arc length exactly through careful choice of the functions. In general, exact answers are much more difficult to come by and numerical approximations are necessary.

Example 7.4.3 **Approximating arc length numerically**
Find the length of the sine curve from $x = 0$ to $x = \pi$.

SOLUTION This is somewhat of a mathematical curiosity; in Example 5.4.3 we found the area under one "hump" of the sine curve is 2 square units; now we are measuring its arc length.
The setup is straightforward: $f(x) = \sin x$ and $f'(x) = \cos x$. Thus

$$L = \int_0^\pi \sqrt{1 + \cos^2 x} \, dx.$$

This integral *cannot* be evaluated in terms of elementary functions so we will approximate it with Simpson's Method with $n = 4$. Figure 7.4.5 gives $\sqrt{1 + \cos^2 x}$ evaluated at 5 evenly spaced points in $[0, \pi]$. Simpson's Rule then states that

$$\int_0^\pi \sqrt{1 + \cos^2 x} \, dx \approx \frac{\pi - 0}{4 \cdot 3} \left(\sqrt{2} + 4\sqrt{3/2} + 2(1) + 4\sqrt{3/2} + \sqrt{2} \right)$$

$$= 3.82918.$$

x	$\sqrt{1 + \cos^2 x}$
0	$\sqrt{2}$
$\pi/4$	$\sqrt{3/2}$
$\pi/2$	1
$3\pi/4$	$\sqrt{3/2}$
π	$\sqrt{2}$

Figure 7.4.5: A table of values of $y = \sqrt{1 + \cos^2 x}$ to evaluate a definite integral in Example 7.4.3.

Using a computer with $n = 100$ the approximation is $L \approx 3.8202$; our approximation with $n = 4$ is quite good.

Notes:

Surface Area of Solids of Revolution

We have already seen how a curve $y = f(x)$ on $[a, b]$ can be revolved around an axis to form a solid. Instead of computing its volume, we now consider its surface area.

We begin as we have in the previous sections: we partition the interval $[a, b]$ with n subintervals, where the ith subinterval is $[x_i, x_{i+1}]$. On each subinterval, we can approximate the curve $y = f(x)$ with a straight line that connects $f(x_i)$ and $f(x_{i+1})$ as shown in Figure 7.4.6(a). Revolving this line segment about the x-axis creates part of a cone (called a *frustum* of a cone) as shown in Figure 7.4.6(b). The surface area of a frustum of a cone is

$$2\pi \cdot \text{ length } \cdot \text{ average of the two radii } R \text{ and } r.$$

The length is given by L; we use the material just covered by arc length to state that

$$L \approx \sqrt{1 + f'(c_i)^2}\,\Delta x_i$$

for some c_i in the ith subinterval. The radii are just the function evaluated at the endpoints of the interval. That is,

$$R = f(x_{i+1}) \quad \text{and} \quad r = f(x_i).$$

Thus the surface area of this sample frustum of the cone is approximately

$$2\pi \frac{f(x_i) + f(x_{i+1})}{2} \sqrt{1 + f'(c_i)^2}\,\Delta x_i.$$

Since f is a continuous function, the Intermediate Value Theorem states there is some d_i in $[x_i, x_{i+1}]$ such that $f(d_i) = \dfrac{f(x_i) + f(x_{i+1})}{2}$; we can use this to rewrite the above equation as

$$2\pi f(d_i) \sqrt{1 + f'(c_i)^2}\,\Delta x_i.$$

Summing over all the subintervals we get the total surface area to be approximately

$$\text{Surface Area} \approx \sum_{i=1}^{n} 2\pi f(d_i) \sqrt{1 + f'(c_i)^2}\,\Delta x_i,$$

which is a Riemann Sum. Taking the limit as the subinterval lengths go to zero gives us the exact surface area, given in the following theorem.

(a)

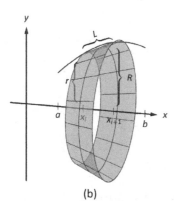

(b)

Figure 7.4.6: Establishing the formula for surface area.

Notes:

Theorem 7.4.2   **Surface Area of a Solid of Revolution**

Let f be differentiable on $[a, b]$, where f' is also continuous on $[a, b]$.

1. The surface area of the solid formed by revolving the graph of $y = f(x)$, where $f(x) \geq 0$, about the x-axis is

$$\text{Surface Area} = 2\pi \int_a^b f(x)\sqrt{1 + f'(x)^2}\, dx.$$

2. The surface area of the solid formed by revolving the graph of $y = f(x)$ about the y-axis, where $a, b \geq 0$, is

$$\text{Surface Area} = 2\pi \int_a^b x\sqrt{1 + f'(x)^2}\, dx.$$

(When revolving $y = f(x)$ about the y-axis, the radii of the resulting frustum are x_i and x_{i+1}; their average value is simply the midpoint of the interval. In the limit, this midpoint is just x. This gives the second part of Theorem 7.4.2.)

Example 7.4.4   **Finding surface area of a solid of revolution**
Find the surface area of the solid formed by revolving $y = \sin x$ on $[0, \pi]$ around the x-axis, as shown in Figure 7.4.7.

SOLUTION   The setup is relatively straightforward. Using Theorem 7.4.2, we have the surface area SA is:

$$\begin{aligned}
SA &= 2\pi \int_0^\pi \sin x \sqrt{1 + \cos^2 x}\, dx \\
&= -2\pi \frac{1}{2} \left(\sinh^{-1}(\cos x) + \cos x \sqrt{1 + \cos^2 x} \right) \Big|_0^\pi \\
&= 2\pi \left(\sqrt{2} + \sinh^{-1} 1 \right) \approx 14.42 \text{ units}^2.
\end{aligned}$$

The integration step above is nontrivial, utilizing an integration method called Trigonometric Substitution.

It is interesting to see that the surface area of a solid, whose shape is defined by a trigonometric function, involves both a square root and an inverse hyperbolic trigonometric function.

Example 7.4.5   **Finding surface area of a solid of revolution**
Find the surface area of the solid formed by revolving the curve $y = x^2$ on $[0, 1]$ about the x-axis and the y-axis.

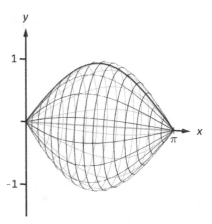

Figure 7.4.7: Revolving $y = \sin x$ on $[0, \pi]$ about the x-axis.

Notes:

(a)

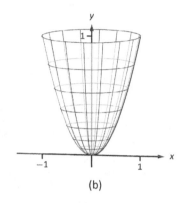

(b)

Figure 7.4.8: The solids used in Example 7.4.5.

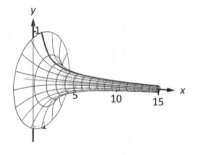

Figure 7.4.9: A graph of Gabriel's Horn.

SOLUTION   About the x-axis: the integral is straightforward to setup:

$$SA = 2\pi \int_0^1 x^2 \sqrt{1 + (2x)^2} \, dx.$$

Like the integral in Example 7.4.4, this requires Trigonometric Substitution.

$$= \frac{\pi}{32} \left(2(8x^3 + x) \sqrt{1 + 4x^2} - \sinh^{-1}(2x) \right) \Big|_0^1$$

$$= \frac{\pi}{32} \left(18\sqrt{5} - \sinh^{-1} 2 \right)$$

$$\approx 3.81 \text{ units}^2.$$

The solid formed by revolving $y = x^2$ around the x-axis is graphed in Figure 7.4.8 (a).

About the y-axis: since we are revolving around the y-axis, the "radius" of the solid is not $f(x)$ but rather x. Thus the integral to compute the surface area is:

$$SA = 2\pi \int_0^1 x \sqrt{1 + (2x)^2} \, dx.$$

This integral can be solved using substitution. Set $u = 1 + 4x^2$; the new bounds are $u = 1$ to $u = 5$. We then have

$$= \frac{\pi}{4} \int_1^5 \sqrt{u} \, du$$

$$= \frac{\pi}{4} \frac{2}{3} u^{3/2} \Big|_1^5$$

$$= \frac{\pi}{6} \left(5\sqrt{5} - 1 \right)$$

$$\approx 5.33 \text{ units}^2.$$

The solid formed by revolving $y = x^2$ about the y-axis is graphed in Figure 7.4.8 (b).

Our final example is a famous mathematical "paradox."

Example 7.4.6   The surface area and volume of Gabriel's Horn
Consider the solid formed by revolving $y = 1/x$ about the x-axis on $[1, \infty)$. Find the volume and surface area of this solid. (This shape, as graphed in Figure 7.4.9, is known as "Gabriel's Horn" since it looks like a very long horn that only a supernatural person, such as an angel, could play.)

Notes:

SOLUTION To compute the volume it is natural to use the Disk Method. We have:

$$V = \pi \int_1^\infty \frac{1}{x^2}\, dx$$

$$= \lim_{b\to\infty} \pi \int_1^b \frac{1}{x^2}\, dx$$

$$= \lim_{b\to\infty} \pi \left(\frac{-1}{x}\right)\Big|_1^b$$

$$= \lim_{b\to\infty} \pi \left(1 - \frac{1}{b}\right)$$

$$= \pi \text{ units}^3.$$

Gabriel's Horn has a finite volume of π cubic units. Since we have already seen that regions with infinite length can have a finite area, this is not too difficult to accept.

We now consider its surface area. The integral is straightforward to setup:

$$SA = 2\pi \int_1^\infty \frac{1}{x}\sqrt{1 + 1/x^4}\, dx.$$

Integrating this expression is not trivial. We can, however, compare it to other improper integrals. Since $1 < \sqrt{1 + 1/x^4}$ on $[1, \infty)$, we can state that

$$2\pi \int_1^\infty \frac{1}{x}\, dx < 2\pi \int_1^\infty \frac{1}{x}\sqrt{1 + 1/x^4}\, dx.$$

By Key Idea 6.8.1, the improper integral on the left diverges. Since the integral on the right is larger, we conclude it also diverges, meaning Gabriel's Horn has infinite surface area.

Hence the "paradox": we can fill Gabriel's Horn with a finite amount of paint, but since it has infinite surface area, we can never paint it.

Somehow this paradox is striking when we think about it in terms of volume and area. However, we have seen a similar paradox before, as referenced above. We know that the area under the curve $y = 1/x^2$ on $[1, \infty)$ is finite, yet the shape has an infinite perimeter. Strange things can occur when we deal with the infinite.

A standard equation from physics is "Work = force $\times$ distance", when the force applied is constant. In the next section we learn how to compute work when the force applied is variable.

Notes:

Exercises 7.4

Terms and Concepts

1. T/F: The integral formula for computing Arc Length was found by first approximating arc length with straight line segments.

2. T/F: The integral formula for computing Arc Length includes a square–root, meaning the integration is probably easy.

Problems

In Exercises 3 – 12, find the arc length of the function on the given interval.

3. $f(x) = x$ on $[0, 1]$.

4. $f(x) = \sqrt{8}x$ on $[-1, 1]$.

5. $f(x) = \frac{1}{3}x^{3/2} - x^{1/2}$ on $[0, 1]$.

6. $f(x) = \frac{1}{12}x^3 + \frac{1}{x}$ on $[1, 4]$.

7. $f(x) = 2x^{3/2} - \frac{1}{6}\sqrt{x}$ on $[0, 9]$.

8. $f(x) = \cosh x$ on $[-\ln 2, \ln 2]$.

9. $f(x) = \frac{1}{2}\left(e^x + e^{-x}\right)$ on $[0, \ln 5]$.

10. $f(x) = \frac{1}{12}x^5 + \frac{1}{5x^3}$ on $[.1, 1]$.

11. $f(x) = \ln\left(\sin x\right)$ on $[\pi/6, \pi/2]$.

12. $f(x) = \ln\left(\cos x\right)$ on $[0, \pi/4]$.

In Exercises 13 – 20, set up the integral to compute the arc length of the function on the given interval. Do not evaluate the integral.

13. $f(x) = x^2$ on $[0, 1]$.

14. $f(x) = x^{10}$ on $[0, 1]$.

15. $f(x) = \sqrt{x}$ on $[0, 1]$.

16. $f(x) = \ln x$ on $[1, e]$.

17. $f(x) = \sqrt{1 - x^2}$ on $[-1, 1]$. (Note: this describes the top half of a circle with radius 1.)

18. $f(x) = \sqrt{1 - x^2/9}$ on $[-3, 3]$. (Note: this describes the top half of an ellipse with a major axis of length 6 and a minor axis of length 2.)

19. $f(x) = \frac{1}{x}$ on $[1, 2]$.

20. $f(x) = \sec x$ on $[-\pi/4, \pi/4]$.

In Exercises 21 – 28, use Simpson's Rule, with $n = 4$, to approximate the arc length of the function on the given interval. Note: these are the same problems as in Exercises 13–20.

21. $f(x) = x^2$ on $[0, 1]$.

22. $f(x) = x^{10}$ on $[0, 1]$.

23. $f(x) = \sqrt{x}$ on $[0, 1]$. (Note: $f'(x)$ is not defined at $x = 0$.)

24. $f(x) = \ln x$ on $[1, e]$.

25. $f(x) = \sqrt{1 - x^2}$ on $[-1, 1]$. (Note: $f'(x)$ is not defined at the endpoints.)

26. $f(x) = \sqrt{1 - x^2/9}$ on $[-3, 3]$. (Note: $f'(x)$ is not defined at the endpoints.)

27. $f(x) = \frac{1}{x}$ on $[1, 2]$.

28. $f(x) = \sec x$ on $[-\pi/4, \pi/4]$.

In Exercises 29 – 33, find the surface area of the described solid of revolution.

29. The solid formed by revolving $y = 2x$ on $[0, 1]$ about the x-axis.

30. The solid formed by revolving $y = x^2$ on $[0, 1]$ about the y-axis.

31. The solid formed by revolving $y = x^3$ on $[0, 1]$ about the x-axis.

32. The solid formed by revolving $y = \sqrt{x}$ on $[0, 1]$ about the x-axis.

33. The sphere formed by revolving $y = \sqrt{1 - x^2}$ on $[-1, 1]$ about the x-axis.

7.5 Work

Work is the scientific term used to describe the action of a force which moves an object. When a constant force F is applied to move an object a distance d, the amount of work performed is $W = F \cdot d$.

The SI unit of force is the Newton, (kg·m/s^2), and the SI unit of distance is a meter (m). The fundamental unit of work is one Newton–meter, or a joule (J). That is, applying a force of one Newton for one meter performs one joule of work. In Imperial units (as used in the United States), force is measured in pounds (lb) and distance is measured in feet (ft), hence work is measured in ft–lb.

When force is constant, the measurement of work is straightforward. For instance, lifting a 200 lb object 5 ft performs $200 \cdot 5 = 1000$ ft–lb of work.

What if the force applied is variable? For instance, imagine a climber pulling a 200 ft rope up a vertical face. The rope becomes lighter as more is pulled in, requiring less force and hence the climber performs less work.

In general, let $F(x)$ be a force function on an interval $[a, b]$. We want to measure the amount of work done applying the force F from $x = a$ to $x = b$. We can approximate the amount of work being done by partitioning $[a, b]$ into subintervals $a = x_1 < x_2 < \cdots < x_{n+1} = b$ and assuming that F is constant on each subinterval. Let c_i be a value in the ith subinterval $[x_i, x_{i+1}]$. Then the work done on this interval is approximately $W_i \approx F(c_i) \cdot (x_{i+1} - x_i) = F(c_i)\Delta x_i$, a constant force $\times$ the distance over which it is applied. The total work is

$$W = \sum_{i=1}^{n} W_i \approx \sum_{i-1}^{n} F(c_i)\Delta x_i.$$

This, of course, is a Riemann sum. Taking a limit as the subinterval lengths go to zero gives an exact value of work which can be evaluated through a definite integral.

Note: *Mass* and *weight* are closely related, yet different, concepts. The mass m of an object is a quantitative measure of that object's resistance to acceleration. The weight w of an object is a measurement of the force applied to the object by the acceleration of gravity g.

Since the two measurements are proportional, $w = m \cdot g$, they are often used interchangeably in everyday conversation. When computing work, one must be careful to note which is being referred to. When mass is given, it must be multiplied by the acceleration of gravity to reference the related force.

> **Key Idea 7.5.1   Work**
>
> Let $F(x)$ be a continuous function on $[a, b]$ describing the amount of force being applied to an object in the direction of travel from distance $x = a$ to distance $x = b$. The total work W done on $[a, b]$ is
>
> $$W = \int_a^b F(x)\, dx.$$

Notes:

Example 7.5.1   Computing work performed: applying variable force
A 60m climbing rope is hanging over the side of a tall cliff. How much work is performed in pulling the rope up to the top, where the rope has a mass of 66g/m?

SOLUTION   We need to create a force function $F(x)$ on the interval $[0, 60]$. To do so, we must first decide what x is measuring: it is the length of the rope still hanging or is it the amount of rope pulled in? As long as we are consistent, either approach is fine. We adopt for this example the convention that x is the amount of rope pulled in. This seems to match intuition better; pulling up the first 10 meters of rope involves $x = 0$ to $x = 10$ instead of $x = 60$ to $x = 50$.

As x is the amount of rope pulled in, the amount of rope still hanging is $60-x$. This length of rope has a mass of 66 g/m, or 0.066 kg/m. The mass of the rope still hanging is $0.066(60 - x)$ kg; multiplying this mass by the acceleration of gravity, 9.8 m/s^2, gives our variable force function

$$F(x) = (9.8)(0.066)(60 - x) = 0.6468(60 - x).$$

Thus the total work performed in pulling up the rope is

$$W = \int_0^{60} 0.6468(60 - x)\, dx = 1,164.24 \text{ J}.$$

By comparison, consider the work done in lifting the entire rope 60 meters. The rope weighs $60 \times 0.066 \times 9.8 = 38.808$ N, so the work applying this force for 60 meters is $60 \times 38.808 = 2,328.48$ J. This is exactly twice the work calculated before (and we leave it to the reader to understand why.)

Example 7.5.2   Computing work performed: applying variable force
Consider again pulling a 60 m rope up a cliff face, where the rope has a mass of 66 g/m. At what point is exactly half the work performed?

SOLUTION   From Example 7.5.1 we know the total work performed is $1,164.24$ J. We want to find a height h such that the work in pulling the rope from a height of $x = 0$ to a height of $x = h$ is 582.12, half the total work. Thus we want to solve the equation

$$\int_0^{h} 0.6468(60 - x)\, dx = 582.12$$

for h.

Notes:

$$\int_0^h 0.6468(60 - x)\, dx = 582.12$$

$$\left(38.808x - 0.3234x^2\right)\Big|_0^h = 582.12$$

$$38.808h - 0.3234h^2 = 582.12$$

$$-0.3234h^2 + 38.808h - 582.12 = 0.$$

Note: In Example 7.5.2, we find that half of the work performed in pulling up a 60 m rope is done in the last 42.43 m. Why is it not coincidental that $60/\sqrt{2} = 42.43$?

Apply the Quadratic Formula:

$$h = 17.57 \text{ and } 102.43$$

As the rope is only 60m long, the only sensible answer is $h = 17.57$. Thus about half the work is done pulling up the first 17.5m the other half of the work is done pulling up the remaining 42.43m.

Example 7.5.3   Computing work performed: applying variable force
A box of 100 lb of sand is being pulled up at a uniform rate a distance of 50 ft over 1 minute. The sand is leaking from the box at a rate of 1 lb/s. The box itself weighs 5 lb and is pulled by a rope weighing .2 lb/ft.

1. How much work is done lifting just the rope?

2. How much work is done lifting just the box and sand?

3. What is the total amount of work performed?

SOLUTION

1. We start by forming the force function $F_r(x)$ for the rope (where the subscript denotes we are considering the rope). As in the previous example, let x denote the amount of rope, in feet, pulled in. (This is the same as saying x denotes the height of the box.) The weight of the rope with x feet pulled in is $F_r(x) = 0.2(50 - x) = 10 - 0.2x$. (Note that we do not have to include the acceleration of gravity here, for the *weight* of the rope per foot is given, not its *mass* per meter as before.) The work performed lifting the rope is

$$W_r = \int_0^{50} (10 - 0.2x)\, dx = 250 \text{ ft–lb.}$$

Notes:

2. The sand is leaving the box at a rate of 1 lb/s. As the vertical trip is to take one minute, we know that 60 lb will have left when the box reaches its final height of 50 ft. Again letting x represent the height of the box, we have two points on the line that describes the weight of the sand: when $x = 0$, the sand weight is 100 lb, producing the point $(0, 100)$; when $x = 50$, the sand in the box weighs 40 lb, producing the point $(50, 40)$. The slope of this line is $\frac{100-40}{0-50} = -1.2$, giving the equation of the weight of the sand at height x as $w(x) = -1.2x + 100$. The box itself weighs a constant 5 lb, so the total force function is $F_b(x) = -1.2x + 105$. Integrating from $x = 0$ to $x = 50$ gives the work performed in lifting box and sand:

$$W_b = \int_0^{50} (-1.2x + 105)\, dx = 3750 \text{ ft–lb.}$$

3. The total work is the sum of W_r and W_b: $250 + 3750 = 4000$ ft–lb. We can also arrive at this via integration:

$$W = \int_0^{50} (F_r(x) + F_b(x))\, dx$$
$$= \int_0^{50} (10 - 0.2x - 1.2x + 105)\, dx$$
$$= \int_0^{50} (-1.4x + 115)\, dx$$
$$= 4000 \text{ ft–lb.}$$

Hooke's Law and Springs

Hooke's Law states that the force required to compress or stretch a spring x units from its natural length is proportional to x; that is, this force is $F(x) = kx$ for some constant k. For example, if a force of 1 N stretches a given spring 2 cm, then a force of 5 N will stretch the spring 10 cm. Converting the distances to meters, we have that stretching this spring 0.02 m requires a force of $F(0.02) = k(0.02) = 1$ N, hence $k = 1/0.02 = 50$ N/m.

Example 7.5.4   Computing work performed: stretching a spring
A force of 20 lb stretches a spring from a natural length of 7 inches to a length of 12 inches. How much work was performed in stretching the spring to this length?

Solution   In many ways, we are not at all concerned with the actual length of the spring, only with the amount of its change. Hence, we do not care

Notes:

that 20 lb of force stretches the spring to a length of 12 inches, but rather that a force of 20 lb stretches the spring by 5 in. This is illustrated in Figure 7.5.1; we only measure the change in the spring's length, not the overall length of the spring.

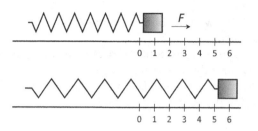

Figure 7.5.1: Illustrating the important aspects of stretching a spring in computing work in Example 7.5.4.

Converting the units of length to feet, we have

$$F(5/12) = 5/12k = 20 \text{ lb}.$$

Thus $k = 48$ lb/ft and $F(x) = 48x$.

We compute the total work performed by integrating $F(x)$ from $x = 0$ to $x = 5/12$:

$$W = \int_0^{5/12} 48x \, dx$$

$$= 24x^2 \Big|_0^{5/12}$$

$$= 25/6 \approx 4.1667 \text{ ft–lb}.$$

Pumping Fluids

Another useful example of the application of integration to compute work comes in the pumping of fluids, often illustrated in the context of emptying a storage tank by pumping the fluid out the top. This situation is different than our previous examples for the forces involved are constant. After all, the force required to move one cubic foot of water (about 62.4 lb) is the same regardless of its location in the tank. What is variable is the distance that cubic foot of water has to travel; water closer to the top travels less distance than water at the bottom, producing less work.

We demonstrate how to compute the total work done in pumping a fluid out of the top of a tank in the next two examples.

Fluid	lb/ft^3	kg/m^3
Concrete	150	2400
Fuel Oil	55.46	890.13
Gasoline	45.93	737.22
Iodine	307	4927
Methanol	49.3	791.3
Mercury	844	13546
Milk	63.6–65.4	1020 – 1050
Water	62.4	1000

Figure 7.5.2: Weight and Mass densities

Notes:

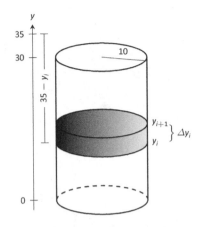

Figure 7.5.3: Illustrating a water tank in order to compute the work required to empty it in Example 7.5.5.

Example 7.5.5   Computing work performed: pumping fluids

A cylindrical storage tank with a radius of 10 ft and a height of 30 ft is filled with water, which weighs approximately 62.4 lb/ft^3. Compute the amount of work performed by pumping the water up to a point 5 feet above the top of the tank.

SOLUTION   We will refer often to Figure 7.5.3 which illustrates the salient aspects of this problem.

We start as we often do: we partition an interval into subintervals. We orient our tank vertically since this makes intuitive sense with the base of the tank at $y = 0$. Hence the top of the water is at $y = 30$, meaning we are interested in subdividing the y-interval $[0, 30]$ into n subintervals as

$$0 = y_1 < y_2 < \cdots < y_{n+1} = 30.$$

Consider the work W_i of pumping only the water residing in the ith subinterval, illustrated in Figure 7.5.3. The force required to move this water is equal to its weight which we calculate as volume $\times$ density. The volume of water in this subinterval is $V_i = 10^2 \pi \Delta y_i$; its density is 62.4 lb/ft^3. Thus the required force is $6240\pi \Delta y_i$ lb.

We approximate the distance the force is applied by using any y-value contained in the ith subinterval; for simplicity, we arbitrarily use y_i for now (it will not matter later on). The water will be pumped to a point 5 feet above the top of the tank, that is, to the height of $y = 35$ ft. Thus the distance the water at height y_i travels is $35 - y_i$ ft.

In all, the approximate work W_i peformed in moving the water in the ith subinterval to a point 5 feet above the tank is

$$W_i \approx 6240\pi \Delta y_i (35 - y_i).$$

To approximate the total work performed in pumping out all the water from the tank, we sum all the work W_i performed in pumping the water from each of the n subintervals of $[0, 30]$:

$$W \approx \sum_{i=1}^{n} W_i = \sum_{i=1}^{n} 6240\pi \Delta y_i (35 - y_i).$$

This is a Riemann sum. Taking the limit as the subinterval length goes to 0 gives

$$W = \int_0^{30} 6240\pi (35 - y)\, dy$$

$$= 6240\pi \left(35y - 1/2 y^2 \right) \Big|_0^{30}$$

$$= 11{,}762{,}123 \text{ ft–lb}$$

$$\approx 1.176 \times 10^7 \text{ ft–lb}.$$

Notes:

We can "streamline" the above process a bit as we may now recognize what the important features of the problem are. Figure 7.5.4 shows the tank from Example 7.5.5 without the ith subinterval identified. Instead, we just draw one differential element. This helps establish the height a small amount of water must travel along with the force required to move it (where the force is volume × density).

We demonstrate the concepts again in the next examples.

Example 7.5.6   Computing work performed: pumping fluids

A conical water tank has its top at ground level and its base 10 feet below ground. The radius of the cone at ground level is 2 ft. It is filled with water weighing 62.4 lb/ft^3 and is to be emptied by pumping the water to a spigot 3 feet above ground level. Find the total amount of work performed in emptying the tank.

SOLUTION   The conical tank is sketched in Figure 7.5.5. We can orient the tank in a variety of ways; we could let $y = 0$ represent the base of the tank and $y = 10$ represent the top of the tank, but we choose to keep the convention of the wording given in the problem and let $y = 0$ represent ground level and hence $y = -10$ represents the bottom of the tank. The actual "height" of the water does not matter; rather, we are concerned with the distance the water travels.

The figure also sketches a differential element, a cross–sectional circle. The radius of this circle is variable, depending on y. When $y = -10$, the circle has radius 0; when $y = 0$, the circle has radius 2. These two points, $(-10, 0)$ and $(0, 2)$, allow us to find the equation of the line that gives the radius of the cross–sectional circle, which is $r(y) = 1/5y + 2$. Hence the volume of water at this height is $V(y) = \pi(1/5y + 2)^2 dy$, where dy represents a very small height of the differential element. The force required to move the water at height y is $F(y) = 62.4 \times V(y)$.

The distance the water at height y travels is given by $h(y) = 3 - y$. Thus the total work done in pumping the water from the tank is

$$W = \int_{-10}^{0} 62.4\pi (1/5y + 2)^2 (3 - y)\, dy$$

$$= 62.4\pi \int_{-10}^{0} \left(-\frac{1}{25}y^3 - \frac{17}{25}y^2 - \frac{8}{5}y + 12 \right) dy$$

$$= 62.2\pi \cdot \frac{220}{3} \approx 14,376 \text{ ft–lb}.$$

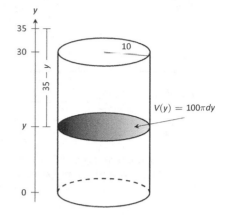

Figure 7.5.4: A simplified illustration for computing work.

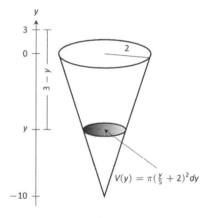

Figure 7.5.5: A graph of the conical water tank in Example 7.5.6.

Notes:

Figure 7.5.6: The cross–section of a swimming pool filled with water in Example 7.5.7.

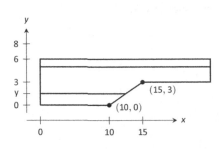

Figure 7.5.7: Orienting the pool and showing differential elements for Example 7.5.7.

Example 7.5.7   **Computing work performed: pumping fluids**

A rectangular swimming pool is 20 ft wide and has a 3 ft "shallow end" and a 6 ft "deep end." It is to have its water pumped out to a point 2 ft above the current top of the water. The cross–sectional dimensions of the water in the pool are given in Figure 7.5.6; note that the dimensions are for the water, not the pool itself. Compute the amount of work performed in draining the pool.

SOLUTION   For the purposes of this problem we choose to set $y = 0$ to represent the bottom of the pool, meaning the top of the water is at $y = 6$. Figure 7.5.7 shows the pool oriented with this y-axis, along with 2 differential elements as the pool must be split into two different regions.

The top region lies in the y-interval of $[3, 6]$, where the length of the differential element is 25 ft as shown. As the pool is 20 ft wide, this differential element represents a thin slice of water with volume $V(y) = 20 \cdot 25 \cdot dy$. The water is to be pumped to a height of $y = 8$, so the height function is $h(y) = 8 - y$. The work done in pumping this top region of water is

$$W_t = 62.4 \int_3^6 500(8 - y)\, dy = 327,600 \text{ ft–lb.}$$

The bottom region lies in the y-interval of $[0, 3]$; we need to compute the length of the differential element in this interval.

One end of the differential element is at $x = 0$ and the other is along the line segment joining the points $(10, 0)$ and $(15, 3)$. The equation of this line is $y = 3/5(x - 10)$; as we will be integrating with respect to y, we rewrite this equation as $x = 5/3y + 10$. So the length of the differential element is a difference of x-values: $x = 0$ and $x = 5/3y + 10$, giving a length of $x = 5/3y + 10$.

Again, as the pool is 20 ft wide, this differential element represents a thin slice of water with volume $V(y) = 20 \cdot (5/3y + 10) \cdot dy$; the height function is the same as before at $h(y) = 8 - y$. The work performed in emptying this part of the pool is

$$W_b = 62.4 \int_0^3 20(5/3y + 10)(8 - y)\, dy = 299,520 \text{ ft–lb.}$$

The total work in emptying the pool is

$$W = W_b + W_t = 327,600 + 299,520 = 627,120 \text{ ft–lb.}$$

Notice how the emptying of the bottom of the pool performs almost as much work as emptying the top. The top portion travels a shorter distance but has more water. In the end, this extra water produces more work.

The next section introduces one final application of the definite integral, the calculation of fluid force on a plate.

Notes:

Exercises 7.5

Terms and Concepts

1. What are the typical units of work?

2. If a man has a mass of 80 kg on Earth, will his mass on the moon be bigger, smaller, or the same?

3. If a woman weighs 130 lb on Earth, will her weight on the moon be bigger, smaller, or the same?

4. Fill in the blanks:

   Some integrals in this section are set up by multiplying a variable _____ by a constant distance; others are set up by multiplying a constant force by a variable _____.

Problems

5. A 100 ft rope, weighing 0.1 lb/ft, hangs over the edge of a tall building.

   (a) How much work is done pulling the entire rope to the top of the building?

   (b) How much rope is pulled in when half of the total work is done?

6. A 50 m rope, with a mass density of 0.2 kg/m, hangs over the edge of a tall building.

   (a) How much work is done pulling the entire rope to the top of the building?

   (b) How much work is done pulling in the first 20 m?

7. A rope of length ℓ ft hangs over the edge of tall cliff. (Assume the cliff is taller than the length of the rope.) The rope has a weight density of d lb/ft.

   (a) How much work is done pulling the entire rope to the top of the cliff?

   (b) What percentage of the total work is done pulling in the first half of the rope?

   (c) How much rope is pulled in when half of the total work is done?

8. A 20 m rope with mass density of 0.5 kg/m hangs over the edge of a 10 m building. How much work is done pulling the rope to the top?

9. A crane lifts a 2,000 lb load vertically 30 ft with a 1″ cable weighing 1.68 lb/ft.

   (a) How much work is done lifting the cable alone?

   (b) How much work is done lifting the load alone?

   (c) Could one conclude that the work done lifting the cable is negligible compared to the work done lifting the load?

10. A 100 lb bag of sand is lifted uniformly 120 ft in one minute. Sand leaks from the bag at a rate of 1/4 lb/s. What is the total work done in lifting the bag?

11. A box weighing 2 lb lifts 10 lb of sand vertically 50 ft. A crack in the box allows the sand to leak out such that 9 lb of sand is in the box at the end of the trip. Assume the sand leaked out at a uniform rate. What is the total work done in lifting the box and sand?

12. A force of 1000 lb compresses a spring 3 in. How much work is performed in compressing the spring?

13. A force of 2 N stretches a spring 5 cm. How much work is performed in stretching the spring?

14. A force of 50 lb compresses a spring from a natural length of 18 in to 12 in. How much work is performed in compressing the spring?

15. A force of 20 lb stretches a spring from a natural length of 6 in to 8 in. How much work is performed in stretching the spring?

16. A force of 7 N stretches a spring from a natural length of 11 cm to 21 cm. How much work is performed in stretching the spring from a length of 16 cm to 21 cm?

17. A force of f N stretches a spring d m from its natural length. How much work is performed in stretching the spring?

18. A 20 lb weight is attached to a spring. The weight rests on the spring, compressing the spring from a natural length of 1 ft to 6 in.

    How much work is done in lifting the box 1.5 ft (i.e, the spring will be stretched 1 ft beyond its natural length)?

19. A 20 lb weight is attached to a spring. The weight rests on the spring, compressing the spring from a natural length of 1 ft to 6 in.

    How much work is done in lifting the box 6 in (i.e, bringing the spring back to its natural length)?

20. A 5 m tall cylindrical tank with radius of 2 m is filled with 3 m of gasoline, with a mass density of 737.22 kg/m^3. Compute the total work performed in pumping all the gasoline to the top of the tank.

21. A 6 ft cylindrical tank with a radius of 3 ft is filled with water, which has a weight density of 62.4 lb/ft^3. The water is to be pumped to a point 2 ft above the top of the tank.

    (a) How much work is performed in pumping all the water from the tank?

    (b) How much work is performed in pumping 3 ft of water from the tank?

    (c) At what point is 1/2 of the total work done?

22. A gasoline tanker is filled with gasoline with a weight density of 45.93 lb/ft^3. The dispensing valve at the base is jammed shut, forcing the operator to empty the tank via pumping the gas to a point 1 ft above the top of the tank. Assume the tank is a perfect cylinder, 20 ft long with a diameter of 7.5 ft. How much work is performed in pumping all the gasoline from the tank?

23. A fuel oil storage tank is 10 ft deep with trapezoidal sides, 5 ft at the top and 2 ft at the bottom, and is 15 ft wide (see diagram below). Given that fuel oil weighs 55.46 lb/ft^3, find the work performed in pumping all the oil from the tank to a point 3 ft above the top of the tank.

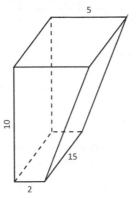

24. A conical water tank is 5 m deep with a top radius of 3 m. (This is similar to Example 7.5.6.) The tank is filled with pure water, with a mass density of 1000 kg/m^3.

    (a) Find the work performed in pumping all the water to the top of the tank.

    (b) Find the work performed in pumping the top 2.5 m of water to the top of the tank.

    (c) Find the work performed in pumping the top half of the water, by volume, to the top of the tank.

25. A water tank has the shape of a truncated cone, with dimensions given below, and is filled with water with a weight density of 62.4 lb/ft^3. Find the work performed in pumping all water to a point 1 ft above the top of the tank.

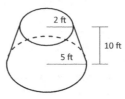

26. A water tank has the shape of an inverted pyramid, with dimensions given below, and is filled with water with a mass density of 1000 kg/m^3. Find the work performed in pumping all water to a point 5 m above the top of the tank.

27. A water tank has the shape of an truncated, inverted pyramid, with dimensions given below, and is filled with water with a mass density of 1000 kg/m^3. Find the work performed in pumping all water to a point 1 m above the top of the tank.

7.6 Fluid Forces

In the unfortunate situation of a car driving into a body of water, the conventional wisdom is that the water pressure on the doors will quickly be so great that they will be effectively unopenable. (Survival techniques suggest immediately opening the door, rolling down or breaking the window, or waiting until the water fills up the interior at which point the pressure is equalized and the door will open. See Mythbusters episode #72 to watch Adam Savage test these options.)

How can this be true? How much force does it take to open the door of a submerged car? In this section we will find the answer to this question by examining the forces exerted by fluids.

We start with **pressure**, which is related to **force** by the following equations:

$$\text{Pressure} = \frac{\text{Force}}{\text{Area}} \quad \Leftrightarrow \quad \text{Force} = \text{Pressure} \times \text{Area}.$$

In the context of fluids, we have the following definition.

Definition 7.6.1   Fluid Pressure

Let w be the weight–density of a fluid. The **pressure** p exerted on an object at depth d in the fluid is $p = w \cdot d$.

We use this definition to find the **force** exerted on a horizontal sheet by considering the sheet's area.

Example 7.6.1   Computing fluid force

1. A cylindrical storage tank has a radius of 2 ft and holds 10 ft of a fluid with a weight–density of 50 lb/ft^3. (See Figure 7.6.1(a).) What is the force exerted on the base of the cylinder by the fluid?

2. A rectangular tank whose base is a 5 ft square has a circular hatch at the bottom with a radius of 2 ft. The tank holds 10 ft of a fluid with a weight–density of 50 lb/ft^3. (See Figure 7.6.1(b).) What is the force exerted on the hatch by the fluid?

SOLUTION

1. Using Definition 7.6.1, we calculate that the pressure exerted on the cylinder's base is $w \cdot d = 50$ lb/ft$^3 \times 10$ ft $= 500$ lb/ft^2. The area of the base is

(a)

(b)

Figure 7.6.1: The cylindrical and rectangular tank in Example 7.6.1.

Notes:

$\pi \cdot 2^2 = 4\pi$ ft^2. So the force exerted by the fluid is

$$F = 500 \times 4\pi = 6283 \text{ lb.}$$

Note that we effectively just computed the *weight* of the fluid in the tank.

2. The dimensions of the tank in this problem are irrelevant. All we are concerned with are the dimensions of the hatch and the depth of the fluid. Since the dimensions of the hatch are the same as the base of the tank in the previous part of this example, as is the depth, we see that the fluid force is the same. That is, $F = 6283$ lb.

   A key concept to understand here is that we are effectively measuring the weight of a 10 ft column of water above the hatch. The size of the tank holding the fluid does not matter.

The previous example demonstrates that computing the force exerted on a horizontally oriented plate is relatively easy to compute. What about a vertically oriented plate? For instance, suppose we have a circular porthole located on the side of a submarine. How do we compute the fluid force exerted on it?

Pascal's Principle states that the pressure exerted by a fluid at a depth is equal in all directions. Thus the pressure on any portion of a plate that is 1 ft below the surface of water is the same no matter how the plate is oriented. (Thus a hollow cube submerged at a great depth will not simply be "crushed" from above, but the sides will also crumple in. The fluid will exert force on *all* sides of the cube.)

So consider a vertically oriented plate as shown in Figure 7.6.2 submerged in a fluid with weight–density w. What is the total fluid force exerted on this plate? We find this force by first approximating the force on small horizontal strips.

Let the top of the plate be at depth b and let the bottom be at depth a. (For now we assume that surface of the fluid is at depth 0, so if the bottom of the plate is 3 ft under the surface, we have $a = -3$. We will come back to this later.) We partition the interval $[a, b]$ into n subintervals

$$a = y_1 < y_2 < \cdots < y_{n+1} = b,$$

with the ith subinterval having length Δy_i. The force F_i exerted on the plate in the ith subinterval is $F_i = \text{Pressure} \times \text{Area}$.

The pressure is depth $\times w$. We approximate the depth of this thin strip by choosing any value d_i in $[y_i, y_{i+1}]$; the depth is approximately $-d_i$. (Our convention has d_i being a negative number, so $-d_i$ is positive.) For convenience, we let d_i be an endpoint of the subinterval; we let $d_i = y_i$.

The area of the thin strip is approximately length $\times$ width. The width is Δy_i. The length is a function of some y-value c_i in the ith subinterval. We state the

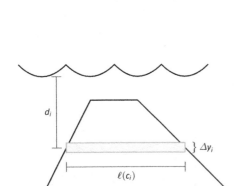

Figure 7.6.2: A thin, vertically oriented plate submerged in a fluid with weight–density w.

length is $\ell(c_i)$. Thus

$$F_i = \text{Pressure} \times \text{Area}$$
$$= -y_i \cdot w \times \ell(c_i) \cdot \Delta y_i.$$

To approximate the total force, we add up the approximate forces on each of the n thin strips:

$$F = \sum_{i=1}^{n} F_i \approx \sum_{i=1}^{n} -w \cdot y_i \cdot \ell(c_i) \cdot \Delta y_i.$$

This is, of course, another Riemann Sum. We can find the exact force by taking a limit as the subinterval lengths go to 0; we evaluate this limit with a definite integral.

Key Idea 7.6.1   Fluid Force on a Vertically Oriented Plate

Let a vertically oriented plate be submerged in a fluid with weight–density w where the top of the plate is at $y = b$ and the bottom is at $y = a$. Let $\ell(y)$ be the length of the plate at y.

1. If $y = 0$ corresponds to the surface of the fluid, then the force exerted on the plate by the fluid is

$$F = \int_a^b w \cdot (-y) \cdot \ell(y) \, dy.$$

2. In general, let $d(y)$ represent the distance between the surface of the fluid and the plate at y. Then the force exerted on the plate by the fluid is

$$F = \int_a^b w \cdot d(y) \cdot \ell(y) \, dy.$$

Example 7.6.2   Finding fluid force

Consider a thin plate in the shape of an isosceles triangle as shown in Figure 7.6.3 submerged in water with a weight–density of 62.4 lb/ft^3. If the bottom of the plate is 10 ft below the surface of the water, what is the total fluid force exerted on this plate?

SOLUTION   We approach this problem in two different ways to illustrate the different ways Key Idea 7.6.1 can be implemented. First we will let $y = 0$ represent the surface of the water, then we will consider an alternate convention.

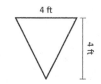

Figure 7.6.3: A thin plate in the shape of an isosceles triangle in Example 7.6.2.

Notes:

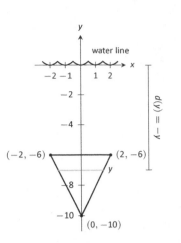

Figure 7.6.4: Sketching the triangular plate in Example 7.6.2 with the convention that the water level is at $y = 0$.

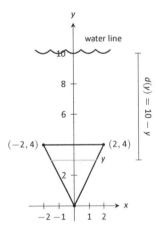

Figure 7.6.5: Sketching the triangular plate in Example 7.6.2 with the convention that the base of the triangle is at $(0, 0)$.

1. We let $y = 0$ represent the surface of the water; therefore the bottom of the plate is at $y = -10$. We center the triangle on the y-axis as shown in Figure 7.6.4. The depth of the plate at y is $-y$ as indicated by the Key Idea. We now consider the length of the plate at y.

   We need to find equations of the left and right edges of the plate. The right hand side is a line that connects the points $(0, -10)$ and $(2, -6)$: that line has equation $x = 1/2(y + 10)$. (Find the equation in the familiar $y = mx + b$ format and solve for x.) Likewise, the left hand side is described by the line $x = -1/2(y + 10)$. The total length is the distance between these two lines: $\ell(y) = 1/2(y + 10) - (-1/2(y + 10)) = y + 10$.

   The total fluid force is then:

   $$F = \int_{-10}^{-6} 62.4(-y)(y + 10) \, dy$$

   $$= 62.4 \cdot \frac{176}{3} \approx 3660.8 \text{ lb.}$$

2. Sometimes it seems easier to orient the thin plate nearer the origin. For instance, consider the convention that the bottom of the triangular plate is at $(0, 0)$, as shown in Figure 7.6.5. The equations of the left and right hand sides are easy to find. They are $y = 2x$ and $y = -2x$, respectively, which we rewrite as $x = 1/2y$ and $x = -1/2y$. Thus the length function is $\ell(y) = 1/2y - (-1/2y) = y$.

   As the surface of the water is 10 ft above the base of the plate, we have that the surface of the water is at $y = 10$. Thus the depth function is the distance between $y = 10$ and y; $d(y) = 10 - y$. We compute the total fluid force as:

   $$F = \int_{0}^{4} 62.4(10 - y)(y) \, dy$$

   $$\approx 3660.8 \text{ lb.}$$

The correct answer is, of course, independent of the placement of the plate in the coordinate plane as long as we are consistent.

Example 7.6.3   Finding fluid force
Find the total fluid force on a car door submerged up to the bottom of its window in water, where the car door is a rectangle 40" long and 27" high (based on the dimensions of a 2005 Fiat Grande Punto.)

SOLUTION   The car door, as a rectangle, is drawn in Figure 7.6.6. Its length is 10/3 ft and its height is 2.25 ft. We adopt the convention that the top

Notes:

of the door is at the surface of the water, both of which are at $y = 0$. Using the weight–density of water of 62.4 lb/ft^3, we have the total force as

$$F = \int_{-2.25}^{0} 62.4(-y)10/3 \, dy$$

$$= \int_{-2.25}^{0} -208y \, dy$$

$$= -104y^2 \Big|_{-2.25}^{0}$$

$$= 526.5 \text{ lb.}$$

Figure 7.6.6: Sketching a submerged car door in Example 7.6.3.

Most adults would find it very difficult to apply over 500 lb of force to a car door while seated inside, making the door effectively impossible to open. This is counter–intuitive as most assume that the door would be relatively easy to open. The truth is that it is not, hence the survival tips mentioned at the beginning of this section.

Example 7.6.4   Finding fluid force

An underwater observation tower is being built with circular viewing portholes enabling visitors to see underwater life. Each vertically oriented porthole is to have a 3 ft diameter whose center is to be located 50 ft underwater. Find the total fluid force exerted on each porthole. Also, compute the fluid force on a horizontally oriented porthole that is under 50 ft of water.

SOLUTION   We place the center of the porthole at the origin, meaning the surface of the water is at $y = 50$ and the depth function will be $d(y) = 50 - y$; see Figure 7.6.7

The equation of a circle with a radius of 1.5 is $x^2 + y^2 = 2.25$; solving for x we have $x = \pm\sqrt{2.25 - y^2}$, where the positive square root corresponds to the right side of the circle and the negative square root corresponds to the left side of the circle. Thus the length function at depth y is $\ell(y) = 2\sqrt{2.25 - y^2}$. Integrating on $[-1.5, 1.5]$ we have:

Figure 7.6.7: Measuring the fluid force on an underwater porthole in Example 7.6.4.

$$F = 62.4 \int_{-1.5}^{1.5} 2(50 - y)\sqrt{2.25 - y^2} \, dy$$

$$= 62.4 \int_{-1.5}^{1.5} \left(100\sqrt{2.25 - y^2} - 2y\sqrt{2.25 - y^2}\right) dy$$

$$= 6240 \int_{-1.5}^{1.5} \left(\sqrt{2.25 - y^2}\right) dy - 62.4 \int_{-1.5}^{1.5} \left(2y\sqrt{2.25 - y^2}\right) dy.$$

Notes:

The second integral above can be evaluated using substitution. Let $u = 2.25 - y^2$ with $du = -2y\,dy$. The new bounds are: $u(-1.5) = 0$ and $u(1.5) = 0$; the new integral will integrate from $u = 0$ to $u = 0$, hence the integral is 0.

The first integral above finds the area of half a circle of radius 1.5, thus the first integral evaluates to $6240 \cdot \pi \cdot 1.5^2/2 = 22{,}054$. Thus the total fluid force on a vertically oriented porthole is 22,054 lb.

Finding the force on a horizontally oriented porthole is more straightforward:

$$F = \text{Pressure} \times \text{Area} = 62.4 \cdot 50 \times \pi \cdot 1.5^2 = 22{,}054 \text{ lb.}$$

That these two forces are equal is not coincidental; it turns out that the fluid force applied to a vertically oriented circle whose center is at depth d is the same as force applied to a horizontally oriented circle at depth d.

We end this chapter with a reminder of the true skills meant to be developed here. We are not truly concerned with an ability to find fluid forces or the volumes of solids of revolution. Work done by a variable force is important, though measuring the work done in pulling a rope up a cliff is probably not.

What we are actually concerned with is the ability to solve certain problems by first approximating the solution, then refining the approximation, then recognizing if/when this refining process results in a definite integral through a limit. Knowing the formulas found inside the special boxes within this chapter is beneficial as it helps solve problems found in the exercises, and other mathematical skills are strengthened by properly applying these formulas. However, more importantly, understand how each of these formulas was constructed. Each is the result of a summation of approximations; each summation was a Riemann sum, allowing us to take a limit and find the exact answer through a definite integral.

The next chapter addresses an entirely different topic: sequences and series. In short, a sequence is a list of numbers, where a series is the summation of a list of numbers. These seemingly–simple ideas lead to very powerful mathematics.

Notes:

Exercises 7.6

Terms and Concepts

1. State in your own words Pascal's Principle.

2. State in your own words how pressure is different from force.

Problems

In Exercises 3 – 12, find the fluid force exerted on the given plate, submerged in water with a weight density of 62.4 lb/ft^3.

3.

4.

5.

6.

7.

8.

9.

10.

11.

12.

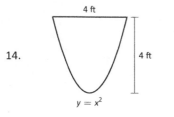

In Exercises 13 – 18, the side of a container is pictured. Find the fluid force exerted on this plate when the container is full of:

    1. water, with a weight density of 62.4 lb/ft^3, and

    2. concrete, with a weight density of 150 lb/ft^3.

13.

5 ft

3 ft

14.

15.

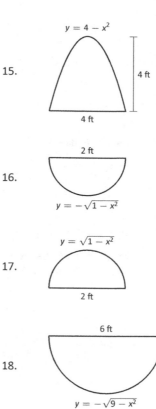

16.

17.

18.

19. How deep must the center of a vertically oriented circular plate with a radius of 1 ft be submerged in water, with a weight density of 62.4 lb/ft^3, for the fluid force on the plate to reach 1,000 lb?

20. How deep must the center of a vertically oriented square plate with a side length of 2 ft be submerged in water, with a weight density of 62.4 lb/ft^3, for the fluid force on the plate to reach 1,000 lb?

8: SEQUENCES AND SERIES

This chapter introduces **sequences** and **series**, important mathematical constructions that are useful when solving a large variety of mathematical problems. The content of this chapter is considerably different from the content of the chapters before it. While the material we learn here definitely falls under the scope of "calculus," we will make very little use of derivatives or integrals. Limits are extremely important, though, especially limits that involve infinity.

One of the problems addressed by this chapter is this: suppose we know information about a function and its derivatives at a point, such as $f(1) = 3$, $f'(1) = 1, f''(1) = -2, f'''(1) = 7$, and so on. What can I say about $f(x)$ itself? Is there any reasonable approximation of the value of $f(2)$? The topic of Taylor Series addresses this problem, and allows us to make excellent approximations of functions when limited knowledge of the function is available.

8.1 Sequences

We commonly refer to a set of events that occur one after the other as a *sequence* of events. In mathematics, we use the word *sequence* to refer to an ordered set of numbers, i.e., a set of numbers that "occur one after the other."

For instance, the numbers 2, 4, 6, 8, ..., form a sequence. The order is important; the first number is 2, the second is 4, etc. It seems natural to seek a formula that describes a given sequence, and often this can be done. For instance, the sequence above could be described by the function $a(n) = 2n$, for the values of $n = 1, 2, \ldots$ To find the 10th term in the sequence, we would compute $a(10)$. This leads us to the following, formal definition of a sequence.

Definition 8.1.1   Sequence

A **sequence** is a function $a(n)$ whose domain is $\mathbb{N}$. The **range** of a sequence is the set of all distinct values of $a(n)$.

The **terms** of a sequence are the values $a(1)$, $a(2)$, ..., which are usually denoted with subscripts as $a_1, a_2, \ldots$.

A sequence $a(n)$ is often denoted as $\{a_n\}$.

Notation: We use $\mathbb{N}$ to describe the set of natural numbers, that is, the integers 1, 2, 3, ...

Factorial: The expression 4! refers to the number $4 \cdot 3 \cdot 2 \cdot 1 = 24$.

In general, $n! = n \cdot (n-1) \cdot (n-2) \cdots 2 \cdot 1$, where n is a natural number.

We define $0! = 1$. While this does not immediately make sense, it makes many mathematical formulas work properly.

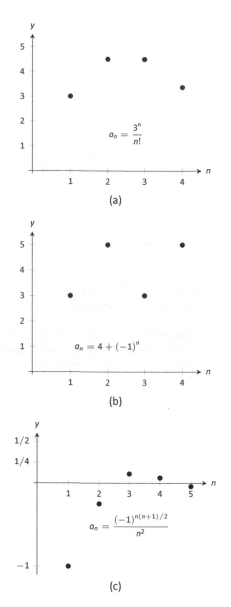

Figure 8.1.1: Plotting sequences in Example 8.1.1.

Example 8.1.1 Listing terms of a sequence

List the first four terms of the following sequences.

1. $\{a_n\} = \left\{\dfrac{3^n}{n!}\right\}$
2. $\{a_n\} = \{4 + (-1)^n\}$
3. $\{a_n\} = \left\{\dfrac{(-1)^{n(n+1)/2}}{n^2}\right\}$

SOLUTION

1. $a_1 = \dfrac{3^1}{1!} = 3;$ $a_2 = \dfrac{3^2}{2!} = \dfrac{9}{2};$ $a_3 = \dfrac{3^3}{3!} = \dfrac{9}{2};$ $a_4 = \dfrac{3^4}{4!} = \dfrac{27}{8}$

We can plot the terms of a sequence with a scatter plot. The "x"-axis is used for the values of n, and the values of the terms are plotted on the y-axis. To visualize this sequence, see Figure 8.1.1(a).

2. $a_1 = 4 + (-1)^1 = 3;$ $a_2 = 4 + (-1)^2 = 5;$

$a_3 = 4 + (-1)^3 = 3;$ $a_4 = 4 + (-1)^4 = 5.$ Note that the range of this sequence is finite, consisting of only the values 3 and 5. This sequence is plotted in Figure 8.1.1(b).

3. $a_1 = \dfrac{(-1)^{1(2)/2}}{1^2} = -1;$ $a_2 = \dfrac{(-1)^{2(3)/2}}{2^2} = -\dfrac{1}{4}$

$a_3 = \dfrac{(-1)^{3(4)/2}}{3^2} = \dfrac{1}{9}$ $a_4 = \dfrac{(-1)^{4(5)/2}}{4^2} = \dfrac{1}{16};$

$a_5 = \dfrac{(-1)^{5(6)/2}}{5^2} = -\dfrac{1}{25}.$

We gave one extra term to begin to show the pattern of signs is "$-, -, +, +, -, -, \ldots$" due to the fact that the exponent of -1 is a special quadratic. This sequence is plotted in Figure 8.1.1(c).

Example 8.1.2 Determining a formula for a sequence

Find the nth term of the following sequences, i.e., find a function that describes each of the given sequences.

1. $2, 5, 8, 11, 14, \ldots$

2. $2, -5, 10, -17, 26, -37, \ldots$

3. $1, 1, 2, 6, 24, 120, 720, \ldots$

4. $\dfrac{5}{2}, \dfrac{5}{2}, \dfrac{15}{8}, \dfrac{5}{4}, \dfrac{25}{32}, \ldots$

Notes:

SOLUTION   We should first note that there is never exactly one function that describes a finite set of numbers as a sequence. There are many sequences that start with 2, then 5, as our first example does. We are looking for a simple formula that describes the terms given, knowing there is possibly more than one answer.

1. Note how each term is 3 more than the previous one. This implies a linear function would be appropriate: $a(n) = a_n = 3n + b$ for some appropriate value of b. As we want $a_1 = 2$, we set $b = -1$. Thus $a_n = 3n - 1$.

2. First notice how the sign changes from term to term. This is most commonly accomplished by multiplying the terms by either $(-1)^n$ or $(-1)^{n+1}$. Using $(-1)^n$ multiplies the odd terms by (-1); using $(-1)^{n+1}$ multiplies the even terms by (-1). As this sequence has negative even terms, we will multiply by $(-1)^{n+1}$.

   After this, we might feel a bit stuck as to how to proceed. At this point, we are just looking for a pattern of some sort: what do the numbers 2, 5, 10, 17, etc., have in common? There are many correct answers, but the one that we'll use here is that each is one more than a perfect square. That is, $2 = 1^2 + 1$, $5 = 2^2 + 1$, $10 = 3^2 + 1$, etc. Thus our formula is $a_n = (-1)^{n+1}(n^2 + 1)$.

3. One who is familiar with the factorial function will readily recognize these numbers. They are $0!$, $1!$, $2!$, $3!$, etc. Since our sequences start with $n = 1$, we cannot write $a_n = n!$, for this misses the $0!$ term. Instead, we shift by 1, and write $a_n = (n - 1)!$.

4. This one may appear difficult, especially as the first two terms are the same, but a little "sleuthing" will help. Notice how the terms in the numerator are always multiples of 5, and the terms in the denominator are always powers of 2. Does something as simple as $a_n = \frac{5n}{2^n}$ work?

   When $n = 1$, we see that we indeed get $5/2$ as desired. When $n = 2$, we get $10/4 = 5/2$. Further checking shows that this formula indeed matches the other terms of the sequence.

A common mathematical endeavor is to create a new mathematical object (for instance, a sequence) and then apply previously known mathematics to the new object. We do so here. The fundamental concept of calculus is the limit, so we will investigate what it means to find the limit of a sequence.

Notes:

Definition 8.1.2   Limit of a Sequence, Convergent, Divergent

Let $\{a_n\}$ be a sequence and let L be a real number. Given any $\varepsilon > 0$, if an m can be found such that $|a_n - L| < \varepsilon$ for all $n > m$, then we say the **limit of $\{a_n\}$, as n approaches infinity, is** L, denoted

$$\lim_{n \to \infty} a_n = L.$$

If $\lim\limits_{n \to \infty} a_n$ exists, we say the sequence **converges**; otherwise, the sequence **diverges**.

This definition states, informally, that if the limit of a sequence is L, then if you go far enough out along the sequence, all subsequent terms will be *really close* to L. Of course, the terms "far enough" and "really close" are subjective terms, but hopefully the intent is clear.

This definition is reminiscent of the ε–δ proofs of Chapter 1. In that chapter we developed other tools to evaluate limits apart from the formal definition; we do so here as well.

Theorem 8.1.1   Limit of a Sequence

Let $\{a_n\}$ be a sequence and let $f(x)$ be a function whose domain contains the positive real numbers where $f(n) = a_n$ for all n in $\mathbb{N}$.

If $\lim\limits_{x \to \infty} f(x) = L$, then $\lim\limits_{n \to \infty} a_n = L$.

Theorem 8.1.1 allows us, in certain cases, to apply the tools developed in Chapter 1 to limits of sequences. Note two things *not* stated by the theorem:

1. If $\lim\limits_{x \to \infty} f(x)$ does not exist, we cannot conclude that $\lim\limits_{n \to \infty} a_n$ does not exist. It may, or may not, exist. For instance, we can define a sequence $\{a_n\} = \{\cos(2\pi n)\}$. Let $f(x) = \cos(2\pi x)$. Since the cosine function oscillates over the real numbers, the limit $\lim\limits_{x \to \infty} f(x)$ does not exist.

   However, for every positive integer n, $\cos(2\pi n) = 1$, so $\lim\limits_{n \to \infty} a_n = 1$.

2. If we cannot find a function $f(x)$ whose domain contains the positive real numbers where $f(n) = a_n$ for all n in $\mathbb{N}$, we cannot conclude $\lim\limits_{n \to \infty} a_n$ does not exist. It may, or may not, exist.

Notes:

Example 8.1.3   Determining convergence/divergence of a sequence

Determine the convergence or divergence of the following sequences.

1. $\{a_n\} = \left\{ \dfrac{3n^2 - 2n + 1}{n^2 - 1000} \right\}$ 2. $\{a_n\} = \{\cos n\}$ 3. $\{a_n\} = \left\{ \dfrac{(-1)^n}{n} \right\}$

SOLUTION

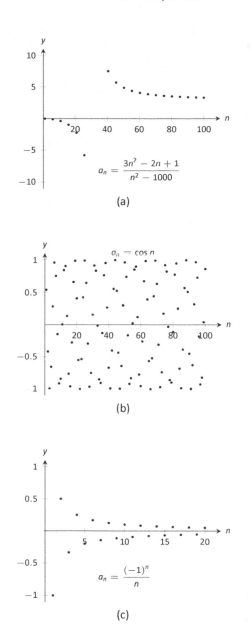

1. Using Theorem 1.6.1, we can state that $\lim\limits_{x \to \infty} \dfrac{3x^2 - 2x + 1}{x^2 - 1000} = 3$. (We could have also directly applied l'Hôpital's Rule.) Thus the sequence $\{a_n\}$ converges, and its limit is 3. A scatter plot of every 5 values of a_n is given in Figure 8.1.2 (a). The values of a_n vary widely near $n = 30$, ranging from about -73 to 125, but as n grows, the values approach 3.

2. The limit $\lim\limits_{x \to \infty} \cos x$ does not exist as $\cos x$ oscillates (and takes on every value in $[-1, 1]$ infinitely many times). Thus we cannot apply Theorem 8.1.1.

   The fact that the cosine function oscillates strongly hints that $\cos n$, when n is restricted to $\mathbb{N}$, will also oscillate. Figure 8.1.2 (b), where the sequence is plotted, implies that this is true. Because only discrete values of cosine are plotted, it does not bear strong resemblance to the familiar cosine wave. The proof of the following statement is beyond the scope of this text, but it is true: there are infinitely many integers n that are arbitrarily (i.e., *very*) close to an even multiple of π, so that $\cos n \approx 1$. Similarly, there are infinitely many integers m that are arbitrarily close to an odd multiple of π, so that $\cos m \approx -1$. As the sequence takes on values near 1 and -1 infinitely many times, we conclude that $\lim\limits_{n \to \infty} a_n$ does not exist.

3. We cannot actually apply Theorem 8.1.1 here, as the function $f(x) = (-1)^x/x$ is not well defined. (What does $(-1)^{\sqrt{2}}$ mean? In actuality, there is an answer, but it involves *complex analysis*, beyond the scope of this text.)

   Instead, we invoke the definition of the limit of a sequence. By looking at the plot in Figure 8.1.2 (c), we would like to conclude that the sequence converges to $L = 0$. Let $\varepsilon > 0$ be given. We can find a natural number m

Figure 8.1.2: Scatter plots of the sequences in Example 8.1.3.

Notes:

such that $1/m < \varepsilon$. Let $n > m$, and consider $|a_n - L|$:

$$|a_n - L| = \left| \frac{(-1)^n}{n} - 0 \right|$$

$$= \frac{1}{n}$$

$$< \frac{1}{m} \quad \text{(since } n > m\text{)}$$

$$< \varepsilon.$$

We have shown that by picking m large enough, we can ensure that a_n is arbitrarily close to our limit, $L = 0$, hence by the definition of the limit of a sequence, we can say $\lim_{n \to \infty} a_n = 0$.

In the previous example we used the definition of the limit of a sequence to determine the convergence of a sequence as we could not apply Theorem 8.1.1. In general, we like to avoid invoking the definition of a limit, and the following theorem gives us tool that we could use in that example instead.

Theorem 8.1.2   Absolute Value Theorem

Let $\{a_n\}$ be a sequence. If $\lim_{n \to \infty} |a_n| = 0$, then $\lim_{n \to \infty} a_n = 0$

Example 8.1.4   Determining the convergence/divergence of a sequence
Determine the convergence or divergence of the following sequences.

$$1. \ \{a_n\} = \left\{ \frac{(-1)^n}{n} \right\} \qquad 2. \ \{a_n\} = \left\{ \frac{(-1)^n(n+1)}{n} \right\}$$

SOLUTION

1. This appeared in Example 8.1.3. We want to apply Theorem 8.1.2, so consider the limit of $\{|a_n|\}$:

$$\lim_{n \to \infty} |a_n| = \lim_{n \to \infty} \left| \frac{(-1)^n}{n} \right|$$

$$= \lim_{n \to \infty} \frac{1}{n}$$

$$= 0.$$

Since this limit is 0, we can apply Theorem 8.1.2 and state that $\lim_{n \to \infty} a_n = 0$.

Notes:

2. Because of the alternating nature of this sequence (i.e., every other term is multiplied by -1), we cannot simply look at the limit $\lim\limits_{x\to\infty} \dfrac{(-1)^x(x+1)}{x}$. We can try to apply the techniques of Theorem 8.1.2:

$$\lim_{n\to\infty} |a_n| = \lim_{n\to\infty} \left| \frac{(-1)^n(n+1)}{n} \right|$$
$$= \lim_{n\to\infty} \frac{n+1}{n}$$
$$= 1.$$

We have concluded that when we ignore the alternating sign, the sequence approaches 1. This means we cannot apply Theorem 8.1.2; it states the the limit must be 0 in order to conclude anything.

Since we know that the signs of the terms alternate *and* we know that the limit of $|a_n|$ is 1, we know that as n approaches infinity, the terms will alternate between values close to 1 and -1, meaning the sequence diverges. A plot of this sequence is given in Figure 8.1.3.

We continue our study of the limits of sequences by considering some of the properties of these limits.

Theorem 8.1.3   Properties of the Limits of Sequences

Let $\{a_n\}$ and $\{b_n\}$ be sequences such that $\lim\limits_{n\to\infty} a_n = L$, $\lim\limits_{n\to\infty} b_n = K$, and let c be a real number.

1. $\lim\limits_{n\to\infty} (a_n \pm b_n) = L \pm K$ 3. $\lim\limits_{n\to\infty} (a_n/b_n) = L/K, K \neq 0$

2. $\lim\limits_{n\to\infty} (a_n \cdot b_n) = L \cdot K$ 4. $\lim\limits_{n\to\infty} c \cdot a_n = c \cdot L$

Example 8.1.5   Applying properties of limits of sequences
Let the following sequences, and their limits, be given:

- $\{a_n\} = \left\{ \dfrac{n+1}{n^2} \right\}$, and $\lim\limits_{n\to\infty} a_n = 0$;

- $\{b_n\} = \left\{ \left(1 + \dfrac{1}{n} \right)^n \right\}$, and $\lim\limits_{n\to\infty} b_n = e$; and

- $\{c_n\} = \left\{ n \cdot \sin(5/n) \right\}$, and $\lim\limits_{n\to\infty} c_n = 5$.

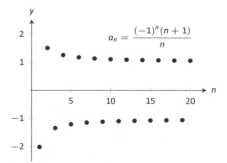

Figure 8.1.3: A plot of a sequence in Example 8.1.4, part 2.

Notes:

Evaluate the following limits.

1. $\lim\limits_{n\to\infty} (a_n + b_n)$ 2. $\lim\limits_{n\to\infty} (b_n \cdot c_n)$ 3. $\lim\limits_{n\to\infty} (1000 \cdot a_n)$

SOLUTION   We will use Theorem 8.1.3 to answer each of these.

1. Since $\lim\limits_{n\to\infty} a_n = 0$ and $\lim\limits_{n\to\infty} b_n = e$, we conclude that $\lim\limits_{n\to\infty} (a_n + b_n) = 0 + e = e$. So even though we are adding something to each term of the sequence b_n, we are adding something so small that the final limit is the same as before.

2. Since $\lim\limits_{n\to\infty} b_n = e$ and $\lim\limits_{n\to\infty} c_n = 5$, we conclude that $\lim\limits_{n\to\infty} (b_n \cdot c_n) = e \cdot 5 = 5e$.

3. Since $\lim\limits_{n\to\infty} a_n = 0$, we have $\lim\limits_{n\to\infty} 1000a_n = 1000 \cdot 0 = 0$. It does not matter that we multiply each term by 1000; the sequence still approaches 0. (It just takes longer to get close to 0.)

There is more to learn about sequences than just their limits. We will also study their range and the relationships terms have with the terms that follow. We start with some definitions describing properties of the range.

Definition 8.1.3   Bounded and Unbounded Sequences

A sequence $\{a_n\}$ is said to be **bounded** if there exist real numbers m and M such that $m < a_n < M$ for all n in $\mathbb{N}$.

A sequence $\{a_n\}$ is said to be **unbounded** if it is not bounded.

A sequence $\{a_n\}$ is said to be **bounded above** if there exists an M such that $a_n < M$ for all n in $\mathbb{N}$; it is **bounded below** if there exists an m such that $m < a_n$ for all n in $\mathbb{N}$.

It follows from this definition that an unbounded sequence may be bounded above or bounded below; a sequence that is both bounded above and below is simply a bounded sequence.

Example 8.1.6   Determining boundedness of sequences
Determine the boundedness of the following sequences.

1. $\{a_n\} = \left\{\dfrac{1}{n}\right\}$ 2. $\{a_n\} = \{2^n\}$

Notes:

SOLUTION

1. The terms of this sequence are always positive but are decreasing, so we have $0 < a_n < 2$ for all n. Thus this sequence is bounded. Figure 8.1.4(a) illustrates this.

2. The terms of this sequence obviously grow without bound. However, it is also true that these terms are all positive, meaning $0 < a_n$. Thus we can say the sequence is unbounded, but also bounded below. Figure 8.1.4(b) illustrates this.

The previous example produces some interesting concepts. First, we can recognize that the sequence $\{1/n\}$ converges to 0. This says, informally, that "most" of the terms of the sequence are "really close" to 0. This implies that the sequence is bounded, using the following logic. First, "most" terms are near 0, so we could find some sort of bound on these terms (using Definition 8.1.2, the bound is ε). That leaves a "few" terms that are not near 0 (i.e., a *finite* number of terms). A finite list of numbers is always bounded.

This logic implies that if a sequence converges, it must be bounded. This is indeed true, as stated by the following theorem.

Theorem 8.1.4   Convergent Sequences are Bounded

Let $\{a_n\}$ be a convergent sequence. Then $\{a_n\}$ is bounded.

In Example 8.1.5 we saw the sequence $\{b_n\} = \left\{(1 + 1/n)^n\right\}$, where it was stated that $\lim\limits_{n\to\infty} b_n = e$. (Note that this is simply restating part of Theorem 1.3.5.) Even though it may be difficult to intuitively grasp the behavior of this sequence, we know immediately that it is bounded.

Another interesting concept to come out of Example 8.1.6 again involves the sequence $\{1/n\}$. We stated, without proof, that the terms of the sequence were decreasing. That is, that $a_{n+1} < a_n$ for all n. (This is easy to show. Clearly $n < n+1$. Taking reciprocals flips the inequality: $1/n > 1/(n+1)$. This is the same as $a_n > a_{n+1}$.) Sequences that either steadily increase or decrease are important, so we give this property a name.

Note: Keep in mind what Theorem 8.1.4 does *not* say. It does not say that bounded sequences must converge, nor does it say that if a sequence does not converge, it is not bounded.

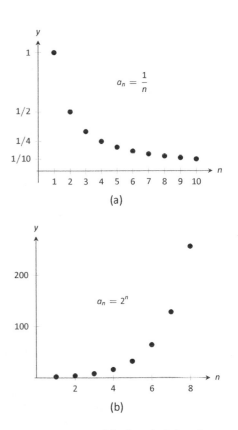

(a)

(b)

Figure 8.1.4: A plot of $\{a_n\} = \{1/n\}$ and $\{a_n\} = \{2^n\}$ from Example 8.1.6.

Notes:

Note: It is sometimes useful to call a monotonically increasing sequence *strictly increasing* if $a_n < a_{n+1}$ for all n; i.e, we remove the possibility that subsequent terms are equal.

A similar statement holds for *strictly decreasing*.

Definition 8.1.4   Monotonic Sequences

1. A sequence $\{a_n\}$ is **monotonically increasing** if $a_n \leq a_{n+1}$ for all n, i.e.,
$$a_1 \leq a_2 \leq a_3 \leq \cdots a_n \leq a_{n+1} \cdots$$

2. A sequence $\{a_n\}$ is **monotonically decreasing** if $a_n \geq a_{n+1}$ for all n, i.e.,
$$a_1 \geq a_2 \geq a_3 \geq \cdots a_n \geq a_{n+1} \cdots$$

3. A sequence is **monotonic** if it is monotonically increasing or monotonically decreasing.

Example 8.1.7   Determining monotonicity

Determine the monotonicity of the following sequences.

1. $\{a_n\} = \left\{ \dfrac{n+1}{n} \right\}$
3. $\{a_n\} = \left\{ \dfrac{n^2 - 9}{n^2 - 10n + 26} \right\}$

2. $\{a_n\} = \left\{ \dfrac{n^2 + 1}{n + 1} \right\}$
4. $\{a_n\} = \left\{ \dfrac{n^2}{n!} \right\}$

Solution   In each of the following, we will examine $a_{n+1} - a_n$. If $a_{n+1} - a_n \geq 0$, we conclude that $a_n \leq a_{n+1}$ and hence the sequence is increasing. If $a_{n+1} - a_n \leq 0$, we conclude that $a_n \geq a_{n+1}$ and the sequence is decreasing. Of course, a sequence need not be monotonic and perhaps neither of the above will apply.

We also give a scatter plot of each sequence. These are useful as they suggest a pattern of monotonicity, but analytic work should be done to confirm a graphical trend.

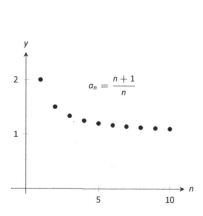

Figure 8.1.5: A plot of $\{a_n\} = \{(n+1)/n\}$ in Example 8.1.7.

1.
$$\begin{aligned}
a_{n+1} - a_n &= \frac{n+2}{n+1} - \frac{n+1}{n} \\
&= \frac{(n+2)(n) - (n+1)^2}{(n+1)n} \\
&= \frac{-1}{n(n+1)} \\
&< 0 \quad \text{for all } n.
\end{aligned}$$

Since $a_{n+1} - a_n < 0$ for all n, we conclude that the sequence is decreasing.

Notes:

2.
$$a_{n+1} - a_n = \frac{(n+1)^2 + 1}{n+2} - \frac{n^2+1}{n+1}$$
$$= \frac{((n+1)^2+1)(n+1) - (n^2+1)(n+2)}{(n+1)(n+2)}$$
$$= \frac{n^2 + 4n + 1}{(n+1)(n+2)}$$
$$> 0 \quad \text{for all } n.$$

Since $a_{n+1} - a_n > 0$ for all n, we conclude the sequence is increasing; see Figure 8.1.6(a).

3. We can clearly see in Figure 8.1.6(b), where the sequence is plotted, that it is not monotonic. However, it does seem that after the first 4 terms it is decreasing. To understand why, perform the same analysis as done before:

$$a_{n+1} - a_n = \frac{(n+1)^2 - 9}{(n+1)^2 - 10(n+1) + 26} - \frac{n^2 - 9}{n^2 - 10n + 26}$$
$$= \frac{n^2 + 2n - 8}{n^2 - 8n + 17} - \frac{n^2 - 9}{n^2 - 10n + 26}$$
$$= \frac{(n^2 + 2n - 8)(n^2 - 10n + 26) - (n^2 - 9)(n^2 - 8n + 17)}{(n^2 - 8n + 17)(n^2 - 10n + 26)}$$
$$= \frac{-10n^2 + 60n - 55}{(n^2 - 8n + 17)(n^2 - 10n + 26)}.$$

We want to know when this is greater than, or less than, 0. The denominator is always positive, therefore we are only concerned with the numerator. For small values of n, the numerator is positive. As n grows large, the numerator is dominated by $-10n^2$, meaning the entire fraction will be negative; i.e., for large enough n, $a_{n+1} - a_n < 0$. Using the quadratic formula we can determine that the numerator is negative for $n \geq 5$.

In short, the sequence is simply not monotonic, though it is useful to note that for $n \geq 5$, the sequence is monotonically decreasing.

4. Again, the plot in Figure 8.1.6(c) shows that the sequence is not monotonic, but it suggests that it is monotonically decreasing after the first term. We perform the usual analysis to confirm this.

$$a_{n+1} - a_n = \frac{(n+1)^2}{(n+1)!} - \frac{n^2}{n!}$$
$$= \frac{(n+1)^2 - n^2(n+1)}{(n+1)!}$$
$$= \frac{-n^3 + 2n + 1}{(n+1)!}$$

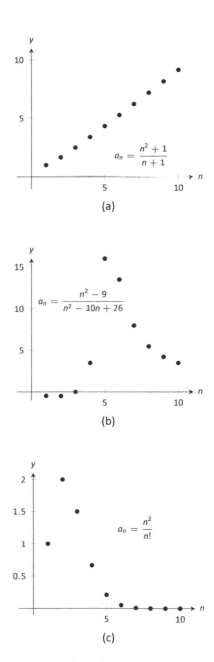

(a)

(b)

(c)

Figure 8.1.6: Plots of sequences in Example 8.1.7.

When $n = 1$, the above expression is > 0; for $n \geq 2$, the above expression is < 0. Thus this sequence is not monotonic, but it is monotonically decreasing after the first term.

Knowing that a sequence is monotonic can be useful. Consider, for example, a sequence that is monotonically decreasing and is bounded below. We know the sequence is always getting smaller, but that there is a bound to how small it can become. This is enough to prove that the sequence will converge, as stated in the following theorem.

Theorem 8.1.5   Bounded Monotonic Sequences are Convergent

1. Let $\{a_n\}$ be a monotonically increasing sequence that is bounded above. Then $\{a_n\}$ converges.

2. Let $\{a_n\}$ be a monotonically decreasing sequence that is bounded below. Then $\{a_n\}$ converges.

Consider once again the sequence $\{a_n\} = \{1/n\}$. It is easy to show it is monotonically decreasing and that it is always positive (i.e., bounded below by 0). Therefore we can conclude by Theorem 8.1.5 that the sequence converges. We already knew this by other means, but in the following section this theorem will become very useful.

We can replace Theorem 8.1.5 with the statement "Let $\{a_n\}$ be a bounded, monotonic sequence. Then $\{a_n\}$ converges; i.e., $\lim\limits_{n\to\infty} a_n$ exists." We leave it to the reader in the exercises to show the theorem and the above statement are equivalent.

Sequences are a great source of mathematical inquiry. The On-Line Encyclopedia of Integer Sequences (http://oeis.org) contains thousands of sequences and their formulae. (As of this writing, there are 297,573 sequences in the database.) Perusing this database quickly demonstrates that a single sequence can represent several different "real life" phenomena.

Interesting as this is, our interest actually lies elsewhere. We are more interested in the *sum* of a sequence. That is, given a sequence $\{a_n\}$, we are very interested in $a_1 + a_2 + a_3 + \cdots$. Of course, one might immediately counter with "Doesn't this just add up to 'infinity'?" Many times, yes, but there are many important cases where the answer is no. This is the topic of *series*, which we begin to investigate in the next section.

Notes:

Exercises 8.1

Terms and Concepts

1. Use your own words to define a *sequence*.

2. The domain of a sequence is the _____ numbers.

3. Use your own words to describe the *range* of a sequence.

4. Describe what it means for a sequence to be *bounded*.

Problems

In Exercises 5 – 8, give the first five terms of the given sequence.

5. $\{a_n\} = \left\{\dfrac{4^n}{(n+1)!}\right\}$

6. $\{b_n\} = \left\{\left(-\dfrac{3}{2}\right)^n\right\}$

7. $\{c_n\} = \left\{-\dfrac{n^{n+1}}{n+2}\right\}$

8. $\{d_n\} = \left\{\dfrac{1}{\sqrt{5}}\left(\left(\dfrac{1+\sqrt{5}}{2}\right)^n - \left(\dfrac{1-\sqrt{5}}{2}\right)^n\right)\right\}$

In Exercises 9 – 12, determine the nth term of the given sequence.

9. $4, 7, 10, 13, 16, \ldots$

10. $3, -\dfrac{3}{2}, \dfrac{3}{4}, -\dfrac{3}{8}, \ldots$

11. $10, 20, 40, 80, 160, \ldots$

12. $1, 1, \dfrac{1}{2}, \dfrac{1}{6}, \dfrac{1}{24}, \dfrac{1}{120}, \ldots$

In Exercises 13 – 16, use the following information to determine the limit of the given sequences.

- $\{a_n\} = \left\{\dfrac{2^n - 20}{2^n}\right\}; \qquad \lim\limits_{n\to\infty} a_n = 1$

- $\{b_n\} = \left\{\left(1 + \dfrac{2}{n}\right)^n\right\}; \qquad \lim\limits_{n\to\infty} b_n = e^2$

- $\{c_n\} = \{\sin(3/n)\}; \qquad \lim\limits_{n\to\infty} c_n = 0$

13. $\{a_n\} = \left\{\dfrac{2^n - 20}{7 \cdot 2^n}\right\}$

14. $\{a_n\} = \{3b_n - a_n\}$

15. $\{a_n\} = \left\{\sin(3/n)\left(1 + \dfrac{2}{n}\right)^n\right\}$

16. $\{a_n\} = \left\{\left(1 + \dfrac{2}{n}\right)^{2n}\right\}$

In Exercises 17 – 28, determine whether the sequence converges or diverges. If convergent, give the limit of the sequence.

17. $\{a_n\} = \left\{(-1)^n \dfrac{n}{n+1}\right\}$

18. $\{a_n\} = \left\{\dfrac{4n^2 - n + 5}{3n^2 + 1}\right\}$

19. $\{a_n\} = \left\{\dfrac{4^n}{5^n}\right\}$

20. $\{a_n\} = \left\{\dfrac{n-1}{n} - \dfrac{n}{n-1}\right\}, n \geq 2$

21. $\{a_n\} = \{\ln(n)\}$

22. $\{a_n\} = \left\{\dfrac{3n}{\sqrt{n^2 + 1}}\right\}$

23. $\{a_n\} = \left\{\left(1 + \dfrac{1}{n}\right)^n\right\}$

24. $\{a_n\} = \left\{5 - \dfrac{1}{n}\right\}$

25. $\{a_n\} = \left\{\dfrac{(-1)^{n+1}}{n}\right\}$

26. $\{a_n\} = \left\{\dfrac{1.1^n}{n}\right\}$

27. $\{a_n\} = \left\{\dfrac{2n}{n+1}\right\}$

28. $\{a_n\} = \left\{(-1)^n \dfrac{n^2}{2^n - 1}\right\}$

In Exercises 29 – 34, determine whether the sequence is bounded, bounded above, bounded below, or none of the above.

29. $\{a_n\} = \{\sin n\}$

30. $\{a_n\} = \{\tan n\}$

31. $\{a_n\} = \left\{(-1)^n \dfrac{3n - 1}{n}\right\}$

32. $\{a_n\} = \left\{\dfrac{3n^2 - 1}{n}\right\}$

33. $\{a_n\} = \{n \cos n\}$

34. $\{a_n\} = \{2^n - n!\}$

0.

In Exercises 35 – 38, determine whether the sequence is monotonically increasing or decreasing. If it is not, determine if there is an m such that it is monotonic for all $n \geq m$.

35. $\{a_n\} = \left\{\dfrac{n}{n+2}\right\}$

36. $\{a_n\} = \left\{\dfrac{n^2 - 6n + 9}{n}\right\}$

37. $\{a_n\} = \left\{(-1)^n \dfrac{1}{n^3}\right\}$

38. $\{a_n\} = \left\{\dfrac{n^2}{2^n}\right\}$

Exercises 39 – 42 explore further the theory of sequences.

39. Prove Theorem 8.1.2; that is, use the definition of the limit of a sequence to show that if $\lim\limits_{n\to\infty} |a_n| = 0$, then $\lim\limits_{n\to\infty} a_n = $

40. Let $\{a_n\}$ and $\{b_n\}$ be sequences such that $\lim\limits_{n\to\infty} a_n = L$ and $\lim\limits_{n\to\infty} b_n = K$.

   (a) Show that if $a_n < b_n$ for all n, then $L \leq K$.

   (b) Give an example where $L = K$.

41. Prove the Squeeze Theorem for sequences: Let $\{a_n\}$ and $\{b_n\}$ be such that $\lim\limits_{n\to\infty} a_n = L$ and $\lim\limits_{n\to\infty} b_n = L$, and let $\{c_n\}$ be such that $a_n \leq c_n \leq b_n$ for all n. Then $\lim\limits_{n\to\infty} c_n = L$

42. Prove the statement "Let $\{a_n\}$ be a bounded, monotonic sequence. Then $\{a_n\}$ converges; i.e., $\lim\limits_{n\to\infty} a_n$ exists." is equivalent to Theorem 8.1.5. That is,

   (a) Show that if Theorem 8.1.5 is true, then above statement is true, and

   (b) Show that if the above statement is true, then Theorem 8.1.5 is true.

8.2 Infinite Series

Given the sequence $\{a_n\} = \{1/2^n\} = 1/2,\ 1/4,\ 1/8,\ \ldots$, consider the following sums:

$$
\begin{array}{rclcl}
a_1 & = & 1/2 & = & 1/2 \\
a_1 + a_2 & = & 1/2 + 1/4 & = & 3/4 \\
a_1 + a_2 + a_3 & = & 1/2 + 1/4 + 1/8 & = & 7/8 \\
a_1 + a_2 + a_3 + a_4 & = & 1/2 + 1/4 + 1/8 + 1/16 & = & 15/16
\end{array}
$$

In general, we can show that

$$a_1 + a_2 + a_3 + \cdots + a_n = \frac{2^n - 1}{2^n} = 1 - \frac{1}{2^n}.$$

Let S_n be the sum of the first n terms of the sequence $\{1/2^n\}$. From the above, we see that $S_1 = 1/2$, $S_2 = 3/4$, etc. Our formula at the end shows that $S_n = 1 - 1/2^n$.

Now consider the following limit: $\lim_{n\to\infty} S_n = \lim_{n\to\infty}\left(1 - 1/2^n\right) = 1$. This limit can be interpreted as saying something amazing: *the sum of* all *the terms of the sequence* $\{1/2^n\}$ *is 1.*

This example illustrates some interesting concepts that we explore in this section. We begin this exploration with some definitions.

Definition 8.2.1   Infinite Series, n^{th} Partial Sums, Convergence, Divergence

Let $\{a_n\}$ be a sequence.

1. The sum $\displaystyle\sum_{n=1}^{\infty} a_n$ is an **infinite series** (or, simply **series**).

2. Let $S_n = \displaystyle\sum_{i=1}^{n} a_i$; the sequence $\{S_n\}$ is the sequence of nth **partial sums** of $\{a_n\}$.

3. If the sequence $\{S_n\}$ converges to L, we say the series $\displaystyle\sum_{n=1}^{\infty} a_n$ **converges** to L, and we write $\displaystyle\sum_{n=1}^{\infty} a_n = L$.

4. If the sequence $\{S_n\}$ diverges, the series $\displaystyle\sum_{n=1}^{\infty} a_n$ **diverges**.

Notes:

Using our new terminology, we can state that the series $\sum_{n=1}^{\infty} 1/2^n$ converges,

and $\sum_{n=1}^{\infty} 1/2^n = 1$.

We will explore a variety of series in this section. We start with two series that diverge, showing how we might discern divergence.

Example 8.2.1   **Showing series diverge**

1. Let $\{a_n\} = \{n^2\}$. Show $\sum_{n=1}^{\infty} a_n$ diverges.

2. Let $\{b_n\} = \{(-1)^{n+1}\}$. Show $\sum_{n=1}^{\infty} b_n$ diverges.

SOLUTION

1. Consider S_n, the nth partial sum.

$$S_n = a_1 + a_2 + a_3 + \cdots + a_n$$
$$= 1^2 + 2^2 + 3^2 \cdots + n^2.$$

By Theorem 5.3.1, this is

$$= \frac{n(n+1)(2n+1)}{6}.$$

Since $\lim_{n \to \infty} S_n = \infty$, we conclude that the series $\sum_{n=1}^{\infty} n^2$ diverges. It is instructive to write $\sum_{n=1}^{\infty} n^2 = \infty$ for this tells us *how* the series diverges: it grows without bound.

A scatter plot of the sequences $\{a_n\}$ and $\{S_n\}$ is given in Figure 8.2.1(a). The terms of $\{a_n\}$ are growing, so the terms of the partial sums $\{S_n\}$ are growing even faster, illustrating that the series diverges.

Notes:

2. The sequence $\{b_n\}$ starts with $1, -1, 1, -1, \ldots$. Consider some of the partial sums S_n of $\{b_n\}$:

$$S_1 = 1$$
$$S_2 = 0$$
$$S_3 = 1$$
$$S_4 = 0$$

This pattern repeats; we find that $S_n = \begin{cases} 1 & n \text{ is odd} \\ 0 & n \text{ is even} \end{cases}$. As $\{S_n\}$ oscillates, repeating $1, 0, 1, 0, \ldots$, we conclude that $\lim\limits_{n\to\infty} S_n$ does not exist, hence $\sum\limits_{n=1}^{\infty} (-1)^{n+1}$ diverges.

A scatter plot of the sequence $\{b_n\}$ and the partial sums $\{S_n\}$ is given in Figure 8.2.1(b). When n is odd, $b_n = S_n$ so the marks for b_n are drawn oversized to show they coincide.

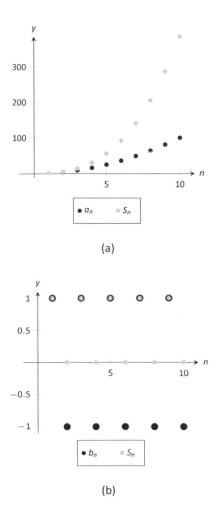

(a)

(b)

Figure 8.2.1: Scatter plots relating to Example 8.2.1.

While it is important to recognize when a series diverges, we are generally more interested in the series that converge. In this section we will demonstrate a few general techniques for determining convergence; later sections will delve deeper into this topic.

Geometric Series

One important type of series is a *geometric series*.

Definition 8.2.2　　**Geometric Series**

A **geometric series** is a series of the form

$$\sum_{n=0}^{\infty} r^n = 1 + r + r^2 + r^3 + \cdots + r^n + \cdots$$

Note that the index starts at $n = 0$, not $n = 1$.

We started this section with a geometric series, although we dropped the first term of 1. One reason geometric series are important is that they have nice convergence properties.

Notes:

> **Theorem 8.2.1   Geometric Series Test**
>
> Consider the geometric series $\displaystyle\sum_{n=0}^{\infty} r^n$.
>
> 1. The nth partial sum is: $S_n = \dfrac{1 - r^{n+1}}{1 - r}, r \neq 1$.
>
> 2. The series converges if, and only if, $|r| < 1$. When $|r| < 1$,
>
> $$\sum_{n=0}^{\infty} r^n = \frac{1}{1 - r}.$$

According to Theorem 8.2.1, the series

$$\sum_{n=0}^{\infty} \frac{1}{2^n} = \sum_{n=0}^{\infty} \left(\frac{1}{2}\right)^2 = 1 + \frac{1}{2} + \frac{1}{4} + \cdots$$

converges as $r = 1/2$, and $\displaystyle\sum_{n=0}^{\infty} \frac{1}{2^n} = \frac{1}{1 - 1/2} = 2$. This concurs with our introductory example; while there we got a sum of 1, we skipped the first term of 1.

Example 8.2.2   Exploring geometric series
Check the convergence of the following series. If the series converges, find its sum.

1. $\displaystyle\sum_{n=2}^{\infty} \left(\frac{3}{4}\right)^n$ 2. $\displaystyle\sum_{n=0}^{\infty} \left(\frac{-1}{2}\right)^n$ 3. $\displaystyle\sum_{n=0}^{\infty} 3^n$

SOLUTION

1. Since $r = 3/4 < 1$, this series converges. By Theorem 8.2.1, we have that

$$\sum_{n=0}^{\infty} \left(\frac{3}{4}\right)^n = \frac{1}{1 - 3/4} = 4.$$

However, note the subscript of the summation in the given series: we are to start with $n = 2$. Therefore we subtract off the first two terms, giving:

$$\sum_{n=2}^{\infty} \left(\frac{3}{4}\right)^n = 4 - 1 - \frac{3}{4} = \frac{9}{4}.$$

This is illustrated in Figure 8.2.2.

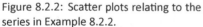

Figure 8.2.2: Scatter plots relating to the series in Example 8.2.2.

Notes:

2. Since $|r| = 1/2 < 1$, this series converges, and by Theorem 8.2.1,

$$\sum_{n=0}^{\infty} \left(\frac{-1}{2}\right)^n = \frac{1}{1-(-1/2)} = \frac{2}{3}.$$

The partial sums of this series are plotted in Figure 8.2.3(a). Note how the partial sums are not purely increasing as some of the terms of the sequence $\{(-1/2)^n\}$ are negative.

3. Since $r > 1$, the series diverges. (This makes "common sense"; we expect the sum

$$1 + 3 + 9 + 27 + 81 + 243 + \cdots$$

to diverge.) This is illustrated in Figure 8.2.3(b).

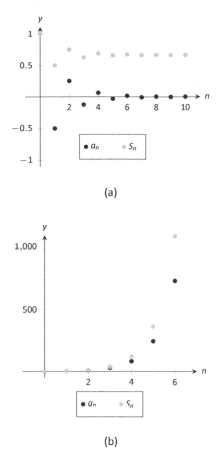

(a)

(b)

Figure 8.2.3: Scatter plots relating to the series in Example 8.2.2.

p–Series

Another important type of series is the *p-series*.

Definition 8.2.3   p–Series, General p–Series

1. A *p–series* is a series of the form

$$\sum_{n=1}^{\infty} \frac{1}{n^p}, \qquad \text{where } p > 0.$$

2. A **general p–series** is a series of the form

$$\sum_{n=1}^{\infty} \frac{1}{(an+b)^p}, \qquad \text{where } p > 0 \text{ and } a, b \text{ are real numbers.}$$

Like geometric series, one of the nice things about p–series is that they have easy to determine convergence properties.

Theorem 8.2.2   p–Series Test

A general p–series $\displaystyle\sum_{n=1}^{\infty} \frac{1}{(an+b)^p}$ will converge if, and only if, $p > 1$.

Note: Theorem 8.2.2 assumes that $an + b \neq 0$ for all n. If $an + b = 0$ for some n, then of course the series does not converge regardless of p as not all of the terms of the sequence are defined.

Notes:

Example 8.2.3   Determining convergence of series
Determine the convergence of the following series.

1. $\displaystyle\sum_{n=1}^{\infty} \frac{1}{n}$

2. $\displaystyle\sum_{n=1}^{\infty} \frac{1}{n^2}$

3. $\displaystyle\sum_{n=1}^{\infty} \frac{1}{\sqrt{n}}$

4. $\displaystyle\sum_{n=1}^{\infty} \frac{(-1)^n}{n}$

5. $\displaystyle\sum_{n=11}^{\infty} \frac{1}{\left(\frac{1}{2}n - 5\right)^3}$

6. $\displaystyle\sum_{n=1}^{\infty} \frac{1}{2^n}$

SOLUTION

1. This is a p–series with $p = 1$. By Theorem 8.2.2, this series diverges.

   This series is a famous series, called the *Harmonic Series*, so named because of its relationship to *harmonics* in the study of music and sound.

2. This is a p–series with $p = 2$. By Theorem 8.2.2, it converges. Note that the theorem does not give a formula by which we can determine *what* the series converges to; we just know it converges. A famous, unexpected result is that this series converges to $\pi^2/6$.

3. This is a p–series with $p = 1/2$; the theorem states that it diverges.

4. This is not a p–series; the definition does not allow for alternating signs. Therefore we cannot apply Theorem 8.2.2. (Another famous result states that this series, the *Alternating Harmonic Series*, converges to ln 2.)

5. This is a general p–series with $p = 3$, therefore it converges.

6. This is not a p–series, but a geometric series with $r = 1/2$. It converges.

   Later sections will provide tests by which we can determine whether or not a given series converges. This, in general, is much easier than determining *what* a given series converges to. There are many cases, though, where the sum can be determined.

Example 8.2.4   Telescoping series
Evaluate the sum $\displaystyle\sum_{n=1}^{\infty} \left(\frac{1}{n} - \frac{1}{n+1}\right)$.

   SOLUTION   It will help to write down some of the first few partial sums

Notes:

of this series.

$$S_1 = \frac{1}{1} - \frac{1}{2} \qquad\qquad\qquad = 1 - \frac{1}{2}$$

$$S_2 = \left(\frac{1}{1} - \frac{1}{2}\right) + \left(\frac{1}{2} - \frac{1}{3}\right) \qquad = 1 - \frac{1}{3}$$

$$S_3 = \left(\frac{1}{1} - \frac{1}{2}\right) + \left(\frac{1}{2} - \frac{1}{3}\right) + \left(\frac{1}{3} - \frac{1}{4}\right) \qquad = 1 - \frac{1}{4}$$

$$S_4 = \left(\frac{1}{1} - \frac{1}{2}\right) + \left(\frac{1}{2} - \frac{1}{3}\right) + \left(\frac{1}{3} - \frac{1}{4}\right) + \left(\frac{1}{4} - \frac{1}{5}\right) \qquad = 1 - \frac{1}{5}$$

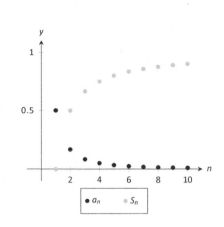

Note how most of the terms in each partial sum are canceled out! In general, we see that $S_n = 1 - \dfrac{1}{n+1}$. The sequence $\{S_n\}$ converges, as $\lim\limits_{n\to\infty} S_n = \lim\limits_{n\to\infty}\left(1 - \dfrac{1}{n+1}\right) = 1$, and so we conclude that $\sum\limits_{n=1}^{\infty}\left(\dfrac{1}{n} - \dfrac{1}{n+1}\right) = 1$. Partial sums of the series are plotted in Figure 8.2.4.

Figure 8.2.4: Scatter plots relating to the series of Example 8.2.4.

The series in Example 8.2.4 is an example of a **telescoping series**. Informally, a telescoping series is one in which most terms cancel with preceding or following terms, reducing the number of terms in each partial sum. The partial sum S_n did not contain n terms, but rather just two: 1 and $1/(n+1)$.

When possible, seek a way to write an explicit formula for the nth partial sum S_n. This makes evaluating the limit $\lim\limits_{n\to\infty} S_n$ much more approachable. We do so in the next example.

Example 8.2.5   Evaluating series
Evaluate each of the following infinite series.

1. $\displaystyle\sum_{n=1}^{\infty} \frac{2}{n^2 + 2n}$ 2. $\displaystyle\sum_{n=1}^{\infty} \ln\left(\frac{n+1}{n}\right)$

SOLUTION

1. We can decompose the fraction $2/(n^2 + 2n)$ as

$$\frac{2}{n^2 + 2n} = \frac{1}{n} - \frac{1}{n+2}.$$

(See Section 6.5, Partial Fraction Decomposition, to recall how this is done, if necessary.)

Notes:

Expressing the terms of $\{S_n\}$ is now more instructive:

$$S_1 = 1 - \frac{1}{3} \qquad\qquad = 1 - \frac{1}{3}$$

$$S_2 = \left(1 - \frac{1}{3}\right) + \left(\frac{1}{2} - \frac{1}{4}\right) \qquad\qquad = 1 + \frac{1}{2} - \frac{1}{3} - \frac{1}{4}$$

$$S_3 = \left(1 - \frac{1}{3}\right) + \left(\frac{1}{2} - \frac{1}{4}\right) + \left(\frac{1}{3} - \frac{1}{5}\right) \qquad = 1 + \frac{1}{2} - \frac{1}{4} - \frac{1}{5}$$

$$S_4 = \left(1 - \frac{1}{3}\right) + \left(\frac{1}{2} - \frac{1}{4}\right) + \left(\frac{1}{3} - \frac{1}{5}\right) + \left(\frac{1}{4} - \frac{1}{6}\right) \qquad = 1 + \frac{1}{2} - \frac{1}{5} - \frac{1}{6}$$

$$S_5 = \left(1 - \frac{1}{3}\right) + \left(\frac{1}{2} - \frac{1}{4}\right) + \left(\frac{1}{3} - \frac{1}{5}\right) + \left(\frac{1}{4} - \frac{1}{6}\right) + \left(\frac{1}{5} - \frac{1}{7}\right) \quad = 1 + \frac{1}{2} - \frac{1}{6} - \frac{1}{7}$$

We again have a telescoping series. In each partial sum, most of the terms cancel and we obtain the formula $S_n = 1 + \dfrac{1}{2} - \dfrac{1}{n+1} - \dfrac{1}{n+2}$. Taking limits allows us to determine the convergence of the series:

$$\lim_{n\to\infty} S_n = \lim_{n\to\infty} \left(1 + \frac{1}{2} - \frac{1}{n+1} - \frac{1}{n+2}\right) = \frac{3}{2}, \quad \text{so} \quad \sum_{n=1}^{\infty} \frac{1}{n^2 + 2n} = \frac{3}{2}.$$

This is illustrated in Figure 8.2.5(a).

2. We begin by writing the first few partial sums of the series:

$$S_1 = \ln(2)$$

$$S_2 = \ln(2) + \ln\left(\frac{3}{2}\right)$$

$$S_3 = \ln(2) + \ln\left(\frac{3}{2}\right) + \ln\left(\frac{4}{3}\right)$$

$$S_4 = \ln(2) + \ln\left(\frac{3}{2}\right) + \ln\left(\frac{4}{3}\right) + \ln\left(\frac{5}{4}\right)$$

At first, this does not seem helpful, but recall the logarithmic identity: $\ln x + \ln y = \ln(xy)$. Applying this to S_4 gives:

$$S_4 = \ln(2) + \ln\left(\frac{3}{2}\right) + \ln\left(\frac{4}{3}\right) + \ln\left(\frac{5}{4}\right) = \ln\left(\frac{2}{1}\cdot\frac{3}{2}\cdot\frac{4}{3}\cdot\frac{5}{4}\right) = \ln(5).$$

We can conclude that $\{S_n\} = \{\ln(n+1)\}$. This sequence does not converge, as $\lim\limits_{n\to\infty} S_n = \infty$. Therefore $\sum\limits_{n=1}^{\infty} \ln\left(\dfrac{n+1}{n}\right) = \infty$; the series diverges. Note in Figure 8.2.5(b) how the sequence of partial sums grows

Figure 8.2.5: Scatter plots relating to the series in Example 8.2.5.

Notes:

slowly; after 100 terms, it is not yet over 5. Graphically we may be fooled into thinking the series converges, but our analysis above shows that it does not.

We are learning about a new mathematical object, the series. As done before, we apply "old" mathematics to this new topic.

Theorem 8.2.3   Properties of Infinite Series

Let $\displaystyle\sum_{n=1}^{\infty} a_n = L$, $\displaystyle\sum_{n=1}^{\infty} b_n = K$, and let c be a constant.

1. Constant Multiple Rule: $\displaystyle\sum_{n=1}^{\infty} c \cdot a_n = c \cdot \sum_{n=1}^{\infty} a_n = c \cdot L$.

2. Sum/Difference Rule: $\displaystyle\sum_{n=1}^{\infty} \left(a_n \pm b_n\right) = \sum_{n=1}^{\infty} a_n \pm \sum_{n=1}^{\infty} b_n = L \pm K$.

Before using this theorem, we provide a few "famous" series.

Key Idea 8.2.1   Important Series

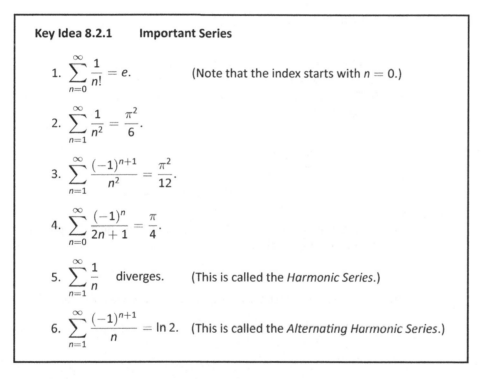

1. $\displaystyle\sum_{n=0}^{\infty} \frac{1}{n!} = e$.   (Note that the index starts with $n = 0$.)

2. $\displaystyle\sum_{n=1}^{\infty} \frac{1}{n^2} = \frac{\pi^2}{6}$.

3. $\displaystyle\sum_{n=1}^{\infty} \frac{(-1)^{n+1}}{n^2} = \frac{\pi^2}{12}$.

4. $\displaystyle\sum_{n=0}^{\infty} \frac{(-1)^n}{2n+1} = \frac{\pi}{4}$.

5. $\displaystyle\sum_{n=1}^{\infty} \frac{1}{n}$ diverges.   (This is called the *Harmonic Series*.)

6. $\displaystyle\sum_{n=1}^{\infty} \frac{(-1)^{n+1}}{n} = \ln 2$.   (This is called the *Alternating Harmonic Series*.)

Notes:

Example 8.2.6   Evaluating series
Evaluate the given series.

1. $\displaystyle\sum_{n=1}^{\infty} \frac{(-1)^{n+1}(n^2 - n)}{n^3}$ 2. $\displaystyle\sum_{n=1}^{\infty} \frac{1000}{n!}$ 3. $\dfrac{1}{16} + \dfrac{1}{25} + \dfrac{1}{36} + \dfrac{1}{49} + \cdots$

SOLUTION

1. We start by using algebra to break the series apart:

$$\sum_{n=1}^{\infty} \frac{(-1)^{n+1}(n^2 - n)}{n^3} = \sum_{n=1}^{\infty}\left(\frac{(-1)^{n+1}n^2}{n^3} - \frac{(-1)^{n+1}n}{n^3}\right)$$

$$= \sum_{n=1}^{\infty}\frac{(-1)^{n+1}}{n} - \sum_{n=1}^{\infty}\frac{(-1)^{n+1}}{n^2}$$

$$= \ln(2) - \frac{\pi^2}{12} \approx -0.1293.$$

This is illustrated in Figure 8.2.6(a).

2. This looks very similar to the series that involves e in Key Idea 8.2.1. Note, however, that the series given in this example starts with $n = 1$ and not $n = 0$. The first term of the series in the Key Idea is $1/0! = 1$, so we will subtract this from our result below:

$$\sum_{n=1}^{\infty} \frac{1000}{n!} = 1000 \cdot \sum_{n=1}^{\infty} \frac{1}{n!}$$

$$= 1000 \cdot (e - 1) \approx 1718.28.$$

This is illustrated in Figure 8.2.6(b). The graph shows how this particular series converges very rapidly.

3. The denominators in each term are perfect squares; we are adding $\displaystyle\sum_{n=4}^{\infty} \frac{1}{n^2}$ (note we start with $n = 4$, not $n = 1$). This series will converge. Using the

(a)

(b)

Figure 8.2.6: Scatter plots relating to the series in Example 8.2.6.

Notes:

formula from Key Idea 8.2.1, we have the following:

$$\sum_{n=1}^{\infty} \frac{1}{n^2} = \sum_{n=1}^{3} \frac{1}{n^2} + \sum_{n=4}^{\infty} \frac{1}{n^2}$$

$$\sum_{n=1}^{\infty} \frac{1}{n^2} - \sum_{n=1}^{3} \frac{1}{n^2} = \sum_{n=4}^{\infty} \frac{1}{n^2}$$

$$\frac{\pi^2}{6} - \left(\frac{1}{1} + \frac{1}{4} + \frac{1}{9} \right) = \sum_{n=4}^{\infty} \frac{1}{n^2}$$

$$\frac{\pi^2}{6} - \frac{49}{36} = \sum_{n=4}^{\infty} \frac{1}{n^2}$$

$$0.2838 \approx \sum_{n=4}^{\infty} \frac{1}{n^2}$$

It may take a while before one is comfortable with this statement, whose truth lies at the heart of the study of infinite series: *it is possible that the sum of an infinite list of nonzero numbers is finite.* We have seen this repeatedly in this section, yet it still may "take some getting used to."

As one contemplates the behavior of series, a few facts become clear.

1. In order to add an infinite list of nonzero numbers and get a finite result, "most" of those numbers must be "very near" 0.

2. If a series diverges, it means that the sum of an infinite list of numbers is not finite (it may approach $\pm\infty$ or it may oscillate), and:

    (a) The series will still diverge if the first term is removed.

    (b) The series will still diverge if the first 10 terms are removed.

    (c) The series will still diverge if the first $1,000,000$ terms are removed.

    (d) The series will still diverge if any finite number of terms from anywhere in the series are removed.

These concepts are very important and lie at the heart of the next two theorems.

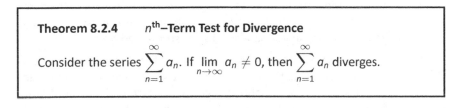

Theorem 8.2.4 n^{th}**–Term Test for Divergence**

Consider the series $\sum_{n=1}^{\infty} a_n$. If $\lim_{n \to \infty} a_n \neq 0$, then $\sum_{n=1}^{\infty} a_n$ diverges.

Notes:

Important! This theorem *does not state* that if $\lim\limits_{n\to\infty} a_n = 0$ then $\sum\limits_{n=1}^{\infty} a_n$ converges. The standard example of this is the Harmonic Series, as given in Key Idea 8.2.1. The Harmonic Sequence, $\{1/n\}$, converges to 0; the Harmonic Series, $\sum\limits_{n=1}^{\infty} \dfrac{1}{n}$, diverges.

Looking back, we can apply this theorem to the series in Example 8.2.1. In that example, the nth terms of both sequences do not converge to 0, therefore we can quickly conclude that each series diverges.

One can rewrite Theorem 8.2.4 to state "If a series converges, then the underlying sequence converges to 0." While it is important to understand the truth of this statement, in practice it is rarely used. It is generally far easier to prove the convergence of a sequence than the convergence of a series.

Theorem 8.2.5   Infinite Nature of Series

The convergence or divergence of an infinite series remains unchanged by the addition or subtraction of any finite number of terms. That is:

1. A divergent series will remain divergent with the addition or subtraction of any finite number of terms.

2. A convergent series will remain convergent with the addition or subtraction of any finite number of terms. (Of course, the *sum* will likely change.)

Consider once more the Harmonic Series $\sum\limits_{n=1}^{\infty} \dfrac{1}{n}$ which diverges; that is, the sequence of partial sums $\{S_n\}$ grows (very, very slowly) without bound. One might think that by removing the "large" terms of the sequence that perhaps the series will converge. This is simply not the case. For instance, the sum of the first 10 million terms of the Harmonic Series is about 16.7. Removing the first 10 million terms from the Harmonic Series changes the nth partial sums, effectively subtracting 16.7 from the sum. However, a sequence that is growing without bound will still grow without bound when 16.7 is subtracted from it.

The equations below illustrate this. The first line shows the infinite sum of the Harmonic Series split into the sum of the first 10 million terms plus the sum of "everything else." The next equation shows us subtracting these first 10 million terms from both sides. The final equation employs a bit of "psuedo–math":

Notes:

subtracting 16.7 from "infinity" still leaves one with "infinity."

$$\sum_{n=1}^{\infty} \frac{1}{n} = \sum_{n=1}^{10,000,000} \frac{1}{n} + \sum_{n=10,000,001}^{\infty} \frac{1}{n}$$

$$\sum_{n=1}^{\infty} \frac{1}{n} - \sum_{n=1}^{10,000,000} \frac{1}{n} = \sum_{n=10,000,001}^{\infty} \frac{1}{n}$$

$$\infty \quad - \quad 16.7 \quad = \quad \infty.$$

This section introduced us to series and defined a few special types of series whose convergence properties are well known: we know when a *p*-series or a geometric series converges or diverges. Most series that we encounter are not one of these types, but we are still interested in knowing whether or not they converge. The next three sections introduce tests that help us determine whether or not a given series converges.

Notes:

Exercises 8.2

Terms and Concepts

1. Use your own words to describe how sequences and series are related.

2. Use your own words to define a *partial sum*.

3. Given a series $\sum\limits_{n=1}^{\infty} a_n$, describe the two sequences related to the series that are important.

4. Use your own words to explain what a geometric series is.

5. T/F: If $\{a_n\}$ is convergent, then $\sum\limits_{n=1}^{\infty} a_n$ is also convergent.

6. T/F: If $\{a_n\}$ converges to 0, then $\sum\limits_{n=0}^{\infty} a_n$ converges.

Problems

In Exercises 7 – 14, a series $\sum\limits_{n=1}^{\infty} a_n$ is given.

  (a) Give the first 5 partial sums of the series.

  (b) Give a graph of the first 5 terms of a_n and S_n on the same axes.

7. $\sum\limits_{n=1}^{\infty} \dfrac{(-1)^n}{n}$

8. $\sum\limits_{n=1}^{\infty} \dfrac{1}{n^2}$

9. $\sum\limits_{n=1}^{\infty} \cos(\pi n)$

10. $\sum\limits_{n=1}^{\infty} n$

11. $\sum\limits_{n=1}^{\infty} \dfrac{1}{n!}$

12. $\sum\limits_{n=1}^{\infty} \dfrac{1}{3^n}$

13. $\sum\limits_{n=1}^{\infty} \left(-\dfrac{9}{10}\right)^n$

14. $\sum\limits_{n=1}^{\infty} \left(\dfrac{1}{10}\right)^n$

In Exercises 15 – 20, use Theorem 8.2.4 to show the given series diverges.

15. $\sum\limits_{n=1}^{\infty} \dfrac{3n^2}{n(n+2)}$

16. $\sum\limits_{n=1}^{\infty} \dfrac{2^n}{n^2}$

17. $\sum\limits_{n=1}^{\infty} \dfrac{n!}{10^n}$

18. $\sum\limits_{n=1}^{\infty} \dfrac{5^n - n^5}{5^n + n^5}$

19. $\sum\limits_{n=1}^{\infty} \dfrac{2^n + 1}{2^{n+1}}$

20. $\sum\limits_{n=1}^{\infty} \left(1 + \dfrac{1}{n}\right)^n$

In Exercises 21 – 30, state whether the given series converges or diverges.

21. $\sum\limits_{n=1}^{\infty} \dfrac{1}{n^5}$

22. $\sum\limits_{n=0}^{\infty} \dfrac{1}{5^n}$

23. $\sum\limits_{n=0}^{\infty} \dfrac{6^n}{5^n}$

24. $\sum\limits_{n=1}^{\infty} n^{-4}$

25. $\sum\limits_{n=1}^{\infty} \sqrt{n}$

26. $\sum\limits_{n=1}^{\infty} \dfrac{10}{n!}$

27. T/F: If $\{a_n\}$ converges to 0, then $\sum\limits_{n=0}^{\infty} a_n$ converges.

28. $\sum\limits_{n=1}^{\infty} \dfrac{2}{(2n+8)^2}$

29. $\sum\limits_{n=1}^{\infty} \dfrac{1}{2n}$

30. $\displaystyle\sum_{n=1}^{\infty} \frac{1}{2n-1}$

In Exercises 31 – 46, a series is given.

    (a) Find a formula for S_n, the nth partial sum of the series.

    (b) Determine whether the series converges or diverges. If it converges, state what it converges to.

31. $\displaystyle\sum_{n=0}^{\infty} \frac{1}{4^n}$

32. $\displaystyle\sum_{n=1}^{\infty} 2$

33. $1^3 + 2^3 + 3^3 + 4^3 + \cdots$

34. $\displaystyle\sum_{n=1}^{\infty} (-1)^n n$

35. $\displaystyle\sum_{n=0}^{\infty} \frac{5}{2^n}$

36. $\displaystyle\sum_{n=1}^{\infty} e^{-n}$

37. $1 - \dfrac{1}{3} + \dfrac{1}{9} - \dfrac{1}{27} + \dfrac{1}{81} + \cdots$

38. $\displaystyle\sum_{n=1}^{\infty} \frac{1}{n(n+1)}$

39. $\displaystyle\sum_{n=1}^{\infty} \frac{3}{n(n+2)}$

40. $\displaystyle\sum_{n=1}^{\infty} \frac{1}{(2n-1)(2n+1)}$

41. $\displaystyle\sum_{n=1}^{\infty} \ln\left(\frac{n}{n+1}\right)$

42. $\displaystyle\sum_{n=1}^{\infty} \frac{2n+1}{n^2(n+1)^2}$

43. $\dfrac{1}{1\cdot 4} + \dfrac{1}{2\cdot 5} + \dfrac{1}{3\cdot 6} + \dfrac{1}{4\cdot 7} + \cdots$

44. $2 + \left(\dfrac{1}{2} + \dfrac{1}{3}\right) + \left(\dfrac{1}{4} + \dfrac{1}{9}\right) + \left(\dfrac{1}{8} + \dfrac{1}{27}\right) + \cdots$

45. $\displaystyle\sum_{n=2}^{\infty} \frac{1}{n^2-1}$

46. $\displaystyle\sum_{n=0}^{\infty} (\sin 1)^n$

47. Break the Harmonic Series into the sum of the odd and even terms:

$$\sum_{n=1}^{\infty} \frac{1}{n} = \sum_{n=1}^{\infty} \frac{1}{2n-1} + \sum_{n=1}^{\infty} \frac{1}{2n}.$$

The goal is to show that each of the series on the right diverge.

    (a) Show why $\displaystyle\sum_{n=1}^{\infty} \frac{1}{2n-1} > \sum_{n=1}^{\infty} \frac{1}{2n}$.

    (Compare each nth partial sum.)

    (b) Show why $\displaystyle\sum_{n=1}^{\infty} \frac{1}{2n-1} < 1 + \sum_{n=1}^{\infty} \frac{1}{2n}$

    (c) Explain why (a) and (b) demonstrate that the series of odd terms is convergent, if, and only if, the series of even terms is also convergent. (That is, show both converge or both diverge.)

    (d) Explain why knowing the Harmonic Series is divergent determines that the even and odd series are also divergent.

48. Show the series $\displaystyle\sum_{n=1}^{\infty} \frac{n}{(2n-1)(2n+1)}$ diverges.

8.3   Integral and Comparison Tests

Knowing whether or not a series converges is very important, especially when we discuss Power Series in Section 8.6. Theorems 8.2.1 and 8.2.2 give criteria for when Geometric and p-series converge, and Theorem 8.2.4 gives a quick test to determine if a series diverges. There are many important series whose convergence cannot be determined by these theorems, though, so we introduce a set of tests that allow us to handle a broad range of series. We start with the Integral Test.

Integral Test

We stated in Section 8.1 that a sequence $\{a_n\}$ is a function $a(n)$ whose domain is $\mathbb{N}$, the set of natural numbers. If we can extend $a(n)$ to $\mathbb{R}$, the real numbers, and it is both positive and decreasing on $[1, \infty)$, then the convergence of $\sum_{n=1}^{\infty} a_n$ is the same as $\int_1^{\infty} a(x)\, dx$.

Note: Theorem 8.3.1 does not state that the integral and the summation have the same value.

Theorem 8.3.1   Integral Test

Let a sequence $\{a_n\}$ be defined by $a_n = a(n)$, where $a(n)$ is continuous, positive and decreasing on $[1, \infty)$. Then $\sum_{n=1}^{\infty} a_n$ converges, if, and only if,

$$\int_1^{\infty} a(x)\, dx \text{ converges.}$$

We can demonstrate the truth of the Integral Test with two simple graphs. In Figure 8.3.1(a), the height of each rectangle is $a(n) = a_n$ for $n = 1, 2, \ldots,$ and clearly the rectangles enclose more area than the area under $y = a(x)$. Therefore we can conclude that

$$\int_1^{\infty} a(x)\, dx < \sum_{n=1}^{\infty} a_n. \tag{8.1}$$

In Figure 8.3.1(b), we draw rectangles under $y = a(x)$ with the Right-Hand rule, starting with $n = 2$. This time, the area of the rectangles is less than the area under $y = a(x)$, so $\sum_{n=2}^{\infty} a_n < \int_1^{\infty} a(x)\, dx$. Note how this summation starts with $n = 2$; adding a_1 to both sides lets us rewrite the summation starting with

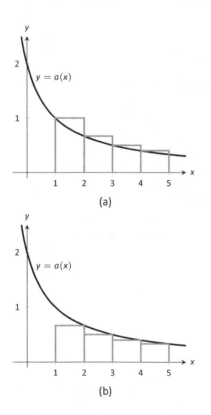

(a)

(b)

Figure 8.3.1: Illustrating the truth of the Integral Test.

Notes:

434

$n = 1$:

$$\sum_{n=1}^{\infty} a_n < a_1 + \int_1^{\infty} a(x)\, dx. \tag{8.2}$$

Combining Equations (8.1) and (8.2), we have

$$\sum_{n=1}^{\infty} a_n < a_1 + \int_1^{\infty} a(x)\, dx < a_1 + \sum_{n=1}^{\infty} a_n. \tag{8.3}$$

From Equation (8.3) we can make the following two statements:

1. If $\displaystyle\sum_{n=1}^{\infty} a_n$ diverges, so does $\displaystyle\int_1^{\infty} a(x)\, dx$ (because $\displaystyle\sum_{n=1}^{\infty} a_n < a_1 + \int_1^{\infty} a(x)\, dx$)

2. If $\displaystyle\sum_{n=1}^{\infty} a_n$ converges, so does $\displaystyle\int_1^{\infty} a(x)\, dx$ (because $\displaystyle\int_1^{\infty} a(x)\, dx < \sum_{n=1}^{\infty} a_n$.)

Therefore the series and integral either both converge or both diverge. Theorem 8.2.5 allows us to extend this theorem to series where $a(n)$ is positive and decreasing on $[b, \infty)$ for some $b > 1$.

Example 8.3.1   Using the Integral Test

Determine the convergence of $\displaystyle\sum_{n=1}^{\infty} \frac{\ln n}{n^2}$. (The terms of the sequence $\{a_n\} = \{\ln n/n^2\}$ and the nth partial sums are given in Figure 8.3.2.)

SOLUTION   Figure 8.3.2 implies that $a(n) = (\ln n)/n^2$ is positive and decreasing on $[2, \infty)$. We can determine this analytically, too. We know $a(n)$ is positive as both $\ln n$ and n^2 are positive on $[2, \infty)$. To determine that $a(n)$ is decreasing, consider $a'(n) = (1 - 2\ln n)/n^3$, which is negative for $n \geq 2$. Since $a'(n)$ is negative, $a(n)$ is decreasing.

Applying the Integral Test, we test the convergence of $\displaystyle\int_1^{\infty} \frac{\ln x}{x^2}\, dx$. Integrating this improper integral requires the use of Integration by Parts, with $u = \ln x$ and $dv = 1/x^2\, dx$.

$$\int_1^{\infty} \frac{\ln x}{x^2}\, dx = \lim_{b \to \infty} \int_1^b \frac{\ln x}{x^2}\, dx$$

$$= \lim_{b \to \infty} -\frac{1}{x} \ln x \Big|_1^b + \int_1^b \frac{1}{x^2}\, dx$$

Figure 8.3.2: Plotting the sequence and series in Example 8.3.1.

Notes:

$$= \lim_{b \to \infty} -\frac{1}{x} \ln x - \frac{1}{x} \Big|_1^b$$

$$= \lim_{b \to \infty} 1 - \frac{1}{b} - \frac{\ln b}{b}. \quad \text{Apply L'Hôpital's Rule:}$$

$$= 1.$$

Since $\int_1^\infty \dfrac{\ln x}{x^2}\, dx$ converges, so does $\displaystyle\sum_{n=1}^\infty \dfrac{\ln n}{n^2}$.

Theorem 8.2.2 was given without justification, stating that the general p-series $\displaystyle\sum_{n=1}^\infty \dfrac{1}{(an+b)^p}$ converges if, and only if, $p > 1$. In the following example, we prove this to be true by applying the Integral Test.

Example 8.3.2   Using the Integral Test to establish Theorem 8.2.2.

Use the Integral Test to prove that $\displaystyle\sum_{n=1}^\infty \dfrac{1}{(an+b)^p}$ converges if, and only if, $p > 1$.

Solution   Consider the integral $\displaystyle\int_1^\infty \dfrac{1}{(ax+b)^p}\, dx$; assuming $p \neq 1$,

$$\int_1^\infty \frac{1}{(ax+b)^p}\, dx = \lim_{c \to \infty} \int_1^c \frac{1}{(ax+b)^p}\, dx$$

$$= \lim_{c \to \infty} \frac{1}{a(1-p)}(ax+b)^{1-p} \Big|_1^c$$

$$= \lim_{c \to \infty} \frac{1}{a(1-p)}\big((ac+b)^{1-p} - (a+b)^{1-p}\big).$$

This limit converges if, and only if, $p > 1$. It is easy to show that the integral also diverges in the case of $p = 1$. (This result is similar to the work preceding Key Idea 6.8.1.)

Therefore $\displaystyle\sum_{n=1}^\infty \dfrac{1}{(an+b)^p}$ converges if, and only if, $p > 1$.

We consider two more convergence tests in this section, both *comparison* tests. That is, we determine the convergence of one series by comparing it to another series with known convergence.

Notes:

Direct Comparison Test

Theorem 8.3.2   Direct Comparison Test

Let $\{a_n\}$ and $\{b_n\}$ be positive sequences where $a_n \leq b_n$ for all $n \geq N$, for some $N \geq 1$.

1. If $\sum_{n=1}^{\infty} b_n$ converges, then $\sum_{n=1}^{\infty} a_n$ converges.

2. If $\sum_{n=1}^{\infty} a_n$ diverges, then $\sum_{n=1}^{\infty} b_n$ diverges.

Note: A sequence $\{a_n\}$ is a **positive sequence** if $a_n > 0$ for all n.

Because of Theorem 8.2.5, any theorem that relies on a positive sequence still holds true when $a_n > 0$ for all but a finite number of values of n.

Example 8.3.3   Applying the Direct Comparison Test

Determine the convergence of $\sum_{n=1}^{\infty} \dfrac{1}{3^n + n^2}$.

SOLUTION   This series is neither a geometric or p-series, but seems related. We predict it will converge, so we look for a series with larger terms that converges. (Note too that the Integral Test seems difficult to apply here.)

Since $3^n < 3^n + n^2$, $\dfrac{1}{3^n} > \dfrac{1}{3^n + n^2}$ for all $n \geq 1$. The series $\sum_{n=1}^{\infty} \dfrac{1}{3^n}$ is a

convergent geometric series; by Theorem 8.3.2, $\sum_{n=1}^{\infty} \dfrac{1}{3^n + n^2}$ converges.

Example 8.3.4   Applying the Direct Comparison Test

Determine the convergence of $\sum_{n=1}^{\infty} \dfrac{1}{n - \ln n}$.

SOLUTION   We know the Harmonic Series $\sum_{n=1}^{\infty} \dfrac{1}{n}$ diverges, and it seems

that the given series is closely related to it, hence we predict it will diverge.

Since $n \geq n - \ln n$ for all $n \geq 1$, $\dfrac{1}{n} \leq \dfrac{1}{n - \ln n}$ for all $n \geq 1$.

The Harmonic Series diverges, so we conclude that $\sum_{n=1}^{\infty} \dfrac{1}{n - \ln n}$ diverges as well.

Notes:

The concept of direct comparison is powerful and often relatively easy to apply. Practice helps one develop the necessary intuition to quickly pick a proper series with which to compare. However, it is easy to construct a series for which it is difficult to apply the Direct Comparison Test.

Consider $\sum_{n=1}^{\infty}\frac{1}{n+\ln n}$. It is very similar to the divergent series given in Example 8.3.4. We suspect that it also diverges, as $\frac{1}{n}\approx\frac{1}{n+\ln n}$ for large n. However, the inequality that we naturally want to use "goes the wrong way": since $n \le n+\ln n$ for all $n \ge 1$, $\frac{1}{n} \ge \frac{1}{n+\ln n}$ for all $n \ge 1$. The given series has terms *less than* the terms of a divergent series, and we cannot conclude anything from this.

Fortunately, we can apply another test to the given series to determine its convergence.

Limit Comparison Test

Theorem 8.3.3   Limit Comparison Test

Let $\{a_n\}$ and $\{b_n\}$ be positive sequences.

1. If $\lim_{n\to\infty}\frac{a_n}{b_n}=L$, where L is a positive real number, then $\sum_{n=1}^{\infty}a_n$ and $\sum_{n=1}^{\infty}b_n$ either both converge or both diverge.

2. If $\lim_{n\to\infty}\frac{a_n}{b_n}=0$, then if $\sum_{n=1}^{\infty}b_n$ converges, then so does $\sum_{n=1}^{\infty}a_n$.

3. If $\lim_{n\to\infty}\frac{a_n}{b_n}=\infty$, then if $\sum_{n=1}^{\infty}b_n$ diverges, then so does $\sum_{n=1}^{\infty}a_n$.

Theorem 8.3.3 is most useful when the convergence of the series from $\{b_n\}$ is known and we are trying to determine the convergence of the series from $\{a_n\}$.

We use the Limit Comparison Test in the next example to examine the series $\sum_{n=1}^{\infty}\frac{1}{n+\ln n}$ which motivated this new test.

Notes:

Example 8.3.5   **Applying the Limit Comparison Test**

Determine the convergence of $\displaystyle\sum_{n=1}^{\infty} \frac{1}{n + \ln n}$ using the Limit Comparison Test.

SOLUTION   We compare the terms of $\displaystyle\sum_{n=1}^{\infty} \frac{1}{n + \ln n}$ to the terms of the

Harmonic Sequence $\displaystyle\sum_{n=1}^{\infty} \frac{1}{n}$:

$$\lim_{n\to\infty} \frac{1/(n + \ln n)}{1/n} = \lim_{n\to\infty} \frac{n}{n + \ln n}$$

$$= 1 \quad \text{(after applying L'Hôpital's Rule).}$$

Since the Harmonic Series diverges, we conclude that $\displaystyle\sum_{n=1}^{\infty} \frac{1}{n + \ln n}$ diverges as well.

Example 8.3.6   **Applying the Limit Comparison Test**

Determine the convergence of $\displaystyle\sum_{n=1}^{\infty} \frac{1}{3^n - n^2}$

SOLUTION   This series is similar to the one in Example 8.3.3, but now we are considering "$3^n - n^2$" instead of "$3^n + n^2$." This difference makes applying the Direct Comparison Test difficult.

Instead, we use the Limit Comparison Test and compare with the series $\displaystyle\sum_{n=1}^{\infty} \frac{1}{3^n}$:

$$\lim_{n\to\infty} \frac{1/(3^n - n^2)}{1/3^n} = \lim_{n\to\infty} \frac{3^n}{3^n - n^2}$$

$$= 1 \quad \text{(after applying L'Hôpital's Rule twice).}$$

We know $\displaystyle\sum_{n=1}^{\infty} \frac{1}{3^n}$ is a convergent geometric series, hence $\displaystyle\sum_{n=1}^{\infty} \frac{1}{3^n - n^2}$ converges as well.

As mentioned before, practice helps one develop the intuition to quickly choose a series with which to compare. A general rule of thumb is to pick a series based on the dominant term in the expression of $\{a_n\}$. It is also helpful to note that factorials dominate exponentials, which dominate algebraic functions (e.g., polynomials), which dominate logarithms. In the previous example,

Notes:

the dominant term of $\dfrac{1}{3^n - n^2}$ was 3^n, so we compared the series to $\displaystyle\sum_{n=1}^{\infty} \dfrac{1}{3^n}$. It is hard to apply the Limit Comparison Test to series containing factorials, though, as we have not learned how to apply L'Hôpital's Rule to $n!$.

Example 8.3.7　Applying the Limit Comparison Test

Determine the convergence of $\displaystyle\sum_{n=1}^{\infty} \dfrac{\sqrt{n}+3}{n^2 - n + 1}$.

SOLUTION　We naïvely attempt to apply the rule of thumb given above and note that the dominant term in the expression of the series is $1/n^2$. Knowing that $\displaystyle\sum_{n=1}^{\infty} \dfrac{1}{n^2}$ converges, we attempt to apply the Limit Comparison Test:

$$\lim_{n\to\infty} \frac{(\sqrt{n}+3)/(n^2-n+1)}{1/n^2} = \lim_{n\to\infty} \frac{n^2(\sqrt{n}+3)}{n^2-n+1}$$
$$= \infty \quad \text{(Apply L'Hôpital's Rule)}.$$

Theorem 8.3.3 part (3) only applies when $\displaystyle\sum_{n=1}^{\infty} b_n$ diverges; in our case, it converges. Ultimately, our test has not revealed anything about the convergence of our series.

The problem is that we chose a poor series with which to compare. Since the numerator and denominator of the terms of the series are both algebraic functions, we should have compared our series to the dominant term of the numerator divided by the dominant term of the denominator.

The dominant term of the numerator is $n^{1/2}$ and the dominant term of the denominator is n^2. Thus we should compare the terms of the given series to $n^{1/2}/n^2 = 1/n^{3/2}$:

$$\lim_{n\to\infty} \frac{(\sqrt{n}+3)/(n^2-n+1)}{1/n^{3/2}} = \lim_{n\to\infty} \frac{n^{3/2}(\sqrt{n}+3)}{n^2-n+1}$$
$$= 1 \quad \text{(Apply L'Hôpital's Rule)}.$$

Since the p-series $\displaystyle\sum_{n=1}^{\infty} \dfrac{1}{n^{3/2}}$ converges, we conclude that $\displaystyle\sum_{n=1}^{\infty} \dfrac{\sqrt{n}+3}{n^2 - n + 1}$ converges as well.

We mentioned earlier that the Integral Test did not work well with series containing factorial terms. The next section introduces the Ratio Test, which does handle such series well. We also introduce the Root Test, which is good for series where each term is raised to a power.

___

Notes:

Exercises 8.3

Terms and Concepts

1. In order to apply the Integral Test to a sequence $\{a_n\}$, the function $a(n) = a_n$ must be _____, _____ and _____.

2. T/F: The Integral Test can be used to determine the sum of a convergent series.

3. What test(s) in this section do not work well with factorials?

4. Suppose $\sum_{n=0}^{\infty} a_n$ is convergent, and there are sequences $\{b_n\}$ and $\{c_n\}$ such that $0 \leq b_n \leq a_n \leq c_n$ for all n. What can be said about the series $\sum_{n=0}^{\infty} b_n$ and $\sum_{n=0}^{\infty} c_n$?

Problems

In Exercises 5 – 12, use the Integral Test to determine the convergence of the given series.

5. $\sum_{n=1}^{\infty} \dfrac{1}{2^n}$

6. $\sum_{n=1}^{\infty} \dfrac{1}{n^4}$

7. $\sum_{n=1}^{\infty} \dfrac{n}{n^2 + 1}$

8. $\sum_{n=2}^{\infty} \dfrac{1}{n \ln n}$

9. $\sum_{n=1}^{\infty} \dfrac{1}{n^2 + 1}$

10. $\sum_{n=2}^{\infty} \dfrac{1}{n(\ln n)^2}$

11. $\sum_{n=1}^{\infty} \dfrac{n}{2^n}$

12. $\sum_{n=1}^{\infty} \dfrac{\ln n}{n^3}$

In Exercises 13 – 22, use the Direct Comparison Test to determine the convergence of the given series; state what series is used for comparison.

13. $\sum_{n=1}^{\infty} \dfrac{1}{n^2 + 3n - 5}$

14. $\sum_{n=1}^{\infty} \dfrac{1}{4^n + n^2 - n}$

15. $\sum_{n=1}^{\infty} \dfrac{\ln n}{n}$

16. $\sum_{n=1}^{\infty} \dfrac{1}{n! + n}$

17. $\sum_{n=2}^{\infty} \dfrac{1}{\sqrt{n^2 - 1}}$

18. $\sum_{n=5}^{\infty} \dfrac{1}{\sqrt{n - 2}}$

19. $\sum_{n=1}^{\infty} \dfrac{n^2 + n + 1}{n^3 \; 5}$

20. $\sum_{n=1}^{\infty} \dfrac{2^n}{5^n + 10}$

21. $\sum_{n=2}^{\infty} \dfrac{n}{n^2 - 1}$

22. $\sum_{n=2}^{\infty} \dfrac{1}{n^2 \ln n}$

In Exercises 23 – 32, use the Limit Comparison Test to determine the convergence of the given series; state what series is used for comparison.

23. $\sum_{n=1}^{\infty} \dfrac{1}{n^2 - 3n + 5}$

24. $\sum_{n=1}^{\infty} \dfrac{1}{4^n - n^2}$

25. $\sum_{n=4}^{\infty} \dfrac{\ln n}{n - 3}$

26. $\sum_{n=1}^{\infty} \dfrac{1}{\sqrt{n^2 + n}}$

27. $\sum_{n=1}^{\infty} \dfrac{1}{n + \sqrt{n}}$

28. $\sum_{n=1}^{\infty} \dfrac{n - 10}{n^2 + 10n + 10}$

29. $\sum_{n=1}^{\infty} \sin\left(1/n\right)$

441

30. $\displaystyle\sum_{n=1}^{\infty} \frac{n+5}{n^3-5}$

31. $\displaystyle\sum_{n=1}^{\infty} \frac{\sqrt{n}+3}{n^2+17}$

32. $\displaystyle\sum_{n=1}^{\infty} \frac{1}{\sqrt{n}+100}$

In Exercises 33 – 40, determine the convergence of the given series. State the test used; more than one test may be appropriate.

33. $\displaystyle\sum_{n=1}^{\infty} \frac{n^2}{2^n}$

34. $\displaystyle\sum_{n=1}^{\infty} \frac{1}{(2n+5)^3}$

35. $\displaystyle\sum_{n=1}^{\infty} \frac{n!}{10^n}$

36. $\displaystyle\sum_{n=1}^{\infty} \frac{\ln n}{n!}$

37. $\displaystyle\sum_{n=1}^{\infty} \frac{1}{3^n+n}$

38. $\displaystyle\sum_{n=1}^{\infty} \frac{n-2}{10n+5}$

39. $\displaystyle\sum_{n=1}^{\infty} \frac{3^n}{n^3}$

40. $\displaystyle\sum_{n=1}^{\infty} \frac{\cos(1/n)}{\sqrt{n}}$

41. Given that $\displaystyle\sum_{n=1}^{\infty} a_n$ converges, state which of the following series converges, may converge, or does not converge.

(a) $\displaystyle\sum_{n=1}^{\infty} \frac{a_n}{n}$

(b) $\displaystyle\sum_{n=1}^{\infty} a_n a_{n+1}$

(c) $\displaystyle\sum_{n=1}^{\infty} (a_n)^2$

(d) $\displaystyle\sum_{n=1}^{\infty} n a_n$

(e) $\displaystyle\sum_{n=1}^{\infty} \frac{1}{a_n}$

8.4 Ratio and Root Tests

The nth–Term Test of Theorem 8.2.4 states that in order for a series $\sum_{n=1}^{\infty} a_n$ to converge, $\lim_{n\to\infty} a_n = 0$. That is, the terms of $\{a_n\}$ must get very small. Not only must the terms approach 0, they must approach 0 "fast enough": while $\lim_{n\to\infty} 1/n = 0$, the Harmonic Series $\sum_{n=1}^{\infty} \frac{1}{n}$ diverges as the terms of $\{1/n\}$ do not approach 0 "fast enough."

The comparison tests of the previous section determine convergence by comparing terms of a series to terms of another series whose convergence is known. This section introduces the Ratio and Root Tests, which determine convergence by analyzing the terms of a series to see if they approach 0 "fast enough."

Ratio Test

Theorem 8.4.1   Ratio Test

Let $\{a_n\}$ be a positive sequence where $\lim_{n\to\infty} \dfrac{a_{n+1}}{a_n} = L$.

1. If $L < 1$, then $\sum_{n=1}^{\infty} a_n$ converges.

2. If $L > 1$ or $L = \infty$, then $\sum_{n=1}^{\infty} a_n$ diverges.

3. If $L = 1$, the Ratio Test is inconclusive.

Note: Theorem 8.2.5 allows us to apply the Ratio Test to series where $\{a_n\}$ is positive for all but a finite number of terms.

The principle of the Ratio Test is this: if $\lim_{n\to\infty} \dfrac{a_{n+1}}{a_n} = L < 1$, then for large n, each term of $\{a_n\}$ is significantly smaller than its previous term which is enough to ensure convergence.

Example 8.4.1   Applying the Ratio Test
Use the Ratio Test to determine the convergence of the following series:

1. $\sum_{n=1}^{\infty} \dfrac{2^n}{n!}$ 2. $\sum_{n=1}^{\infty} \dfrac{3^n}{n^3}$ 3. $\sum_{n=1}^{\infty} \dfrac{1}{n^2 + 1}$.

Notes:

SOLUTION

1. $\displaystyle\sum_{n=1}^{\infty} \frac{2^n}{n!}$:

$$\lim_{n\to\infty} \frac{2^{n+1}/(n+1)!}{2^n/n!} = \lim_{n\to\infty} \frac{2^{n+1}n!}{2^n(n+1)!}$$
$$= \lim_{n\to\infty} \frac{2}{n+1}$$
$$= 0.$$

Since the limit is $0 < 1$, by the Ratio Test $\displaystyle\sum_{n=1}^{\infty} \frac{2^n}{n!}$ converges.

2. $\displaystyle\sum_{n=1}^{\infty} \frac{3^n}{n^3}$:

$$\lim_{n\to\infty} \frac{3^{n+1}/(n+1)^3}{3^n/n^3} = \lim_{n\to\infty} \frac{3^{n+1}n^3}{3^n(n+1)^3}$$
$$= \lim_{n\to\infty} \frac{3n^3}{(n+1)^3}$$
$$= 3.$$

Since the limit is $3 > 1$, by the Ratio Test $\displaystyle\sum_{n=1}^{\infty} \frac{3^n}{n^3}$ diverges.

3. $\displaystyle\sum_{n=1}^{\infty} \frac{1}{n^2+1}$:

$$\lim_{n\to\infty} \frac{1/\left((n+1)^2+1\right)}{1/(n^2+1)} = \lim_{n\to\infty} \frac{n^2+1}{(n+1)^2+1}$$
$$= 1.$$

Since the limit is 1, the Ratio Test is inconclusive. We can easily show this series converges using the Direct or Limit Comparison Tests, with each comparing to the series $\displaystyle\sum_{n=1}^{\infty} \frac{1}{n^2}$.

Notes:

The Ratio Test is not effective when the terms of a series *only* contain algebraic functions (e.g., polynomials). It is most effective when the terms contain some factorials or exponentials. The previous example also reinforces our developing intuition: factorials dominate exponentials, which dominate algebraic functions, which dominate logarithmic functions. In Part 1 of the example, the factorial in the denominator dominated the exponential in the numerator, causing the series to converge. In Part 2, the exponential in the numerator dominated the algebraic function in the denominator, causing the series to diverge.

While we have used factorials in previous sections, we have not explored them closely and one is likely to not yet have a strong intuitive sense for how they behave. The following example gives more practice with factorials.

Example 8.4.2   **Applying the Ratio Test**

Determine the convergence of $\displaystyle\sum_{n=1}^{\infty} \frac{n!n!}{(2n)!}$.

SOLUTION   Before we begin, be sure to note the difference between $(2n)!$ and $2n!$. When $n = 4$, the former is $8! = 8 \cdot 7 \cdot \ldots \cdot 2 \cdot 1 = 40,320$, whereas the latter is $2(4 \cdot 3 \cdot 2 \cdot 1) = 48$.

Applying the Ratio Test:

$$\lim_{n\to\infty} \frac{(n+1)!(n+1)!/\big(2(n+1)\big)!}{n!n!/(2n)!} = \lim_{n\to\infty} \frac{(n+1)!(n+1)!(2n)!}{n!n!(2n+2)!}$$

Noting that $(2n+2)! = (2n+2) \cdot (2n+1) \cdot (2n)!$, we have

$$= \lim_{n\to\infty} \frac{(n+1)(n+1)}{(2n+2)(2n+1)}$$
$$= 1/4.$$

Since the limit is $1/4 < 1$, by the Ratio Test we conclude $\displaystyle\sum_{n=1}^{\infty} \frac{n!n!}{(2n)!}$ converges.

Root Test

The final test we introduce is the Root Test, which works particularly well on series where each term is raised to a power, and does not work well with terms containing factorials.

Notes:

Note: Theorem 8.2.5 allows us to apply the Root Test to series where $\{a_n\}$ is positive for all but a finite number of terms.

Theorem 8.4.2   Root Test

Let $\{a_n\}$ be a positive sequence, and let $\lim\limits_{n\to\infty} (a_n)^{1/n} = L$.

1. If $L < 1$, then $\sum\limits_{n=1}^{\infty} a_n$ converges.

2. If $L > 1$ or $L = \infty$, then $\sum\limits_{n=1}^{\infty} a_n$ diverges.

3. If $L = 1$, the Root Test is inconclusive.

Example 8.4.3   Applying the Root Test

Determine the convergence of the following series using the Root Test:

1. $\sum\limits_{n=1}^{\infty} \left(\dfrac{3n+1}{5n-2}\right)^n$ 2. $\sum\limits_{n=1}^{\infty} \dfrac{n^4}{(\ln n)^n}$ 3. $\sum\limits_{n=1}^{\infty} \dfrac{2^n}{n^2}$.

SOLUTION

1. $\lim\limits_{n\to\infty} \left(\left(\dfrac{3n+1}{5n-2}\right)^n\right)^{1/n} = \lim\limits_{n\to\infty} \dfrac{3n+1}{5n-2} = \dfrac{3}{5}$.

   Since the limit is less than 1, we conclude the series converges. Note: it is difficult to apply the Ratio Test to this series.

2. $\lim\limits_{n\to\infty} \left(\dfrac{n^4}{(\ln n)^n}\right)^{1/n} = \lim\limits_{n\to\infty} \dfrac{\left(n^{1/n}\right)^4}{\ln n}$.

   As n grows, the numerator approaches 1 (apply L'Hôpital's Rule) and the denominator grows to infinity. Thus

   $$\lim\limits_{n\to\infty} \dfrac{\left(n^{1/n}\right)^4}{\ln n} = 0.$$

   Since the limit is less than 1, we conclude the series converges.

3. $\lim\limits_{n\to\infty} \left(\dfrac{2^n}{n^2}\right)^{1/n} = \lim\limits_{n\to\infty} \dfrac{2}{\left(n^{1/n}\right)^2} = 2$.

   Since this is greater than 1, we conclude the series diverges.

Each of the tests we have encountered so far has required that we analyze series from *positive* sequences. The next section relaxes this restriction by considering *alternating series*, where the underlying sequence has terms that alternate between being positive and negative.

Notes:

Exercises 8.4

Terms and Concepts

1. The Ratio Test is not effective when the terms of a sequence only contain _____ functions.

2. The Ratio Test is most effective when the terms of a sequence contains _____ and/or _____ functions.

3. What three convergence tests do not work well with terms containing factorials?

4. The Root Test works particularly well on series where each term is _____ to a _____.

Problems

In Exercises 5 – 14, determine the convergence of the given series using the Ratio Test. If the Ratio Test is inconclusive, state so and determine convergence with another test.

5. $\sum\limits_{n=0}^{\infty} \dfrac{2n}{n!}$

6. $\sum\limits_{n=0}^{\infty} \dfrac{5^n - 3n}{4^n}$

7. $\sum\limits_{n=0}^{\infty} \dfrac{n!10^n}{(2n)!}$

8. $\sum\limits_{n=1}^{\infty} \dfrac{5^n + n^4}{7^n + n^2}$

9. $\sum\limits_{n=1}^{\infty} \dfrac{1}{n}$

10. $\sum\limits_{n=1}^{\infty} \dfrac{1}{3n^3 + 7}$

11. $\sum\limits_{n=1}^{\infty} \dfrac{10 \cdot 5^n}{7^n - 3}$

12. $\sum\limits_{n=1}^{\infty} n \cdot \left(\dfrac{3}{5}\right)^n$

13. $\sum\limits_{n=1}^{\infty} \dfrac{2 \cdot 4 \cdot 6 \cdot 8 \cdots 2n}{3 \cdot 6 \cdot 9 \cdot 12 \cdots 3n}$

14. $\sum\limits_{n=1}^{\infty} \dfrac{n!}{5 \cdot 10 \cdot 15 \cdots (5n)}$

In Exercises 15 – 24, determine the convergence of the given series using the Root Test. If the Root Test is inconclusive, state so and determine convergence with another test.

15. $\sum\limits_{n=1}^{\infty} \left(\dfrac{2n + 5}{3n + 11}\right)^n$

16. $\sum\limits_{n=1}^{\infty} \left(\dfrac{.9n^2 - n - 3}{n^2 + n + 3}\right)^n$

17. $\sum\limits_{n=1}^{\infty} \dfrac{2^n n^2}{3^n}$

18. $\sum\limits_{n=1}^{\infty} \dfrac{1}{n^n}$

19. $\sum\limits_{n=1}^{\infty} \dfrac{3^n}{n^2 2^{n+1}}$

20. $\sum\limits_{n=1}^{\infty} \dfrac{4^{n+7}}{7^n}$

21. $\sum\limits_{n=1}^{\infty} \left(\dfrac{n^2 - n}{n^2 + n}\right)^n$

22. $\sum\limits_{n=1}^{\infty} \left(\dfrac{1}{n} - \dfrac{1}{n^2}\right)^n$

23. $\sum\limits_{n=1}^{\infty} \dfrac{1}{(\ln n)^n}$

24. $\sum\limits_{n=1}^{\infty} \dfrac{n^2}{(\ln n)^n}$

In Exercises 25 – 34, determine the convergence of the given series. State the test used; more than one test may be appropriate.

25. $\sum\limits_{n=1}^{\infty} \dfrac{n^2 + 4n - 2}{n^3 + 4n^2 - 3n + 7}$

26. $\sum\limits_{n=1}^{\infty} \dfrac{n^4 4^n}{n!}$

27. $\sum\limits_{n=1}^{\infty} \dfrac{n^2}{3^n + n}$

28. $\sum\limits_{n=1}^{\infty} \dfrac{3^n}{n^n}$

29. $\sum\limits_{n=1}^{\infty} \dfrac{n}{\sqrt{n^2 + 4n + 1}}$

30. $\displaystyle\sum_{n=1}^{\infty} \frac{n!n!n!}{(3n)!}$

31. $\displaystyle\sum_{n=1}^{\infty} \frac{1}{\ln n}$

32. $\displaystyle\sum_{n=1}^{\infty} \left(\frac{n+2}{n+1}\right)^{n}$

33. $\displaystyle\sum_{n=1}^{\infty} \frac{n^3}{(\ln n)^{n}}$

34. $\displaystyle\sum_{n=1}^{\infty} \left(\frac{1}{n} - \frac{1}{n+2}\right)$

8.5 Alternating Series and Absolute Convergence

All of the series convergence tests we have used require that the underlying sequence $\{a_n\}$ be a positive sequence. (We can relax this with Theorem 8.2.5 and state that there must be an $N > 0$ such that $a_n > 0$ for all $n > N$; that is, $\{a_n\}$ is positive for all but a finite number of values of n.)

In this section we explore series whose summation includes negative terms. We start with a very specific form of series, where the terms of the summation alternate between being positive and negative.

Definition 8.5.1   Alternating Series

Let $\{a_n\}$ be a positive sequence. An **alternating series** is a series of either the form

$$\sum_{n=1}^{\infty}(-1)^n a_n \qquad \text{or} \qquad \sum_{n=1}^{\infty}(-1)^{n+1} a_n.$$

Recall the terms of Harmonic Series come from the Harmonic Sequence $\{a_n\} = \{1/n\}$. An important alternating series is the **Alternating Harmonic Series**:

$$\sum_{n=1}^{\infty}(-1)^{n+1}\frac{1}{n} = 1 - \frac{1}{2} + \frac{1}{3} - \frac{1}{4} + \frac{1}{5} - \frac{1}{6} + \cdots$$

Geometric Series can also be alternating series when $r < 0$. For instance, if $r = -1/2$, the geometric series is

$$\sum_{n=0}^{\infty}\left(\frac{-1}{2}\right)^n = 1 - \frac{1}{2} + \frac{1}{4} - \frac{1}{8} + \frac{1}{16} - \frac{1}{32} + \cdots$$

Theorem 8.2.1 states that geometric series converge when $|r| < 1$ and gives the sum: $\sum_{n=0}^{\infty} r^n = \dfrac{1}{1-r}$. When $r = -1/2$ as above, we find

$$\sum_{n=0}^{\infty}\left(\frac{-1}{2}\right)^n = \frac{1}{1-(-1/2)} = \frac{1}{3/2} = \frac{2}{3}.$$

A powerful convergence theorem exists for other alternating series that meet a few conditions.

Notes:

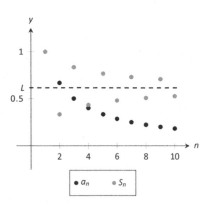

Figure 8.5.1: Illustrating convergence with the Alternating Series Test.

> **Theorem 8.5.1 Alternating Series Test**
>
> Let $\{a_n\}$ be a positive, decreasing sequence where $\lim\limits_{n\to\infty} a_n = 0$. Then
>
> $$\sum_{n=1}^{\infty}(-1)^n a_n \quad \text{and} \quad \sum_{n=1}^{\infty}(-1)^{n+1} a_n$$
>
> converge.

The basic idea behind Theorem 8.5.1 is illustrated in Figure 8.5.1. A positive, decreasing sequence $\{a_n\}$ is shown along with the partial sums

$$S_n = \sum_{i=1}^{n}(-1)^{i+1}a_i = a_1 - a_2 + a_3 - a_4 + \cdots + (-1)^{n+1}a_n.$$

Because $\{a_n\}$ is decreasing, the amount by which S_n bounces up/down decreases. Moreover, the odd terms of S_n form a decreasing, bounded sequence, while the even terms of S_n form an increasing, bounded sequence. Since bounded, monotonic sequences converge (see Theorem 8.1.5) and the terms of $\{a_n\}$ approach 0, one can show the odd and even terms of S_n converge to the same common limit L, the sum of the series.

Example 8.5.1 Applying the Alternating Series Test
Determine if the Alternating Series Test applies to each of the following series.

1. $\displaystyle\sum_{n=1}^{\infty}(-1)^{n+1}\frac{1}{n}$ 2. $\displaystyle\sum_{n=1}^{\infty}(-1)^{n}\frac{\ln n}{n}$ 3. $\displaystyle\sum_{n=1}^{\infty}(-1)^{n+1}\frac{|\sin n|}{n^2}$

SOLUTION

1. This is the Alternating Harmonic Series as seen previously. The underlying sequence is $\{a_n\} = \{1/n\}$, which is positive, decreasing, and approaches 0 as $n \to \infty$. Therefore we can apply the Alternating Series Test and conclude this series converges.

   While the test does not state what the series converges to, we will see later that $\displaystyle\sum_{n=1}^{\infty}(-1)^{n+1}\frac{1}{n} = \ln 2$.

2. The underlying sequence is $\{a_n\} = \{\ln n/n\}$. This is positive and approaches 0 as $n \to \infty$ (use L'Hôpital's Rule). However, the sequence is not decreasing for all n. It is straightforward to compute $a_1 = 0$, $a_2 \approx 0.347$,

Notes:

$a_3 \approx 0.366$, and $a_4 \approx 0.347$: the sequence is increasing for at least the first 3 terms.

We do not immediately conclude that we cannot apply the Alternating Series Test. Rather, consider the long–term behavior of $\{a_n\}$. Treating $a_n = a(n)$ as a continuous function of n defined on $[1, \infty)$, we can take its derivative:

$$a'(n) = \frac{1 - \ln n}{n^2}.$$

The derivative is negative for all $n \geq 3$ (actually, for all $n > e$), meaning $a(n) = a_n$ is decreasing on $[3, \infty)$. We can apply the Alternating Series Test to the series when we start with $n = 3$ and conclude that $\sum_{n=3}^{\infty} (-1)^n \frac{\ln n}{n}$ converges; adding the terms with $n = 1$ and $n = 2$ do not change the convergence (i.e., we apply Theorem 8.2.5).

The important lesson here is that as before, if a series fails to meet the criteria of the Alternating Series Test on only a finite number of terms, we can still apply the test.

3. The underlying sequence is $\{a_n\} = |\sin n|/n^2$. This sequence is positive and approaches 0 as $n \to \infty$. However, it is not a decreasing sequence; the value of $|\sin n|$ oscillates between 0 and 1 as $n \to \infty$. We cannot remove a finite number of terms to make $\{a_n\}$ decreasing, therefore we cannot apply the Alternating Series Test.

Keep in mind that this does not mean we conclude the series diverges; in fact, it does converge. We are just unable to conclude this based on Theorem 8.5.1.

Key Idea 8.2.1 gives the sum of some important series. Two of these are

$$\sum_{n=1}^{\infty} \frac{1}{n^2} = \frac{\pi^2}{6} \approx 1.64493 \quad \text{and} \quad \sum_{n=1}^{\infty} \frac{(-1)^{n+1}}{n^2} = \frac{\pi^2}{12} \approx 0.82247.$$

These two series converge to their sums at different rates. To be accurate to two places after the decimal, we need 202 terms of the first series though only 13 of the second. To get 3 places of accuracy, we need 1069 terms of the first series though only 33 of the second. Why is it that the second series converges so much faster than the first?

While there are many factors involved when studying rates of convergence, the alternating structure of an alternating series gives us a powerful tool when approximating the sum of a convergent series.

Notes:

> **Theorem 8.5.2   The Alternating Series Approximation Theorem**
>
> Let $\{a_n\}$ be a sequence that satisfies the hypotheses of the Alternating Series Test, and let S_n and L be the nth partial sums and sum, respectively, of either $\sum_{n=1}^{\infty}(-1)^n a_n$ or $\sum_{n=1}^{\infty}(-1)^{n+1}a_n$. Then
>
> 1. $|S_n - L| < a_{n+1}$, and
>
> 2. L is between S_n and S_{n+1}.

Part 1 of Theorem 8.5.2 states that the nth partial sum of a convergent alternating series will be within a_{n+1} of its total sum. Consider the alternating series we looked at before the statement of the theorem, $\sum_{n=1}^{\infty}\dfrac{(-1)^{n+1}}{n^2}$. Since $a_{14} = 1/14^2 \approx 0.0051$, we know that S_{13} is within 0.0051 of the total sum.

Moreover, Part 2 of the theorem states that since $S_{13} \approx 0.8252$ and $S_{14} \approx 0.8201$, we know the sum L lies between 0.8201 and 0.8252. One use of this is the knowledge that S_{14} is accurate to two places after the decimal.

Some alternating series converge slowly. In Example 8.5.1 we determined the series $\sum_{n=1}^{\infty}(-1)^{n+1}\dfrac{\ln n}{n}$ converged. With $n = 1001$, we find $\ln n/n \approx 0.0069$, meaning that $S_{1000} \approx 0.1633$ is accurate to one, maybe two, places after the decimal. Since $S_{1001} \approx 0.1564$, we know the sum L is $0.1564 \le L \le 0.1633$.

Example 8.5.2   Approximating the sum of convergent alternating series
Approximate the sum of the following series, accurate to within 0.001.

1. $\sum_{n=1}^{\infty}(-1)^{n+1}\dfrac{1}{n^3}$ 2. $\sum_{n=1}^{\infty}(-1)^{n+1}\dfrac{\ln n}{n}$.

Solution

1. Using Theorem 8.5.2, we want to find n where $1/n^3 \le 0.001$:

$$\frac{1}{n^3} \le 0.001 = \frac{1}{1000}$$
$$n^3 \ge 1000$$
$$n \ge \sqrt[3]{1000}$$
$$n \ge 10.$$

Notes:

Let L be the sum of this series. By Part 1 of the theorem, $|S_9 - L| < a_{10} = 1/1000$. We can compute $S_9 = 0.902116$, which our theorem states is within 0.001 of the total sum.

We can use Part 2 of the theorem to obtain an even more accurate result. As we know the 10th term of the series is $-1/1000$, we can easily compute $S_{10} = 0.901116$. Part 2 of the theorem states that L is between S_9 and S_{10}, so $0.901116 < L < 0.902116$.

2. We want to find n where $\ln(n)/n < 0.001$. We start by solving $\ln(n)/n = 0.001$ for n. This cannot be solved algebraically, so we will use Newton's Method to approximate a solution.

Let $f(x) = \ln(x)/x - 0.001$; we want to know where $f(x) = 0$. We make a guess that x must be "large," so our initial guess will be $x_1 = 1000$. Recall how Newton's Method works: given an approximate solution x_n, our next approximation x_{n+1} is given by

$$x_{n+1} = x_n - \frac{f(x_n)}{f'(x_n)}.$$

We find $f'(x) = \left(1 - \ln(x)\right)/x^2$. This gives

$$x_2 = 1000 - \frac{\ln(1000)/1000 - 0.001}{\left(1 - \ln(1000)\right)/1000^2}$$

$$= 2000.$$

Using a computer, we find that Newton's Method seems to converge to a solution $x = 9118.01$ after 8 iterations. Taking the next integer higher, we have $n = 9119$, where $\ln(9119)/9119 = 0.000999903 < 0.001$.

Again using a computer, we find $S_{9118} = -0.160369$. Part 1 of the theorem states that this is within 0.001 of the actual sum L. Already knowing the 9,119th term, we can compute $S_{9119} = -0.159369$, meaning $-0.159369 < L < -0.160369$.

Notice how the first series converged quite quickly, where we needed only 10 terms to reach the desired accuracy, whereas the second series took over 9,000 terms.

One of the famous results of mathematics is that the Harmonic Series, $\displaystyle\sum_{n=1}^{\infty} \frac{1}{n}$

diverges, yet the Alternating Harmonic Series, $\displaystyle\sum_{n=1}^{\infty} (-1)^{n+1}\frac{1}{n}$, converges. The

Notes:

notion that alternating the signs of the terms in a series can make a series converge leads us to the following definitions.

Note: In Definition 8.5.2, $\sum_{n=1}^{\infty} a_n$ is not necessarily an alternating series; it just may have some negative terms.

Definition 8.5.2   Absolute and Conditional Convergence

1. A series $\sum_{n=1}^{\infty} a_n$ **converges absolutely** if $\sum_{n=1}^{\infty} |a_n|$ converges.

2. A series $\sum_{n=1}^{\infty} a_n$ **converges conditionally** if $\sum_{n=1}^{\infty} a_n$ converges but $\sum_{n=1}^{\infty} |a_n|$ diverges.

Thus we say the Alternating Harmonic Series converges conditionally.

Example 8.5.3   Determining absolute and conditional convergence.
Determine if the following series converge absolutely, conditionally, or diverge.

1. $\displaystyle\sum_{n=1}^{\infty} (-1)^n \frac{n+3}{n^2+2n+5}$ 2. $\displaystyle\sum_{n=1}^{\infty} (-1)^n \frac{n^2+2n+5}{2^n}$ 3. $\displaystyle\sum_{n=3}^{\infty} (-1)^n \frac{3n-3}{5n-10}$

SOLUTION

1. We can show the series

$$\sum_{n=1}^{\infty} \left| (-1)^n \frac{n+3}{n^2+2n+5} \right| = \sum_{n=1}^{\infty} \frac{n+3}{n^2+2n+5}$$

diverges using the Limit Comparison Test, comparing with $1/n$.

The series $\displaystyle\sum_{n=1}^{\infty} (-1)^n \frac{n+3}{n^2+2n+5}$ converges using the Alternating Series Test; we conclude it converges conditionally.

2. We can show the series

$$\sum_{n=1}^{\infty} \left| (-1)^n \frac{n^2+2n+5}{2^n} \right| = \sum_{n=1}^{\infty} \frac{n^2+2n+5}{2^n}$$

converges using the Ratio Test.

Notes:

Therefore we conclude $\sum_{n=1}^{\infty}(-1)^n \dfrac{n^2 + 2n + 5}{2^n}$ converges absolutely.

3. The series

$$\sum_{n=3}^{\infty}\left|(-1)^n \frac{3n - 3}{5n - 10}\right| = \sum_{n=3}^{\infty}\frac{3n - 3}{5n - 10}$$

diverges using the nth Term Test, so it does not converge absolutely.

The series $\sum_{n=3}^{\infty}(-1)^n \dfrac{3n - 3}{5n - 10}$ fails the conditions of the Alternating Series Test as $(3n - 3)/(5n - 10)$ does not approach 0 as $n \to \infty$. We can state further that this series diverges; as $n \to \infty$, the series effectively adds and subtracts $3/5$ over and over. This causes the sequence of partial sums to oscillate and not converge.

Therefore the series $\sum_{n=1}^{\infty}(-1)^n \dfrac{3n - 3}{5n - 10}$ diverges.

Knowing that a series converges absolutely allows us to make two important statements, given in the following theorem. The first is that absolute convergence is "stronger" than regular convergence. That is, just because $\sum_{n=1}^{\infty} a_n$ converges, we cannot conclude that $\sum_{n=1}^{\infty} |a_n|$ will converge, but knowing a series converges absolutely tells us that $\sum_{n=1}^{\infty} a_n$ will converge.

One reason this is important is that our convergence tests all require that the underlying sequence of terms be positive. By taking the absolute value of the terms of a series where not all terms are positive, we are often able to apply an appropriate test and determine absolute convergence. This, in turn, determines that the series we are given also converges.

The second statement relates to **rearrangements** of series. When dealing with a finite set of numbers, the sum of the numbers does not depend on the order which they are added. (So $1+2+3 = 3+1+2$.) One may be surprised to find out that when dealing with an infinite set of numbers, the same statement does not always hold true: some infinite lists of numbers may be rearranged in different orders to achieve different sums. The theorem states that the terms of an absolutely convergent series can be rearranged in any way without affecting the sum.

Notes:

In Example 8.5.3, we determined the series in part 2 converges absolutely. Theorem 8.5.3 tells us the series converges (which we could also determine using the Alternating Series Test).

The theorem states that rearranging the terms of an absolutely convergent series does not affect its sum. This implies that perhaps the sum of a conditionally convergent series can change based on the arrangement of terms. Indeed, it can. The Riemann Rearrangement Theorem (named after Bernhard Riemann) states that any conditionally convergent series can have its terms rearranged so that the sum is any desired value, including ∞!

As an example, consider the Alternating Harmonic Series once more. We have stated that

$$\sum_{n=1}^{\infty}(-1)^{n+1}\frac{1}{n} = 1 - \frac{1}{2} + \frac{1}{3} - \frac{1}{4} + \frac{1}{5} - \frac{1}{6} + \frac{1}{7}\cdots = \ln 2,$$

(see Key Idea 8.2.1 or Example 8.5.1).

Consider the rearrangement where every positive term is followed by two negative terms:

$$1 - \frac{1}{2} - \frac{1}{4} + \frac{1}{3} - \frac{1}{6} - \frac{1}{8} + \frac{1}{5} - \frac{1}{10} - \frac{1}{12}\cdots$$

(Convince yourself that these are exactly the same numbers as appear in the Alternating Harmonic Series, just in a different order.) Now group some terms

Notes:

and simplify:

$$\left(1 - \frac{1}{2}\right) - \frac{1}{4} + \left(\frac{1}{3} - \frac{1}{6}\right) - \frac{1}{8} + \left(\frac{1}{5} - \frac{1}{10}\right) - \frac{1}{12} + \cdots =$$

$$\frac{1}{2} - \frac{1}{4} + \frac{1}{6} - \frac{1}{8} + \frac{1}{10} - \frac{1}{12} + \cdots =$$

$$\frac{1}{2}\left(1 - \frac{1}{2} + \frac{1}{3} - \frac{1}{4} + \frac{1}{5} - \frac{1}{6} + \cdots\right) = \frac{1}{2}\ln 2.$$

By rearranging the terms of the series, we have arrived at a different sum! (One could *try* to argue that the Alternating Harmonic Series does not actually converge to ln 2, because rearranging the terms of the series *shouldn't* change the sum. However, the Alternating Series Test proves this series converges to L, for some number L, and if the rearrangement does not change the sum, then $L = L/2$, implying $L = 0$. But the Alternating Series Approximation Theorem quickly shows that $L > 0$. The only conclusion is that the rearrangement *did* change the sum.) This is an incredible result.

We end here our study of tests to determine convergence. The end of this text contains a table summarizing the tests that one may find useful.

While series are worthy of study in and of themselves, our ultimate goal within calculus is the study of Power Series, which we will consider in the next section. We will use power series to create functions where the output is the result of an infinite summation.

Notes:

Exercises 8.5

Terms and Concepts

1. Why is $\sum_{n=1}^{\infty} \sin n$ not an alternating series?

2. A series $\sum_{n=1}^{\infty} (-1)^n a_n$ converges when $\{a_n\}$ is _____, _____ and $\lim_{n \to \infty} a_n = $ _____.

3. Give an example of a series where $\sum_{n=0}^{\infty} a_n$ converges but $\sum_{n=0}^{\infty} |a_n|$ does not.

4. The sum of a _____ convergent series can be changed by rearranging the order of its terms.

Problems

In Exercises 5 – 20, an alternating series $\sum_{n=i}^{\infty} a_n$ is given.

  (a) Determine if the series converges or diverges.

  (b) Determine if $\sum_{n=0}^{\infty} |a_n|$ converges or diverges.

  (c) If $\sum_{n=0}^{\infty} a_n$ converges, determine if the convergence is conditional or absolute.

5. $\sum_{n=1}^{\infty} \dfrac{(-1)^{n+1}}{n^2}$

6. $\sum_{n=1}^{\infty} \dfrac{(-1)^{n+1}}{\sqrt{n!}}$

7. $\sum_{n=0}^{\infty} (-1)^n \dfrac{n+5}{3n-5}$

8. $\sum_{n=1}^{\infty} (-1)^n \dfrac{2^n}{n^2}$

9. $\sum_{n=0}^{\infty} (-1)^{n+1} \dfrac{3n+5}{n^2-3n+1}$

10. $\sum_{n=1}^{\infty} \dfrac{(-1)^n}{\ln n + 1}$

11. $\sum_{n=2}^{\infty} (-1)^n \dfrac{n}{\ln n}$

12. $\sum_{n=1}^{\infty} \dfrac{(-1)^{n+1}}{1+3+5+\cdots+(2n-1)}$

13. $\sum_{n=1}^{\infty} \cos(\pi n)$

14. $\sum_{n=2}^{\infty} \dfrac{\sin((n+1/2)\pi)}{n \ln n}$

15. $\sum_{n=0}^{\infty} \left(-\dfrac{2}{3}\right)^n$

16. $\sum_{n=0}^{\infty} (-e)^{-n}$

17. $\sum_{n=0}^{\infty} \dfrac{(-1)^n n^2}{n!}$

18. $\sum_{n=0}^{\infty} (-1)^n 2^{-n^2}$

19. $\sum_{n=1}^{\infty} \dfrac{(-1)^n}{\sqrt{n}}$

20. $\sum_{n=1}^{\infty} \dfrac{(-1000)^n}{n!}$

Let S_n be the nth partial sum of a series. In Exercises 21 – 24, a convergent alternating series is given and a value of n. Compute S_n and S_{n+1} and use these values to find bounds on the sum of the series.

21. $\sum_{n=1}^{\infty} \dfrac{(-1)^n}{\ln(n+1)}, \quad n = 5$

22. $\sum_{n=1}^{\infty} \dfrac{(-1)^{n+1}}{n^4}, \quad n = 4$

23. $\sum_{n=0}^{\infty} \dfrac{(-1)^n}{n!}, \quad n = 6$

24. $\sum_{n=0}^{\infty} \left(-\dfrac{1}{2}\right)^n, \quad n = 9$

In Exercises 25 – 28, a convergent alternating series is given along with its sum and a value of ε. Use Theorem 8.5.2 to find n such that the nth partial sum of the series is within ε of the sum of the series.

25. $\sum_{n=1}^{\infty} \dfrac{(-1)^{n+1}}{n^4} = \dfrac{7\pi^4}{720}, \quad \varepsilon = 0.001$

26. $\displaystyle\sum_{n=0}^{\infty} \frac{(-1)^n}{n!} = \frac{1}{e}$, $\varepsilon = 0.0001$

28. $\displaystyle\sum_{n=0}^{\infty} \frac{(-1)^n}{(2n)!} = \cos 1$, $\varepsilon = 10^{-8}$

27. $\displaystyle\sum_{n=0}^{\infty} \frac{(-1)^n}{2n+1} = \frac{\pi}{4}$, $\varepsilon = 0.001$

8.6   Power Series

So far, our study of series has examined the question of "Is the sum of these infinite terms finite?," i.e., "Does the series converge?" We now approach series from a different perspective: as a function. Given a value of x, we evaluate $f(x)$ by finding the sum of a particular series that depends on x (assuming the series converges). We start this new approach to series with a definition.

Definition 8.6.1   Power Series

Let $\{a_n\}$ be a sequence, let x be a variable, and let c be a real number.

1. The **power series in** x is the series

$$\sum_{n=0}^{\infty} a_n x^n = a_0 + a_1 x + a_2 x^2 + a_3 x^3 + \ldots$$

2. The **power series in** x **centered at** c is the series

$$\sum_{n=0}^{\infty} a_n (x - c)^n = a_0 + a_1(x - c) + a_2(x - c)^2 + a_3(x - c)^3 + \ldots$$

Example 8.6.1   Examples of power series
Write out the first five terms of the following power series:

1. $\displaystyle\sum_{n=0}^{\infty} x^n$ 2. $\displaystyle\sum_{n=1}^{\infty}(-1)^{n+1}\frac{(x+1)^n}{n}$ 3. $\displaystyle\sum_{n=0}^{\infty}(-1)^{n+1}\frac{(x-\pi)^{2n}}{(2n)!}$.

SOLUTION

1. One of the conventions we adopt is that $x^0 = 1$ regardless of the value of x. Therefore

$$\sum_{n=0}^{\infty} x^n = 1 + x + x^2 + x^3 + x^4 + \ldots$$

This is a geometric series in x.

2. This series is centered at $c = -1$. Note how this series starts with $n = 1$. We could rewrite this series starting at $n = 0$ with the understanding that $a_0 = 0$, and hence the first term is 0.

$$\sum_{n=1}^{\infty}(-1)^{n+1}\frac{(x+1)^n}{n} = (x+1) - \frac{(x+1)^2}{2} + \frac{(x+1)^3}{3} - \frac{(x+1)^4}{4} + \frac{(x+1)^5}{5} \cdots$$

Notes:

3. This series is centered at $c = \pi$. Recall that $0! = 1$.

$$\sum_{n=0}^{\infty}(-1)^{n+1}\frac{(x-\pi)^{2n}}{(2n)!} = -1 + \frac{(x-\pi)^2}{2} - \frac{(x-\pi)^4}{24} + \frac{(x-\pi)^6}{6!} - \frac{(x-\pi)^8}{8!}\cdots$$

We introduced power series as a type of function, where a value of x is given and the sum of a series is returned. Of course, not every series converges. For instance, in part 1 of Example 8.6.1, we recognized the series $\sum_{n=0}^{\infty}x^n$ as a geometric series in x. Theorem 8.2.1 states that this series converges only when $|x| < 1$.

This raises the question: "For what values of x will a given power series converge?," which leads us to a theorem and definition.

Theorem 8.6.1   Convergence of Power Series

Let a power series $\sum_{n=0}^{\infty}a_n(x-c)^n$ be given. Then one of the following is true:

1. The series converges only at $x = c$.

2. There is an $R > 0$ such that the series converges for all x in $(c-R, c+R)$ and diverges for all $x < c - R$ and $x > c + R$.

3. The series converges for all x.

The value of R is important when understanding a power series, hence it is given a name in the following definition. Also, note that part 2 of Theorem 8.6.1 makes a statement about the interval $(c-R, c+R)$, but the not the endpoints of that interval. A series may/may not converge at these endpoints.

Notes:

> **Definition 8.6.2   Radius and Interval of Convergence**
>
> 1. The number R given in Theorem 8.6.1 is the **radius of convergence** of a given series. When a series converges for only $x = c$, we say the radius of convergence is 0, i.e., $R = 0$. When a series converges for all x, we say the series has an infinite radius of convergence, i.e., $R = \infty$.
>
> 2. The **interval of convergence** is the set of all values of x for which the series converges.

To find the values of x for which a given series converges, we will use the convergence tests we studied previously (especially the Ratio Test). However, the tests all required that the terms of a series be positive. The following theorem gives us a work–around to this problem.

> **Theorem 8.6.2   The Radius of Convergence of a Series and Absolute Convergence**
>
> The series $\displaystyle\sum_{n=0}^{\infty} a_n(x - c)^n$ and $\displaystyle\sum_{n=0}^{\infty} \left|a_n(x - c)^n\right|$ have the same radius of convergence R.

Theorem 8.6.2 allows us to find the radius of convergence R of a series by applying the Ratio Test (or any applicable test) to the absolute value of the terms of the series. We practice this in the following example.

Example 8.6.2   Determining the radius and interval of convergence.
Find the radius and interval of convergence for each of the following series:

1. $\displaystyle\sum_{n=0}^{\infty} \frac{x^n}{n!}$ 2. $\displaystyle\sum_{n=1}^{\infty} (-1)^{n+1}\frac{x^n}{n}$ 3. $\displaystyle\sum_{n=0}^{\infty} 2^n(x - 3)^n$ 4. $\displaystyle\sum_{n=0}^{\infty} n!x^n$

Solution

Notes:

1. We apply the Ratio Test to the series $\sum_{n=0}^{\infty} \left| \dfrac{x^n}{n!} \right|$:

$$\lim_{n\to\infty} \frac{\left| x^{n+1}/(n+1)! \right|}{\left| x^n/n! \right|} = \lim_{n\to\infty} \left| \frac{x^{n+1}}{x^n} \cdot \frac{n!}{(n+1)!} \right|$$

$$= \lim_{n\to\infty} \left| \frac{x}{n+1} \right|$$

$$= 0 \text{ for all } x.$$

The Ratio Test shows us that regardless of the choice of x, the series converges. Therefore the radius of convergence is $R = \infty$, and the interval of convergence is $(-\infty, \infty)$.

2. We apply the Ratio Test to the series $\sum_{n=1}^{\infty} \left| (-1)^{n+1} \dfrac{x^n}{n} \right| = \sum_{n=1}^{\infty} \left| \dfrac{x^n}{n} \right|$:

$$\lim_{n\to\infty} \frac{\left| x^{n+1}/(n+1) \right|}{\left| x^n/n \right|} = \lim_{n\to\infty} \left| \frac{x^{n+1}}{x^n} \cdot \frac{n}{n+1} \right|$$

$$= \lim_{n\to\infty} |x| \frac{n}{n+1}$$

$$= |x|.$$

The Ratio Test states a series converges if the limit of $|a_{n+1}/a_n| = L < 1$. We found the limit above to be $|x|$; therefore, the power series converges when $|x| < 1$, or when x is in $(-1, 1)$. Thus the radius of convergence is $R = 1$.

To determine the interval of convergence, we need to check the endpoints of $(-1, 1)$. When $x = -1$, we have the opposite of the Harmonic Series:

$$\sum_{n=1}^{\infty} (-1)^{n+1} \frac{(-1)^n}{n} = \sum_{n=1}^{\infty} \frac{-1}{n}$$

$$= -\infty.$$

The series diverges when $x = -1$.

When $x = 1$, we have the series $\sum_{n=1}^{\infty} (-1)^{n+1} \dfrac{(1)^n}{n}$, which is the Alternating Harmonic Series, which converges. Therefore the interval of convergence is $(-1, 1]$.

Notes:

3. We apply the Ratio Test to the series $\displaystyle\sum_{n=0}^{\infty}\left|2^n(x-3)^n\right|$:

$$\lim_{n\to\infty}\frac{\left|2^{n+1}(x-3)^{n+1}\right|}{\left|2^n(x-3)^n\right|}=\lim_{n\to\infty}\left|\frac{2^{n+1}}{2^n}\cdot\frac{(x-3)^{n+1}}{(x-3)^n}\right|$$

$$=\lim_{n\to\infty}\left|2(x-3)\right|.$$

According to the Ratio Test, the series converges when $\left|2(x-3)\right|<1\implies$ $\left|x-3\right|<1/2$. The series is centered at 3, and x must be within $1/2$ of 3 in order for the series to converge. Therefore the radius of convergence is $R=1/2$, and we know that the series converges absolutely for all x in $(3-1/2,3+1/2)=(2.5,3.5)$.

We check for convergence at the endpoints to find the interval of convergence. When $x=2.5$, we have:

$$\sum_{n=0}^{\infty}2^n(2.5-3)^n=\sum_{n=0}^{\infty}2^n(-1/2)^n$$

$$=\sum_{n=0}^{\infty}(-1)^n,$$

which diverges. A similar process shows that the series also diverges at $x=3.5$. Therefore the interval of convergence is $(2.5,3.5)$.

4. We apply the Ratio Test to $\displaystyle\sum_{n=0}^{\infty}\left|n!x^n\right|$:

$$\lim_{n\to\infty}\frac{\left|(n+1)!x^{n+1}\right|}{\left|n!x^n\right|}=\lim_{n\to\infty}\left|(n+1)x\right|$$

$$=\infty\text{ for all }x,\text{ except }x=0.$$

The Ratio Test shows that the series diverges for all x except $x=0$. Therefore the radius of convergence is $R=0$.

We can use a power series to define a function:

$$f(x)=\sum_{n=0}^{\infty}a_nx^n$$

where the domain of f is a subset of the interval of convergence of the power series. One can apply calculus techniques to such functions; in particular, we can find derivatives and antiderivatives.

Notes:

Theorem 8.6.3   **Derivatives and Indefinite Integrals of Power Series Functions**

Let $f(x) = \sum_{n=0}^{\infty} a_n(x - c)^n$ be a function defined by a power series, with radius of convergence R.

1. $f(x)$ is continuous and differentiable on $(c - R, c + R)$.

2. $f'(x) = \sum_{n=1}^{\infty} a_n \cdot n \cdot (x - c)^{n-1}$, with radius of convergence R.

3. $\int f(x)\, dx = C + \sum_{n=0}^{\infty} a_n \frac{(x - c)^{n+1}}{n + 1}$, with radius of convergence R.

A few notes about Theorem 8.6.3:

1. The theorem states that differentiation and integration do not change the radius of convergence. It does not state anything about the *interval* of convergence. They are not always the same.

2. Notice how the summation for $f'(x)$ starts with $n = 1$. This is because the constant term a_0 of $f(x)$ goes to 0.

3. Differentiation and integration are simply calculated term–by–term using the Power Rules.

Example 8.6.3   **Derivatives and indefinite integrals of power series**

Let $f(x) = \sum_{n=0}^{\infty} x^n$. Find $f'(x)$ and $F(x) = \int f(x)\, dx$, along with their respective intervals of convergence.

SOLUTION   We find the derivative and indefinite integral of $f(x)$, following Theorem 8.6.3.

1. $f'(x) = \sum_{n=1}^{\infty} nx^{n-1} = 1 + 2x + 3x^2 + 4x^3 + \cdots$.

   In Example 8.6.1, we recognized that $\sum_{n=0}^{\infty} x^n$ is a geometric series in x. We know that such a geometric series converges when $|x| < 1$; that is, the interval of convergence is $(-1, 1)$.

Notes:

To determine the interval of convergence of $f'(x)$, we consider the endpoints of $(-1, 1)$:

$$f'(-1) = 1 - 2 + 3 - 4 + \cdots, \quad \text{which diverges.}$$

$$f'(1) = 1 + 2 + 3 + 4 + \cdots, \quad \text{which diverges.}$$

Therefore, the interval of convergence of $f'(x)$ is $(-1, 1)$.

2. $F(x) = \displaystyle\int f(x)\, dx = C + \sum_{n=0}^{\infty} \frac{x^{n+1}}{n+1} = C + x + \frac{x^2}{2} + \frac{x^3}{3} + \cdots$

To find the interval of convergence of $F(x)$, we again consider the endpoints of $(-1, 1)$:

$$F(-1) = C - 1 + 1/2 - 1/3 + 1/4 + \cdots$$

The value of C is irrelevant; notice that the rest of the series is an Alternating Series that whose terms converge to 0. By the Alternating Series Test, this series converges. (In fact, we can recognize that the terms of the series after C are the opposite of the Alternating Harmonic Series. We can thus say that $F(-1) = C - \ln 2$.)

$$F(1) = C + 1 + 1/2 + 1/3 + 1/4 + \cdots$$

Notice that this summation is $C +$ the Harmonic Series, which diverges. Since F converges for $x = -1$ and diverges for $x = 1$, the interval of convergence of $F(x)$ is $[-1, 1)$.

The previous example showed how to take the derivative and indefinite integral of a power series without motivation for why we care about such operations. We may care for the sheer mathematical enjoyment "that we can", which is motivation enough for many. However, we would be remiss to not recognize that we can learn a great deal from taking derivatives and indefinite integrals.

Recall that $f(x) = \displaystyle\sum_{n=0}^{\infty} x^n$ in Example 8.6.3 is a geometric series. According to Theorem 8.2.1, this series converges to $1/(1-x)$ when $|x| < 1$. Thus we can say

$$f(x) = \sum_{n=0}^{\infty} x^n = \frac{1}{1-x}, \quad \text{on} \quad (-1, 1).$$

Integrating the power series, (as done in Example 8.6.3,) we find

$$F(x) = C_1 + \sum_{n=0}^{\infty} \frac{x^{n+1}}{n+1}, \tag{8.4}$$

Notes:

while integrating the function $f(x) = 1/(1-x)$ gives

$$F(x) - -\ln|1-x| + C_2. \qquad (8.5)$$

Equating Equations (8.4) and (8.5), we have

$$F(x) = C_1 + \sum_{n=0}^{\infty} \frac{x^{n+1}}{n+1} = -\ln|1-x| + C_2.$$

Letting $x = 0$, we have $F(0) = C_1 = C_2$. This implies that we can drop the constants and conclude

$$\sum_{n=0}^{\infty} \frac{x^{n+1}}{n+1} = -\ln|1-x|.$$

We established in Example 8.6.3 that the series on the left converges at $x = -1$; substituting $x - -1$ on both sides of the above equality gives

$$-1 + \frac{1}{2} - \frac{1}{3} + \frac{1}{4} - \frac{1}{5} + \cdots = -\ln 2.$$

On the left we have the opposite of the Alternating Harmonic Series; on the right, we have $-\ln 2$. We conclude that

$$1 - \frac{1}{2} + \frac{1}{3} - \frac{1}{4} + \cdots = \ln 2.$$

Important: We stated in Key Idea 8.2.1 (in Section 8.2) that the Alternating Harmonic Series converges to $\ln 2$, and referred to this fact again in Example 8.5.1 of Section 8.5. However, we never gave an argument for why this was the case. The work above finally shows how we conclude that the Alternating Harmonic Series converges to $\ln 2$.

We use this type of analysis in the next example.

Example 8.6.4   Analyzing power series functions

Let $f(x) = \sum_{n=0}^{\infty} \frac{x^n}{n!}$. Find $f'(x)$ and $\int f(x)\, dx$, and use these to analyze the behavior of $f(x)$.

   Solution   We start by making two notes: first, in Example 8.6.2, we found the interval of convergence of this power series is $(-\infty, \infty)$. Second, we will find it useful later to have a few terms of the series written out:

$$\sum_{n=0}^{\infty} \frac{x^n}{n!} = 1 + x + \frac{x^2}{2} + \frac{x^3}{6} + \frac{x^4}{24} + \cdots \qquad (8.6)$$

Notes:

We now find the derivative:

$$f'(x) = \sum_{n=1}^{\infty} n\frac{x^{n-1}}{n!}$$

$$= \sum_{n=1}^{\infty} \frac{x^{n-1}}{(n-1)!} = 1 + x + \frac{x^2}{2!} + \cdots .$$

Since the series starts at $n = 1$ and each term refers to $(n - 1)$, we can re-index the series starting with $n = 0$:

$$= \sum_{n=0}^{\infty} \frac{x^n}{n!}$$

$$= f(x).$$

We found the derivative of $f(x)$ is $f(x)$. The only functions for which this is true are of the form $y = ce^x$ for some constant c. As $f(0) = 1$ (see Equation (8.6)), c must be 1. Therefore we conclude that

$$f(x) = \sum_{n=0}^{\infty} \frac{x^n}{n!} = e^x$$

for all x.

We can also find $\int f(x)\, dx$:

$$\int f(x)\, dx = C + \sum_{n=0}^{\infty} \frac{x^{n+1}}{n!(n+1)}$$

$$= C + \sum_{n=0}^{\infty} \frac{x^{n+1}}{(n+1)!}$$

We write out a few terms of this last series:

$$C + \sum_{n=0}^{\infty} \frac{x^{n+1}}{(n+1)!} = C + x + \frac{x^2}{2} + \frac{x^3}{6} + \frac{x^4}{24} + \cdots$$

The integral of $f(x)$ differs from $f(x)$ only by a constant, again indicating that $f(x) = e^x$.

Example 8.6.4 and the work following Example 8.6.3 established relationships between a power series function and "regular" functions that we have dealt with in the past. In general, given a power series function, it is difficult (if

---

Notes:

not impossible) to express the function in terms of elementary functions. We chose examples where things worked out nicely.

In this section's last example, we show how to solve a simple differential equation with a power series.

Example 8.6.5   **Solving a differential equation with a power series.**
Give the first 4 terms of the power series solution to $y' = 2y$, where $y(0) = 1$.

SOLUTION   The differential equation $y' = 2y$ describes a function $y = f(x)$ where the derivative of y is twice y and $y(0) = 1$. This is a rather simple differential equation; with a bit of thought one should realize that if $y = Ce^{2x}$, then $y' = 2Ce^{2x}$, and hence $y' = 2y$. By letting $C = 1$ we satisfy the initial condition of $y(0) = 1$.

Let's ignore the fact that we already know the solution and find a power series function that satisfies the equation. The solution we seek will have the form

$$f(x) = \sum_{n=0}^{\infty} a_n x^n = a_0 + a_1 x + a_2 x^2 + a_3 x^3 + \cdots$$

for unknown coefficients a_n. We can find $f'(x)$ using Theorem 8.6.3:

$$f'(x) = \sum_{n=1}^{\infty} a_n \cdot n \cdot x^{n-1} = a_1 + 2a_2 x + 3a_3 x^2 + 4a_4 x^3 \cdots.$$

Since $f'(x) = 2f(x)$, we have

$$a_1 + 2a_2 x + 3a_3 x^2 + 4a_4 x^3 \cdots = 2\big(a_0 + a_1 x + a_2 x^2 + a_3 x^3 + \cdots\big)$$
$$= 2a_0 + 2a_1 x + 2a_2 x^2 + 2a_3 x^3 + \cdots$$

The coefficients of like powers of x must be equal, so we find that

$$a_1 = 2a_0, \quad 2a_2 = 2a_1, \quad 3a_3 = 2a_2, \quad 4a_4 = 2a_3, \quad \text{etc.}$$

The initial condition $y(0) = f(0) = 1$ indicates that $a_0 = 1$; with this, we can find the values of the other coefficients:

$$a_0 = 1 \text{ and } a_1 = 2a_0 \Rightarrow a_1 = 2;$$
$$a_1 = 2 \text{ and } 2a_2 = 2a_1 \Rightarrow a_2 = 4/2 = 2;$$
$$a_2 = 2 \text{ and } 3a_3 = 2a_2 \Rightarrow a_3 = 8/(2 \cdot 3) = 4/3;$$
$$a_3 = 4/3 \text{ and } 4a_4 = 2a_3 \Rightarrow a_4 = 16/(2 \cdot 3 \cdot 4) = 2/3.$$

Thus the first 5 terms of the power series solution to the differential equation $y' = 2y$ is

$$f(x) = 1 + 2x + 2x^2 + \frac{4}{3}x^3 + \frac{2}{3}x^4 + \cdots$$

Notes:

In Section 8.8, as we study Taylor Series, we will learn how to recognize this series as describing $y = e^{2x}$.

Our last example illustrates that it can be difficult to recognize an elementary function by its power series expansion. It is far easier to start with a known function, expressed in terms of elementary functions, and represent it as a power series function. One may wonder why we would bother doing so, as the latter function probably seems more complicated. In the next two sections, we show both *how* to do this and *why* such a process can be beneficial.

Notes:

Exercises 8.6

Terms and Concepts

1. We adopt the convention that $x^0 =$ _____, regardless of the value of x.

2. What is the difference between the radius of convergence and the interval of convergence?

3. If the radius of convergence of $\sum_{n=0}^{\infty} a_n x^n$ is 5, what is the radius of convergence of $\sum_{n=1}^{\infty} n \cdot a_n x^{n-1}$?

4. If the radius of convergence of $\sum_{n=0}^{\infty} a_n x^n$ is 5, what is the radius of convergence of $\sum_{n=0}^{\infty} (-1)^n a_n x^n$?

Problems

In Exercises 5 – 8, write out the sum of the first 5 terms of the given power series.

5. $\sum_{n=0}^{\infty} 2^n x^n$

6. $\sum_{n=1}^{\infty} \frac{1}{n^2} x^n$

7. $\sum_{n=0}^{\infty} \frac{1}{n!} x^n$

8. $\sum_{n=0}^{\infty} \frac{(-1)^n}{(2n)!} x^{2n}$

In Exercises 9 – 24, a power series is given.

  (a) Find the radius of convergence.

  (b) Find the interval of convergence.

9. $\sum_{n=0}^{\infty} \frac{(-1)^{n+1}}{n!} x^n$

10. $\sum_{n=0}^{\infty} n x^n$

11. $\sum_{n=1}^{\infty} \frac{(-1)^n (x-3)^n}{n}$

12. $\sum_{n=0}^{\infty} \frac{(x+4)^n}{n!}$

13. $\sum_{n=0}^{\infty} \frac{x^n}{2^n}$

14. $\sum_{n=0}^{\infty} \frac{(-1)^n (x-5)^n}{10^n}$

15. $\sum_{n=0}^{\infty} 5^n (x-1)^n$

16. $\sum_{n=0}^{\infty} (-2)^n x^n$

17. $\sum_{n=0}^{\infty} \sqrt{n} x^n$

18. $\sum_{n=0}^{\infty} \frac{n}{3^n} x^n$

19. $\sum_{n=0}^{\infty} \frac{3^n}{n!} (x-5)^n$

20. $\sum_{n=0}^{\infty} (-1)^n n! (x-10)^n$

21. $\sum_{n=1}^{\infty} \frac{x^n}{n^2}$

22. $\sum_{n=1}^{\infty} \frac{(x+2)^n}{n^3}$

23. $\sum_{n=0}^{\infty} n! \left(\frac{x}{10}\right)^n$

24. $\sum_{n=0}^{\infty} n^2 \left(\frac{x+4}{4}\right)^n$

In Exercises 25 – 30, a function $f(x) = \sum_{n=0}^{\infty} a_n x^n$ is given.

  (a) Give a power series for $f'(x)$ and its interval of convergence.

  (b) Give a power series for $\int f(x)\, dx$ and its interval of convergence.

25. $\sum_{n=0}^{\infty} n x^n$

26. $\sum_{n=1}^{\infty} \frac{x^n}{n}$

27. $\sum_{n=0}^{\infty} \left(\frac{x}{2}\right)^n$

28. $\displaystyle\sum_{n=0}^{\infty}(-3x)^n$

29. $\displaystyle\sum_{n=0}^{\infty}\frac{(-1)^n x^{2n}}{(2n)!}$

30. $\displaystyle\sum_{n=0}^{\infty}\frac{(-1)^n x^n}{n!}$

In Exercises 31 – 36, give the first 5 terms of the series that is a solution to the given differential equation.

31. $y' = 3y, \quad y(0) = 1$

32. $y' = 5y, \quad y(0) = 5$

33. $y' = y^2, \quad y(0) = 1$

34. $y' = y + 1, \quad y(0) = 1$

35. $y'' = -y, \quad y(0) = 0, y'(0) = 1$

36. $y'' = 2y, \quad y(0) = 1, y'(0) = 1$

8.7 Taylor Polynomials

Consider a function $y = f(x)$ and a point $(c, f(c))$. The derivative, $f'(c)$, gives the instantaneous rate of change of f at $x = c$. Of all lines that pass through the point $(c, f(c))$, the line that best approximates f at this point is the tangent line; that is, the line whose slope (rate of change) is $f'(c)$.

In Figure 8.7.1, we see a function $y = f(x)$ graphed. The table below the graph shows that $f(0) = 2$ and $f'(0) = 1$; therefore, the tangent line to f at $x = 0$ is $p_1(x) = 1(x - 0) + 2 = x + 2$. The tangent line is also given in the figure. Note that "near" $x = 0$, $p_1(x) \approx f(x)$; that is, the tangent line approximates f well.

One shortcoming of this approximation is that the tangent line only matches the slope of f; it does not, for instance, match the concavity of f. We can find a polynomial, $p_2(x)$, that does match the concavity without much difficulty, though. The table in Figure 8.7.1 gives the following information:

$$f(0) = 2 \qquad f'(0) = 1 \qquad f''(0) = 2.$$

Therefore, we want our polynomial $p_2(x)$ to have these same properties. That is, we need

$$p_2(0) = 2 \qquad p_2'(0) = 1 \qquad p_2''(0) = 2.$$

This is simply an initial–value problem. We can solve this using the techniques first described in Section 5.1. To keep $p_2(x)$ as simple as possible, we'll assume that not only $p_2''(0) = 2$, but that $p_2''(x) = 2$. That is, the second derivative of p_2 is constant.

If $p_2''(x) = 2$, then $p_2'(x) = 2x + C$ for some constant C. Since we have determined that $p_2'(0) = 1$, we find that $C = 1$ and so $p_2'(x) = 2x + 1$. Finally, we can compute $p_2(x) = x^2 + x + C$. Using our initial values, we know $p_2(0) = 2$ so $C = 2$. We conclude that $p_2(x) = x^2 + x + 2$. This function is plotted with f in Figure 8.7.2.

We can repeat this approximation process by creating polynomials of higher degree that match more of the derivatives of f at $x = 0$. In general, a polynomial of degree n can be created to match the first n derivatives of f. Figure 8.7.2 also shows $p_4(x) = -x^4/2 - x^3/6 + x^2 + x + 2$, whose first four derivatives at 0 match those of f. (Using the table in Figure 8.7.1, start with $p_4^{(4)}(x) = -12$ and solve the related initial–value problem.)

As we use more and more derivatives, our polynomial approximation to f gets better and better. In this example, the interval on which the approximation is "good" gets bigger and bigger. Figure 8.7.3 shows $p_{13}(x)$; we can visually affirm that this polynomial approximates f very well on $[-2, 3]$. (The polynomial $p_{13}(x)$ is not particularly "nice". It is

$$\frac{16901 x^{13}}{6227020800} + \frac{13 x^{12}}{1209600} - \frac{1321 x^{11}}{39916800} - \frac{779 x^{10}}{1814400} - \frac{359 x^9}{362880} + \frac{x^8}{240} + \frac{139 x^7}{5040} + \frac{11 x^6}{360} + \frac{19 x^5}{120} - \frac{x^4}{2} - \frac{x^3}{6} + x^2 + x + 2.)$$

Notes:

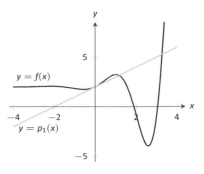

$$
\begin{array}{ll}
f(0) = 2 & f'''(0) = -1 \\
f'(0) = 1 & f^{(4)}(0) = -12 \\
f''(0) = 2 & f^{(5)}(0) = -19
\end{array}
$$

Figure 8.7.1: Plotting $y = f(x)$ and a table of derivatives of f evaluated at 0.

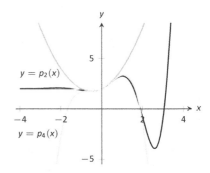

Figure 8.7.2: Plotting f, p_2 and p_4.

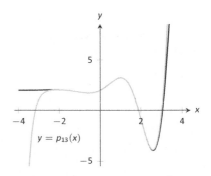

Figure 8.7.3: Plotting f and p_{13}.

The polynomials we have created are examples of *Taylor polynomials,* named after the British mathematician Brook Taylor who made important discoveries about such functions. While we created the above Taylor polynomials by solving initial–value problems, it can be shown that Taylor polynomials follow a general pattern that make their formation much more direct. This is described in the following definition.

Definition 8.7.1   Taylor Polynomial, Maclaurin Polynomial

Let f be a function whose first n derivatives exist at $x = c$.

1. The **Taylor polynomial of degree n of f at $x = c$** is

$$p_n(x) = f(c) + f'(c)(x-c) + \frac{f''(c)}{2!}(x-c)^2 + \frac{f'''(c)}{3!}(x-c)^3 + \cdots + \frac{f^{(n)}(c)}{n!}(x-c)^n.$$

2. A special case of the Taylor polynomial is the Maclaurin polynomial, where $c = 0$. That is, the **Maclaurin polynomial of degree n of f** is

$$p_n(x) = f(0) + f'(0)x + \frac{f''(0)}{2!}x^2 + \frac{f'''(0)}{3!}x^3 + \cdots + \frac{f^{(n)}(0)}{n!}x^n.$$

We will practice creating Taylor and Maclaurin polynomials in the following examples.

Example 8.7.1   Finding and using Maclaurin polynomials

1. Find the nth Maclaurin polynomial for $f(x) = e^x$.

2. Use $p_5(x)$ to approximate the value of e.

SOLUTION

1. We start with creating a table of the derivatives of e^x evaluated at $x = 0$. In this particular case, this is relatively simple, as shown in Figure 8.7.4. By the definition of the Maclaurin series, we have

$$p_n(x) = f(0) + f'(0)x + \frac{f''(0)}{2!}x^2 + \frac{f'''(0)}{3!}x^3 + \cdots + \frac{f^{(n)}(0)}{n!}x^n$$
$$= 1 + x + \frac{1}{2}x^2 + \frac{1}{6}x^3 + \frac{1}{24}x^4 + \cdots + \frac{1}{n!}x^n.$$

$$f(x) = e^x \quad \Rightarrow \quad f(0) = 1$$
$$f'(x) = e^x \quad \Rightarrow \quad f'(0) = 1$$
$$f''(x) = e^x \quad \Rightarrow \quad f''(0) = 1$$
$$\vdots \qquad\qquad \vdots$$
$$f^{(n)}(x) = e^x \quad \Rightarrow \quad f^{(n)}(0) = 1$$

Figure 8.7.4: The derivatives of $f(x) = e^x$ evaluated at $x = 0$.

Notes:

2. Using our answer from part 1, we have

$$p_5 = 1 + x + \frac{1}{2}x^2 + \frac{1}{6}x^3 + \frac{1}{24}x^4 + \frac{1}{120}x^5.$$

To approximate the value of e, note that $e = e^1 = f(1) \approx p_5(1)$. It is very straightforward to evaluate $p_5(1)$:

$$p_5(1) = 1 + 1 + \frac{1}{2} + \frac{1}{6} + \frac{1}{24} + \frac{1}{120} = \frac{163}{60} \approx 2.71667.$$

A plot of $f(x) = e^x$ and $p_5(x)$ is given in Figure 8.7.5.

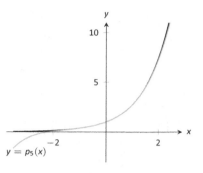

Figure 8.7.5: A plot of $f(x) = e^x$ and its 5th degree Maclaurin polynomial $p_5(x)$.

Example 8.7.2   Finding and using Taylor polynomials

1. Find the nth Taylor polynomial of $y = \ln x$ at $x = 1$.

2. Use $p_6(x)$ to approximate the value of $\ln 1.5$.

3. Use $p_6(x)$ to approximate the value of $\ln 2$.

SOLUTION

1. We begin by creating a table of derivatives of $\ln x$ evaluated at $x = 1$. While this is not as straightforward as it was in the previous example, a pattern does emerge, as shown in Figure 8.7.6.

   Using Definition 8.7.1, we have

$$p_n(x) = f(c) + f'(c)(x - c) + \frac{f''(c)}{2!}(x - c)^2 + \frac{f'''(c)}{3!}(x - c)^3 + \cdots + \frac{f^{(n)}(c)}{n!}(x - c)^n$$

$$= 0 + (x - 1) - \frac{1}{2}(x - 1)^2 + \frac{1}{3}(x - 1)^3 - \frac{1}{4}(x - 1)^4 + \cdots + \frac{(-1)^{n+1}}{n}(x - 1)^n.$$

$$\begin{aligned}
f(x) &= \ln x & \Rightarrow & \quad f(1) = 0 \\
f'(x) &= 1/x & \Rightarrow & \quad f'(1) = 1 \\
f''(x) &= -1/x^2 & \Rightarrow & \quad f''(1) = -1 \\
f'''(x) &= 2/x^3 & \Rightarrow & \quad f'''(1) = 2 \\
f^{(4)}(x) &= -6/x^4 & \Rightarrow & \quad f^{(4)}(1) = -6 \\
&\vdots & & \quad \vdots \\
f^{(n)}(x) &= & \Rightarrow & \quad f^{(n)}(1) = \\
\frac{(-1)^{n+1}(n-1)!}{x^n} & & & \quad (-1)^{n+1}(n-1)!
\end{aligned}$$

Figure 8.7.6: Derivatives of $\ln x$ evaluated at $x = 1$.

   Note how the coefficients of the $(x - 1)$ terms turn out to be "nice."

2. We can compute $p_6(x)$ using our work above:

$$p_6(x) = (x-1) - \frac{1}{2}(x-1)^2 + \frac{1}{3}(x-1)^3 - \frac{1}{4}(x-1)^4 + \frac{1}{5}(x-1)^5 - \frac{1}{6}(x-1)^6.$$

   Since $p_6(x)$ approximates $\ln x$ well near $x = 1$, we approximate $\ln 1.5 \approx p_6(1.5)$:

Notes:

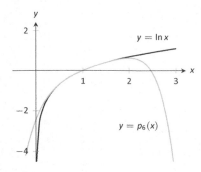

Figure 8.7.7: A plot of $y = \ln x$ and its 6th degree Taylor polynomial at $x = 1$.

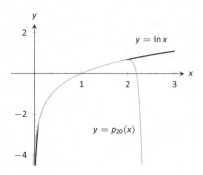

Figure 8.7.8: A plot of $y = \ln x$ and its 20th degree Taylor polynomial at $x = 1$.

$$p_6(1.5) = (1.5 - 1) - \frac{1}{2}(1.5 - 1)^2 + \frac{1}{3}(1.5 - 1)^3 - \frac{1}{4}(1.5 - 1)^4 + \cdots$$
$$\cdots + \frac{1}{5}(1.5 - 1)^5 - \frac{1}{6}(1.5 - 1)^6$$
$$= \frac{259}{640}$$
$$\approx 0.404688.$$

This is a good approximation as a calculator shows that $\ln 1.5 \approx 0.4055$. Figure 8.7.7 plots $y = \ln x$ with $y = p_6(x)$. We can see that $\ln 1.5 \approx p_6(1.5)$.

3. We approximate $\ln 2$ with $p_6(2)$:

$$p_6(2) = (2 - 1) - \frac{1}{2}(2 - 1)^2 + \frac{1}{3}(2 - 1)^3 - \frac{1}{4}(2 - 1)^4 + \cdots$$
$$\cdots + \frac{1}{5}(2 - 1)^5 - \frac{1}{6}(2 - 1)^6$$
$$= 1 - \frac{1}{2} + \frac{1}{3} - \frac{1}{4} + \frac{1}{5} - \frac{1}{6}$$
$$= \frac{37}{60}$$
$$\approx 0.616667.$$

This approximation is not terribly impressive: a hand held calculator shows that $\ln 2 \approx 0.693147$. The graph in Figure 8.7.7 shows that $p_6(x)$ provides less accurate approximations of $\ln x$ as x gets close to 0 or 2.

Surprisingly enough, even the 20th degree Taylor polynomial fails to approximate $\ln x$ for $x > 2$, as shown in Figure 8.7.8. We'll soon discuss why this is.

Taylor polynomials are used to approximate functions $f(x)$ in mainly two situations:

1. When $f(x)$ is known, but perhaps "hard" to compute directly. For instance, we can define $y = \cos x$ as either the ratio of sides of a right triangle ("adjacent over hypotenuse") or with the unit circle. However, neither of these provides a convenient way of computing $\cos 2$. A Taylor polynomial of sufficiently high degree can provide a reasonable method of computing such values using only operations usually hard–wired into a computer ($+$, $-$, $\times$ and $\div$).

Notes:

2. When $f(x)$ is not known, but information about its derivatives is known. This occurs more often than one might think, especially in the study of differential equations.

In both situations, a critical piece of information to have is "How good is my approximation?" If we use a Taylor polynomial to compute $\cos 2$, how do we know how accurate the approximation is?

We had the same problem when studying Numerical Integration. Theorem 5.5.1 provided bounds on the error when using, say, Simpson's Rule to approximate a definite integral. These bounds allowed us to determine that, for instance, using 10 subintervals provided an approximation within $\pm.01$ of the exact value. The following theorem gives similar bounds for Taylor (and hence Maclaurin) polynomials.

Note: Even though Taylor polynomials *could* be used in calculators and computers to calculate values of trigonometric functions, in practice they generally aren't. Other more efficient and accurate methods have been developed, such as the CORDIC algorithm.

Theorem 8.7.1   Taylor's Theorem

1. Let f be a function whose $n + 1^{\text{th}}$ derivative exists on an interval I and let c be in I. Then, for each x in I, there exists z_x between x and c such that

$$f(x) = f(c) + f'(c)(x - c) + \frac{f''(c)}{2!}(x - c)^2 + \cdots + \frac{f^{(n)}(c)}{n!}(x - c)^n + R_n(x),$$

where $R_n(x) = \frac{f^{(n+1)}(z_x)}{(n + 1)!}(x - c)^{(n+1)}$.

2. $\left| R_n(x) \right| \leq \frac{\max \left| f^{(n+1)}(z) \right|}{(n + 1)!} \left| (x - c)^{(n+1)} \right|$, where z is in I.

The first part of Taylor's Theorem states that $f(x) = p_n(x) + R_n(x)$, where $p_n(x)$ is the nth order Taylor polynomial and $R_n(x)$ is the remainder, or error, in the Taylor approximation. The second part gives bounds on how big that error can be. If the $(n + 1)^{\text{th}}$ derivative is large on I, the error may be large; if x is far from c, the error may also be large. However, the $(n + 1)!$ term in the denominator tends to ensure that the error gets smaller as n increases.

The following example computes error estimates for the approximations of $\ln 1.5$ and $\ln 2$ made in Example 8.7.2.

Example 8.7.3   Finding error bounds of a Taylor polynomial
Use Theorem 8.7.1 to find error bounds when approximating $\ln 1.5$ and $\ln 2$ with $p_6(x)$, the Taylor polynomial of degree 6 of $f(x) = \ln x$ at $x = 1$, as calculated in Example 8.7.2.

Notes:

SOLUTION

1. We start with the approximation of ln 1.5 with $p_6(1.5)$. The theorem references an open interval I that contains both x and c. The smaller the interval we use the better; it will give us a more accurate (and smaller!) approximation of the error. We let $I = (0.9, 1.6)$, as this interval contains both $c = 1$ and $x = 1.5$.

   The theorem references max $|f^{(n+1)}(z)|$. In our situation, this is asking "How big can the 7th derivative of $y = \ln x$ be on the interval $(0.9, 1.6)$?" The seventh derivative is $y = -6!/x^7$. The largest value it attains on I is about 1506. Thus we can bound the error as:

   $$\left|R_6(1.5)\right| \leq \frac{\max\left|f^{(7)}(z)\right|}{7!}\left|(1.5-1)^7\right|$$
   $$\leq \frac{1506}{5040} \cdot \frac{1}{2^7}$$
   $$\approx 0.0023.$$

   We computed $p_6(1.5) = 0.404688$; using a calculator, we find $\ln 1.5 \approx 0.405465$, so the actual error is about 0.000778, which is less than our bound of 0.0023. This affirms Taylor's Theorem; the theorem states that our approximation would be within about 2 thousandths of the actual value, whereas the approximation was actually closer.

2. We again find an interval I that contains both $c = 1$ and $x = 2$; we choose $I = (0.9, 2.1)$. The maximum value of the seventh derivative of f on this interval is again about 1506 (as the largest values come near $x = 0.9$). Thus

   $$\left|R_6(2)\right| \leq \frac{\max\left|f^{(7)}(z)\right|}{7!}\left|(2-1)^7\right|$$
   $$\leq \frac{1506}{5040} \cdot 1^7$$
   $$\approx 0.30.$$

   This bound is not as nearly as good as before. Using the degree 6 Taylor polynomial at $x = 1$ will bring us within 0.3 of the correct answer. As $p_6(2) \approx 0.61667$, our error estimate guarantees that the actual value of ln 2 is somewhere between 0.31667 and 0.91667. These bounds are not particularly useful.

   In reality, our approximation was only off by about 0.07. However, we are approximating ostensibly because we do not know the real answer. In order to be assured that we have a good approximation, we would have to resort to using a polynomial of higher degree.

Notes:

We practice again. This time, we use Taylor's theorem to find n that guarantees our approximation is within a certain amount.

Example 8.7.4   Finding sufficiently accurate Taylor polynomials

Find n such that the nth Taylor polynomial of $f(x) = \cos x$ at $x = 0$ approximates $\cos 2$ to within 0.001 of the actual answer. What is $p_n(2)$?

SOLUTION   Following Taylor's theorem, we need bounds on the size of the derivatives of $f(x) = \cos x$. In the case of this trigonometric function, this is easy. All derivatives of cosine are $\pm \sin x$ or $\pm \cos x$. In all cases, these functions are never greater than 1 in absolute value. We want the error to be less than 0.001. To find the appropriate n, consider the following inequalities:

$$\frac{\max \left|f^{(n+1)}(z)\right|}{(n+1)!}\left|(2-0)^{(n+1)}\right| \leq 0.001$$

$$\frac{1}{(n+1)!} \cdot 2^{(n+1)} \leq 0.001$$

We find an n that satisfies this last inequality with trial–and–error. When $n = 8$, we have $\dfrac{2^{8+1}}{(8+1)!} \approx 0.0014$; when $n = 9$, we have $\dfrac{2^{9+1}}{(9+1)!} \approx 0.000282 < 0.001$. Thus we want to approximate $\cos 2$ with $p_9(2)$.

We now set out to compute $p_9(x)$. We again need a table of the derivatives of $f(x) = \cos x$ evaluated at $x = 0$. A table of these values is given in Figure 8.7.9. Notice how the derivatives, evaluated at $x = 0$, follow a certain pattern. All the odd powers of x in the Taylor polynomial will disappear as their coefficient is 0. While our error bounds state that we need $p_9(x)$, our work shows that this will be the same as $p_8(x)$.

Since we are forming our polynomial at $x = 0$, we are creating a Maclaurin polynomial, and:

$$p_8(x) = f(0) + f'(0)x + \frac{f''(0)}{2!}x^2 + \frac{f'''(0)}{3!}x^3 + \cdots + \frac{f^{(8)}(0)}{8!}x^8$$

$$= 1 - \frac{1}{2!}x^2 + \frac{1}{4!}x^4 - \frac{1}{6!}x^6 + \frac{1}{8!}x^8$$

We finally approximate $\cos 2$:

$$\cos 2 \approx p_8(2) = -\frac{131}{315} \approx -0.41587.$$

Our error bound guarantee that this approximation is within 0.001 of the correct answer. Technology shows us that our approximation is actually within about 0.0003 of the correct answer.

Figure 8.7.10 shows a graph of $y = p_8(x)$ and $y = \cos x$. Note how well the two functions agree on about $(-\pi, \pi)$.

$$
\begin{aligned}
f(x) &= \cos x & &\Rightarrow & f(0) &= 1 \\
f'(x) &= -\sin x & &\Rightarrow & f'(0) &= 0 \\
f''(x) &= -\cos x & &\Rightarrow & f''(0) &= -1 \\
f'''(x) &= \sin x & &\Rightarrow & f'''(0) &= 0 \\
f^{(4)}(x) &= \cos x & &\Rightarrow & f^{(4)}(0) &= 1 \\
f^{(5)}(x) &= -\sin x & &\Rightarrow & f^{(5)}(0) &= 0 \\
f^{(6)}(x) &= -\cos x & &\Rightarrow & f^{(6)}(0) &= -1 \\
f^{(7)}(x) &= \sin x & &\Rightarrow & f^{(7)}(0) &= 0 \\
f^{(8)}(x) &= \cos x & &\Rightarrow & f^{(8)}(0) &= 1 \\
f^{(9)}(x) &= -\sin x & &\Rightarrow & f^{(9)}(0) &= 0
\end{aligned}
$$

Figure 8.7.9: A table of the derivatives of $f(x) = \cos x$ evaluated at $x = 0$.

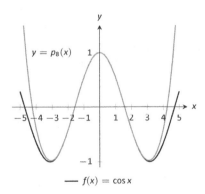

Figure 8.7.10: A graph of $f(x) = \cos x$ and its degree 8 Maclaurin polynomial.

Notes:

$$f(x) = \sqrt{x} \qquad \Rightarrow \qquad f(4) = 2$$

$$f'(x) = \frac{1}{2\sqrt{x}} \qquad \Rightarrow \qquad f'(4) = \frac{1}{4}$$

$$f''(x) = \frac{-1}{4x^{3/2}} \qquad \Rightarrow \qquad f''(4) = \frac{-1}{32}$$

$$f'''(x) = \frac{3}{8x^{5/2}} \qquad \Rightarrow \qquad f'''(4) = \frac{3}{256}$$

$$f^{(4)}(x) = \frac{-15}{16x^{7/2}} \qquad \Rightarrow \qquad f^{(4)}(4) = \frac{-15}{2048}$$

Figure 8.7.11: A table of the derivatives of $f(x) = \sqrt{x}$ evaluated at $x = 4$.

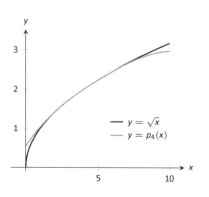

Figure 8.7.12: A graph of $f(x) = \sqrt{x}$ and its degree 4 Taylor polynomial at $x = 4$.

Example 8.7.5   Finding and using Taylor polynomials

1. Find the degree 4 Taylor polynomial, $p_4(x)$, for $f(x) = \sqrt{x}$ at $x = 4$.

2. Use $p_4(x)$ to approximate $\sqrt{3}$.

3. Find bounds on the error when approximating $\sqrt{3}$ with $p_4(3)$.

SOLUTION

1. We begin by evaluating the derivatives of f at $x = 4$. This is done in Figure 8.7.11. These values allow us to form the Taylor polynomial $p_4(x)$:

$$p_4(x) = 2 + \frac{1}{4}(x-4) + \frac{-1/32}{2!}(x-4)^2 + \frac{3/256}{3!}(x-4)^3 + \frac{-15/2048}{4!}(x-4)^4.$$

2. As $p_4(x) \approx \sqrt{x}$ near $x = 4$, we approximate $\sqrt{3}$ with $p_4(3) = 1.73212$.

3. To find a bound on the error, we need an open interval that contains $x = 3$ and $x = 4$. We set $I = (2.9, 4.1)$. The largest value the fifth derivative of $f(x) = \sqrt{x}$ takes on this interval is near $x = 2.9$, at about 0.0273. Thus

$$\big|R_4(3)\big| \le \frac{0.0273}{5!}\big|(3-4)^5\big| \approx 0.00023.$$

This shows our approximation is accurate to at least the first 2 places after the decimal. (It turns out that our approximation is actually accurate to 4 places after the decimal.) A graph of $f(x) = \sqrt{x}$ and $p_4(x)$ is given in Figure 8.7.12. Note how the two functions are nearly indistinguishable on $(2, 7)$.

Our final example gives a brief introduction to using Taylor polynomials to solve differential equations.

Example 8.7.6   Approximating an unknown function

A function $y = f(x)$ is unknown save for the following two facts.

1. $y(0) = f(0) = 1$, and

2. $y' = y^2$

(This second fact says that amazingly, the derivative of the function is actually the function squared!)

Find the degree 3 Maclaurin polynomial $p_3(x)$ of $y = f(x)$.

Notes:

SOLUTION   One might initially think that not enough information is given to find $p_3(x)$. However, note how the second fact above actually lets us know what $y'(0)$ is:

$$y' = y^2 \Rightarrow y'(0) = y^2(0).$$

Since $y(0) = 1$, we conclude that $y'(0) = 1$.

Now we find information about y''. Starting with $y' = y^2$, take derivatives of both sides, *with respect to x*. That means we must use implicit differentiation.

$$y' = y^2$$
$$\frac{d}{dx}(y') = \frac{d}{dx}(y^2)$$
$$y'' = 2y \cdot y'.$$

Now evaluate both sides at $x = 0$:

$$y''(0) = 2y(0) \cdot y'(0)$$
$$y''(0) = 2$$

We repeat this once more to find $y'''(0)$. We again use implicit differentiation; this time the Product Rule is also required.

$$\frac{d}{dx}(y'') = \frac{d}{dx}(2yy')$$
$$y''' = 2y' \cdot y' + 2y \cdot y''.$$

Now evaluate both sides at $x = 0$:

$$y'''(0) = 2y'(0)^2 + 2y(0)y''(0)$$
$$y'''(0) = 2 + 4 = 6$$

In summary, we have:

$$y(0) = 1 \qquad y'(0) = 1 \qquad y''(0) = 2 \qquad y'''(0) = 6.$$

We can now form $p_3(x)$:

$$p_3(x) = 1 + x + \frac{2}{2!}x^2 + \frac{6}{3!}x^3$$
$$= 1 + x + x^2 + x^3.$$

It turns out that the differential equation we started with, $y' = y^2$, where $y(0) = 1$, can be solved without too much difficulty: $y = \dfrac{1}{1-x}$. Figure 8.7.13 shows this function plotted with $p_3(x)$. Note how similar they are near $x = 0$.

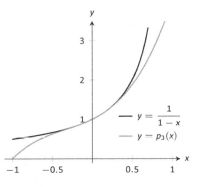

Figure 8.7.13: A graph of $y = -1/(x-1)$ and $y = p_3(x)$ from Example 8.7.6.

Notes:

It is beyond the scope of this text to pursue error analysis when using Taylor polynomials to approximate solutions to differential equations. This topic is often broached in introductory Differential Equations courses and usually covered in depth in Numerical Analysis courses. Such an analysis is very important; one needs to know how good their approximation is. We explored this example simply to demonstrate the usefulness of Taylor polynomials.

Most of this chapter has been devoted to the study of infinite series. This section has taken a step back from this study, focusing instead on finite summation of terms. In the next section, we explore **Taylor Series**, where we represent a function with an infinite series.

Notes:

Exercises 8.7

Terms and Concepts

1. What is the difference between a Taylor polynomial and a Maclaurin polynomial?

2. T/F: In general, $p_n(x)$ approximates $f(x)$ better and better as n gets larger.

3. For some function $f(x)$, the Maclaurin polynomial of degree 4 is $p_4(x) = 6 + 3x - 4x^2 + 5x^3 - 7x^4$. What is $p_2(x)$?

4. For some function $f(x)$, the Maclaurin polynomial of degree 4 is $p_4(x) = 6 + 3x - 4x^2 + 5x^3 - 7x^4$. What is $f'''(0)$?

Problems

In Exercises 5 – 12, find the Maclaurin polynomial of degree n for the given function.

5. $f(x) = e^{-x}, \quad n = 3$

6. $f(x) = \sin x, \quad n = 8$

7. $f(x) = x \cdot e^x, \quad n = 5$

8. $f(x) = \tan x, \quad n = 6$

9. $f(x) = e^{2x}, \quad n = 4$

10. $f(x) = \dfrac{1}{1 - x}, \quad n = 4$

11. $f(x) = \dfrac{1}{1 + x}, \quad n = 4$

12. $f(x) = \dfrac{1}{1 + x}, \quad n = 7$

In Exercises 13 – 20, find the Taylor polynomial of degree n, at $x = c$, for the given function.

13. $f(x) = \sqrt{x}, \quad n = 4, \quad c = 1$

14. $f(x) = \ln(x + 1), \quad n = 4, \quad c = 1$

15. $f(x) = \cos x, \quad n = 6, \quad c = \pi/4$

16. $f(x) = \sin x, \quad n = 5, \quad c = \pi/6$

17. $f(x) = \dfrac{1}{x}, \quad n = 5, \quad c = 2$

18. $f(x) = \dfrac{1}{x^2}, \quad n = 8, \quad c = 1$

19. $f(x) = \dfrac{1}{x^2 + 1}, \quad n = 3, \quad c = -1$

20. $f(x) = x^2 \cos x, \quad n = 2, \quad c = \pi$

In Exercises 21 – 24, approximate the function value with the indicated Taylor polynomial and give approximate bounds on the error.

21. Approximate $\sin 0.1$ with the Maclaurin polynomial of degree 3.

22. Approximate $\cos 1$ with the Maclaurin polynomial of degree 4.

23. Approximate $\sqrt{10}$ with the Taylor polynomial of degree 2 centered at $x = 9$.

24. Approximate $\ln 1.5$ with the Taylor polynomial of degree 3 centered at $x = 1$.

Exercises 25 – 28 ask for an n to be found such that $p_n(x)$ approximates $f(x)$ within a certain bound of accuracy.

25. Find n such that the Maclaurin polynomial of degree n of $f(x) = e^x$ approximates e within 0.0001 of the actual value.

26. Find n such that the Taylor polynomial of degree n of $f(x) = \sqrt{x}$, centered at $x = 4$, approximates $\sqrt{3}$ within 0.0001 of the actual value.

27. Find n such that the Maclaurin polynomial of degree n of $f(x) = \cos x$ approximates $\cos \pi/3$ within 0.0001 of the actual value.

28. Find n such that the Maclaurin polynomial of degree n of $f(x) = \sin x$ approximates $\cos \pi$ within 0.0001 of the actual value.

In Exercises 29 – 34, find the nth term of the indicated Taylor polynomial.

29. Find a formula for the nth term of the Maclaurin polynomial for $f(x) = e^x$.

30. Find a formula for the nth term of the Maclaurin polynomial for $f(x) = \cos x$.

31. Find a formula for the nth term of the Maclaurin polynomial for $f(x) = \sin x$.

32. Find a formula for the nth term of the Maclaurin polynomial for $f(x) = \dfrac{1}{1 - x}$.

33. Find a formula for the nth term of the Maclaurin polynomial for $f(x) = \dfrac{1}{1 + x}$.

34. Find a formula for the nth term of the Taylor polynomial for $f(x) = \ln x$ centered at $x = 1$.

In Exercises 35 – 37, approximate the solution to the given differential equation with a degree 4 Maclaurin polynomial.

35. $y' = y,$ $y(0) = 1$

36. $y' = 5y,$ $y(0) = 3$

37. $y' = \dfrac{2}{y},$ $y(0) = 1$

8.8 Taylor Series

In Section 8.6, we showed how certain functions can be represented by a power series function. In Section 8.7, we showed how we can approximate functions with polynomials, given that enough derivative information is available. In this section we combine these concepts: if a function $f(x)$ is infinitely differentiable, we show how to represent it with a power series function.

Definition 8.8.1   Taylor and Maclaurin Series

Let $f(x)$ have derivatives of all orders at $x = c$.

1. The **Taylor Series of $f(x)$, centered at c** is

$$\sum_{n=0}^{\infty} \frac{f^{(n)}(c)}{n!}(x-c)^n.$$

2. Setting $c = 0$ gives the **Maclaurin Series of $f(x)$**:

$$\sum_{n=0}^{\infty} \frac{f^{(n)}(0)}{n!}x^n.$$

If $p_n(x)$ is the nth degree Taylor polynomial for $f(x)$ centered at $x = c$, we saw how $f(x)$ is *approximately equal* to $p_n(x)$ near $x = c$. We also saw how increasing the degree of the polynomial generally reduced the error.

We are now considering *series*, where we sum an infinite set of terms. Our ultimate hope is to see the error vanish and claim a function is *equal* to its Taylor series.

When creating the Taylor polynomial of degree n for a function $f(x)$ at $x = c$, we needed to evaluate f, and the first n derivatives of f, at $x = c$. When creating the Taylor series of f, it helps to find a pattern that describes the nth derivative of f at $x = c$. We demonstrate this in the next two examples.

Example 8.8.1   The Maclaurin series of $f(x) = \cos x$
Find the Maclaurin series of $f(x) = \cos x$.

SOLUTION   In Example 8.7.4 we found the 8th degree Maclaurin polynomial of $\cos x$. In doing so, we created the table shown in Figure 8.8.1. Notice how $f^{(n)}(0) = 0$ when n is odd, $f^{(n)}(0) = 1$ when n is divisible by 4, and $f^{(n)}(0) = -1$ when n is even but not divisible by 4. Thus the Maclaurin series

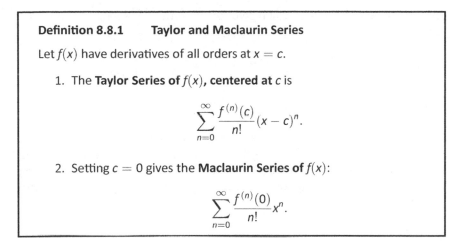

$$f(x) = \cos x \qquad \Rightarrow \qquad f(0) = 1$$
$$f'(x) = -\sin x \qquad \Rightarrow \qquad f'(0) = 0$$
$$f''(x) = -\cos x \qquad \Rightarrow \qquad f''(0) = -1$$
$$f'''(x) = \sin x \qquad \Rightarrow \qquad f'''(0) = 0$$
$$f^{(4)}(x) = \cos x \qquad \Rightarrow \qquad f^{(4)}(0) = 1$$
$$f^{(5)}(x) = -\sin x \qquad \Rightarrow \qquad f^{(5)}(0) = 0$$
$$f^{(6)}(x) = -\cos x \qquad \Rightarrow \qquad f^{(6)}(0) = -1$$
$$f^{(7)}(x) = \sin x \qquad \Rightarrow \qquad f^{(7)}(0) = 0$$
$$f^{(8)}(x) = \cos x \qquad \Rightarrow \qquad f^{(8)}(0) = 1$$
$$f^{(9)}(x) = -\sin x \qquad \Rightarrow \qquad f^{(9)}(0) = 0$$

Figure 8.8.1: A table of the derivatives of $f(x) = \cos x$ evaluated at $x = 0$.

Notes:

of $\cos x$ is

$$1 - \frac{x^2}{2} + \frac{x^4}{4!} - \frac{x^6}{6!} + \frac{x^8}{8!} - \cdots$$

We can go further and write this as a summation. Since we only need the terms where the power of x is even, we write the power series in terms of x^{2n}:

$$\sum_{n=0}^{\infty} (-1)^n \frac{x^{2n}}{(2n)!}.$$

This Maclaurin series is a special type of power series. As such, we should determine its interval of convergence. Applying the Ratio Test, we have

$$\lim_{n\to\infty} \left| (-1)^{n+1} \frac{x^{2(n+1)}}{(2(n+1))!} \right| \Bigg/ \left| (-1)^n \frac{x^{2n}}{(2n)!} \right| = \lim_{n\to\infty} \left| \frac{x^{2n+2}}{x^{2n}} \right| \frac{(2n)!}{(2n+2)!}$$

$$= \lim_{n\to\infty} \frac{x^2}{(2n+2)(2n+1)}.$$

For any fixed x, this limit is 0. Therefore this power series has an infinite radius of convergence, converging for all x. It is important to note what we have, and have not, determined: we have determined the Maclaurin series for $\cos x$ along with its interval of convergence. We *have not* shown that $\cos x$ is *equal* to this power series.

Example 8.8.2   **The Taylor series of $f(x) = \ln x$ at $x = 1$**
Find the Taylor series of $f(x) = \ln x$ centered at $x = 1$.

SOLUTION   Figure 8.8.2 shows the nth derivative of $\ln x$ evaluated at $x = 1$ for $n = 0, \ldots, 5$, along with an expression for the nth term:

$$f^{(n)}(1) = (-1)^{n+1}(n-1)! \quad \text{for } n \geq 1.$$

Remember that this is what distinguishes Taylor series from Taylor polynomials; we are very interested in finding a pattern for the nth term, not just finding a finite set of coefficients for a polynomial. Since $f(1) = \ln 1 = 0$, we skip the first term and start the summation with $n = 1$, giving the Taylor series for $\ln x$, centered at $x = 1$, as

$$\sum_{n=1}^{\infty} (-1)^{n+1}(n-1)!\frac{1}{n!}(x-1)^n = \sum_{n=1}^{\infty} (-1)^{n+1}\frac{(x-1)^n}{n}.$$

We now determine the interval of convergence, using the Ratio Test.

$$\lim_{n\to\infty} \left| (-1)^{n+2}\frac{(x-1)^{n+1}}{n+1} \right| \Bigg/ \left| (-1)^{n+1}\frac{(x-1)^n}{n} \right| = \lim_{n\to\infty} \left| \frac{(x-1)^{n+1}}{(x-1)^n} \right| \frac{n}{n+1}$$

$$= \left| (x-1) \right|.$$

$$\begin{aligned}
f(x) &= \ln x &&\Rightarrow& f(1) &= 0 \\
f'(x) &= 1/x &&\Rightarrow& f'(1) &= 1 \\
f''(x) &= -1/x^2 &&\Rightarrow& f''(1) &= -1 \\
f'''(x) &= 2/x^3 &&\Rightarrow& f'''(1) &= 2 \\
f^{(4)}(x) &= -6/x^4 &&\Rightarrow& f^{(4)}(1) &= -6 \\
f^{(5)}(x) &= 24/x^5 &&\Rightarrow& f^{(5)}(1) &= 24 \\
&\vdots &&& &\vdots \\
f^{(n)}(x) &= &&\Rightarrow& f^{(n)}(1) &= \\
\frac{(-1)^{n+1}(n-1)!}{x^n} &&&& (-1)^{n+1}(n-1)!
\end{aligned}$$

Figure 8.8.2: Derivatives of $\ln x$ evaluated at $x = 1$.

Notes:

By the Ratio Test, we have convergence when $\left|(x-1)\right| < 1$: the radius of convergence is 1, and we have convergence on $(0, 2)$. We now check the endpoints.

At $x = 0$, the series is

$$\sum_{n=1}^{\infty}(-1)^{n+1}\frac{(-1)^n}{n} = -\sum_{n=1}^{\infty}\frac{1}{n},$$

which diverges (it is the Harmonic Series times (-1).)

At $x = 2$, the series is

$$\sum_{n=1}^{\infty}(-1)^{n+1}\frac{(1)^n}{n} = \sum_{n=1}^{\infty}(-1)^{n+1}\frac{1}{n},$$

the Alternating Harmonic Series, which converges.

We have found the Taylor series of $\ln x$ centered at $x = 1$, and have determined the series converges on $(0, 2]$. We cannot (yet) say that $\ln x$ is equal to this Taylor series on $(0, 2]$.

It is important to note that Definition 8.8.1 defines a Taylor series given a function $f(x)$, but makes no claim about their equality. We will find that "most of the time" they are equal, but we need to consider the conditions that allow us to conclude this.

Theorem 8.7.1 states that the error between a function $f(x)$ and its $n^{\text{th}}-$degree Taylor polynomial $p_n(x)$ is $R_n(x)$, where

$$\left|R_n(x)\right| \le \frac{\max\left|f^{(n+1)}(z)\right|}{(n+1)!}\left|(x-c)^{(n+1)}\right|.$$

If $R_n(x)$ goes to 0 for each x in an interval I as n approaches infinity, we conclude that the function is equal to its Taylor series expansion.

Note: It can be shown that $\ln x$ is equal to this Taylor series on $(0, 2]$. From the work in Example 8.8.2, this justifies our previous declaration that the Alternating Harmonic Series converges to $\ln 2$.

Theorem 8.8.1   Function and Taylor Series Equality

Let $f(x)$ have derivatives of all orders at $x = c$, let $R_n(x)$ be as stated in Theorem 8.7.1, and let I be an interval on which the Taylor series of $f(x)$ converges. If $\lim_{n\to\infty} R_n(x) = 0$ for all x in I, then

$$f(x) = \sum_{n=0}^{\infty}\frac{f^{(n)}(c)}{n!}(x-c)^n \text{ on } I.$$

Notes:

We demonstrate the use of this theorem in an example.

Example 8.8.3   **Establishing equality of a function and its Taylor series**
Show that $f(x) = \cos x$ is equal to its Maclaurin series, as found in Example 8.8.1, for all x.

SOLUTION   Given a value x, the magnitude of the error term $R_n(x)$ is bounded by

$$\left| R_n(x) \right| \leq \frac{\max \left| f^{(n+1)}(z) \right|}{(n+1)!} \left| x^{n+1} \right|.$$

Since all derivatives of $\cos x$ are $\pm \sin x$ or $\pm \cos x$, whose magnitudes are bounded by 1, we can state

$$\left| R_n(x) \right| \leq \frac{1}{(n+1)!} \left| x^{n+1} \right|$$

which implies

$$-\frac{\left| x^{n+1} \right|}{(n+1)!} \leq R_n(x) \leq \frac{\left| x^{n+1} \right|}{(n+1)!}. \tag{8.7}$$

For any x, $\displaystyle\lim_{n \to \infty} \frac{x^{n+1}}{(n+1)!} = 0$. Applying the Squeeze Theorem to Equation (8.7), we conclude that $\displaystyle\lim_{n \to \infty} R_n(x) = 0$ for all x, and hence

$$\cos x = \sum_{n=0}^{\infty} (-1)^n \frac{x^{2n}}{(2n)!} \quad \text{for all } x.$$

It is natural to assume that a function is equal to its Taylor series on the series' interval of convergence, but this is not always the case. In order to properly establish equality, one must use Theorem 8.8.1. This is a bit disappointing, as we developed beautiful techniques for determining the interval of convergence of a power series, and proving that $R_n(x) \to 0$ can be difficult. For instance, it is not a simple task to show that $\ln x$ equals its Taylor series on $(0, 2]$ as found in Example 8.8.2; in the Exercises, the reader is only asked to show equality on $(1, 2)$, which is simpler.

There is good news. A function $f(x)$ that is equal to its Taylor series, centered at any point the domain of $f(x)$, is said to be an **analytic function**, and most, if not all, functions that we encounter within this course are analytic functions. Generally speaking, any function that one creates with elementary functions (polynomials, exponentials, trigonometric functions, etc.) that is not piecewise defined is probably analytic. For most functions, we assume the function is equal to its Taylor series on the series' interval of convergence and only use Theorem 8.8.1 when we suspect something may not work as expected.

Notes:

We develop the Taylor series for one more important function, then give a table of the Taylor series for a number of common functions.

Example 8.8.4   The Binomial Series

Find the Maclaurin series of $f(x) = (1 + x)^k$, $k \neq 0$.

SOLUTION   When k is a positive integer, the Maclaurin series is finite. For instance, when $k = 4$, we have

$$f(x) = (1 + x)^4 = 1 + 4x + 6x^2 + 4x^3 + x^4.$$

The coefficients of x when k is a positive integer are known as the *binomial coefficients*, giving the series we are developing its name.

When $k = 1/2$, we have $f(x) = \sqrt{1 + x}$. Knowing a series representation of this function would give a useful way of approximating $\sqrt{1.3}$, for instance.

To develop the Maclaurin series for $f(x) = (1 + x)^k$ for any value of $k \neq 0$, we consider the derivatives of f evaluated at $x = 0$:

$$f(x) = (1 + x)^k \qquad\qquad f(0) = 1$$
$$f'(x) = k(1 + x)^{k-1} \qquad\qquad f'(0) = k$$
$$f''(x) = k(k - 1)(1 + x)^{k-2} \qquad\qquad f''(0) = k(k - 1)$$
$$f'''(x) = k(k - 1)(k - 2)(1 + x)^{k-3} \qquad\qquad f'''(0) = k(k - 1)(k - 2)$$
$$\vdots \qquad\qquad\qquad\qquad \vdots$$
$$f^{(n)}(x) = k(k - 1)\cdots\big(k - (n - 1)\big)(1 + x)^{k-n} \qquad f^{(n)}(0) = k(k - 1)\cdots\big(k - (n - 1)\big)$$

Thus the Maclaurin series for $f(x) = (1 + x)^k$ is

$$1 + kx + \frac{k(k - 1)}{2!}x^2 + \frac{k(k - 1)(k - 2)}{3!}x^3 + \ldots + \frac{k(k - 1)\cdots\big(k - (n - 1)\big)}{n!}x^n + \ldots$$

It is important to determine the interval of convergence of this series. With

$$a_n = \frac{k(k - 1)\cdots\big(k - (n - 1)\big)}{n!}x^n,$$

we apply the Ratio Test:

$$\lim_{n\to\infty} \frac{|a_{n+1}|}{|a_n|} = \lim_{n\to\infty} \left| \frac{k(k - 1)\cdots(k - n)}{(n + 1)!}x^{n+1} \right| \bigg/ \left| \frac{k(k - 1)\cdots\big(k - (n - 1)\big)}{n!}x^n \right|$$

$$= \lim_{n\to\infty} \left| \frac{k - n}{n + 1}x \right|$$

$$= |x|.$$

Notes:

The series converges absolutely when the limit of the Ratio Test is less than 1; therefore, we have absolute convergence when $|x| < 1$.

While outside the scope of this text, the interval of convergence depends on the value of k. When $k > 0$, the interval of convergence is $[-1, 1]$. When $-1 < k < 0$, the interval of convergence is $[-1, 1)$. If $k \leq -1$, the interval of convergence is $(-1, 1)$.

We learned that Taylor polynomials offer a way of approximating a "difficult to compute" function with a polynomial. Taylor series offer a way of exactly representing a function with a series. One probably can see the use of a good approximation; is there any use of representing a function exactly as a series?

While we should not overlook the mathematical beauty of Taylor series (which is reason enough to study them), there are practical uses as well. They provide a valuable tool for solving a variety of problems, including problems relating to integration and differential equations.

In Key Idea 8.8.1 (on the following page) we give a table of the Taylor series of a number of common functions. We then give a theorem about the "algebra of power series," that is, how we can combine power series to create power series of new functions. This allows us to find the Taylor series of functions like $f(x) = e^x \cos x$ by knowing the Taylor series of e^x and $\cos x$.

Before we investigate combining functions, consider the Taylor series for the arctangent function (see Key Idea 8.8.1). Knowing that $\tan^{-1}(1) = \pi/4$, we can use this series to approximate the value of π:

$$\frac{\pi}{4} = \tan^{-1}(1) = 1 - \frac{1}{3} + \frac{1}{5} - \frac{1}{7} + \frac{1}{9} - \cdots$$
$$\pi = 4\left(1 - \frac{1}{3} + \frac{1}{5} - \frac{1}{7} + \frac{1}{9} - \cdots\right)$$

Unfortunately, this particular expansion of π converges very slowly. The first 100 terms approximate π as 3.13159, which is not particularly good.

Notes:

Key Idea 8.8.1   Important Taylor Series Expansions

Function and Series	First Few Terms	Interval of Convergence
$e^x = \sum_{n=0}^{\infty} \dfrac{x^n}{n!}$	$1 + x + \dfrac{x^2}{2!} + \dfrac{x^3}{3!} + \cdots$	$(-\infty, \infty)$
$\sin x = \sum_{n=0}^{\infty} (-1)^n \dfrac{x^{2n+1}}{(2n+1)!}$	$x - \dfrac{x^3}{3!} + \dfrac{x^5}{5!} - \dfrac{x^7}{7!} + \cdots$	$(-\infty, \infty)$
$\cos x = \sum_{n=0}^{\infty} (-1)^n \dfrac{x^{2n}}{(2n)!}$	$1 - \dfrac{x^2}{2!} + \dfrac{x^4}{4!} - \dfrac{x^6}{6!} + \cdots$	$(-\infty, \infty)$
$\ln x = \sum_{n=1}^{\infty} (-1)^{n+1} \dfrac{(x-1)^n}{n}$	$(x-1) - \dfrac{(x-1)^2}{2} + \dfrac{(x-1)^3}{3} - \cdots$	$(0, 2]$
$\dfrac{1}{1-x} = \sum_{n=0}^{\infty} x^n$	$1 + x + x^2 + x^3 + \cdots$	$(-1, 1)$
$(1+x)^k = \sum_{n=0}^{\infty} \dfrac{k(k-1)\cdots(k-(n-1))}{n!} x^n$	$1 + kx + \dfrac{k(k-1)}{2!} x^2 + \cdots$	$(-1, 1)^a$
$\tan^{-1} x = \sum_{n=0}^{\infty} (-1)^n \dfrac{x^{2n+1}}{2n+1}$	$x - \dfrac{x^3}{3} + \dfrac{x^5}{5} - \dfrac{x^7}{7} + \cdots$	$[-1, 1]$

^{a}Convergence at $x = \pm 1$ depends on the value of k.

Theorem 8.8.2   Algebra of Power Series

Let $f(x) = \sum_{n=0}^{\infty} a_n x^n$ and $g(x) = \sum_{n=0}^{\infty} b_n x^n$ converge absolutely for $|x| < R$, and let $h(x)$ be continuous.

1. $f(x) \pm g(x) = \sum_{n=0}^{\infty} (a_n \pm b_n) x^n \quad$ for $|x| < R$.

2. $f(x)g(x) = \left(\sum_{n=0}^{\infty} a_n x^n \right) \left(\sum_{n=0}^{\infty} b_n x^n \right) = \sum_{n=0}^{\infty} \left(a_0 b_n + a_1 b_{n-1} + \ldots a_n b_0 \right) x^n$ for $|x| < R$.

3. $f\big(h(x)\big) = \sum_{n=0}^{\infty} a_n \big(h(x)\big)^n \quad$ for $|h(x)| < R$.

Notes:

Example 8.8.5   Combining Taylor series

Write out the first 3 terms of the Taylor Series for $f(x) = e^x \cos x$ using Key Idea 8.8.1 and Theorem 8.8.2.

SOLUTION   Key Idea 8.8.1 informs us that

$$e^x = 1 + x + \frac{x^2}{2!} + \frac{x^3}{3!} + \cdots \quad \text{and} \quad \cos x = 1 - \frac{x^2}{2!} + \frac{x^4}{4!} + \cdots .$$

Applying Theorem 8.8.2, we find that

$$e^x \cos x = \left(1 + x + \frac{x^2}{2!} + \frac{x^3}{3!} + \cdots\right)\left(1 - \frac{x^2}{2!} + \frac{x^4}{4!} + \cdots\right).$$

Distribute the right hand expression across the left:

$$= 1\left(1 - \frac{x^2}{2!} + \frac{x^4}{4!} + \cdots\right) + x\left(1 - \frac{x^2}{2!} + \frac{x^4}{4!} + \cdots\right) + \frac{x^2}{2!}\left(1 - \frac{x^2}{2!} + \frac{x^4}{4!} + \cdots\right)$$
$$+ \frac{x^3}{3!}\left(1 - \frac{x^2}{2!} + \frac{x^4}{4!} + \cdots\right) + \frac{x^4}{4!}\left(1 - \frac{x^2}{2!} + \frac{x^4}{4!} + \cdots\right) + \cdots$$

Distribute again and collect like terms.

$$= 1 + x - \frac{x^3}{3} - \frac{x^4}{6} - \frac{x^5}{30} + \frac{x^7}{630} + \cdots$$

While this process is a bit tedious, it is much faster than evaluating all the necessary derivatives of $e^x \cos x$ and computing the Taylor series directly.

Because the series for e^x and $\cos x$ both converge on $(-\infty, \infty)$, so does the series expansion for $e^x \cos x$.

Example 8.8.6   Creating new Taylor series

Use Theorem 8.8.2 to create series for $y = \sin(x^2)$ and $y = \ln(\sqrt{x})$.

SOLUTION   Given that

$$\sin x = \sum_{n=0}^{\infty} (-1)^n \frac{x^{2n+1}}{(2n+1)!} = x - \frac{x^3}{3!} + \frac{x^5}{5!} - \frac{x^7}{7!} + \cdots ,$$

we simply substitute x^2 for x in the series, giving

$$\sin(x^2) = \sum_{n=0}^{\infty} (-1)^n \frac{(x^2)^{2n+1}}{(2n+1)!} = x^2 - \frac{x^6}{3!} + \frac{x^{10}}{5!} - \frac{x^{14}}{7!} \cdots .$$

Notes:

Since the Taylor series for $\sin x$ has an infinite radius of convergence, so does the Taylor series for $\sin(x^2)$.

The Taylor expansion for $\ln x$ given in Key Idea 8.8.1 is centered at $x = 1$, so we will center the series for $\ln(\sqrt{x})$ at $x = 1$ as well. With

$$\ln x = \sum_{n=1}^{\infty}(-1)^{n+1}\frac{(x-1)^n}{n} = (x-1) - \frac{(x-1)^2}{2} + \frac{(x-1)^3}{3} - \cdots,$$

we substitute $\sqrt{x}$ for x to obtain

$$\ln(\sqrt{x}) = \sum_{n=1}^{\infty}(-1)^{n+1}\frac{(\sqrt{x}-1)^n}{n} = (\sqrt{x}-1) - \frac{(\sqrt{x}-1)^2}{2} + \frac{(\sqrt{x}-1)^3}{3} - \cdots.$$

While this is not strictly a power series, it is a series that allows us to study the function $\ln(\sqrt{x})$. Since the interval of convergence of $\ln x$ is $(0, 2]$, and the range of $\sqrt{x}$ on $(0, 4]$ is $(0, 2]$, the interval of convergence of this series expansion of $\ln(\sqrt{x})$ is $(0, 4]$.

Note: In Example 8.8.6, one could create a series for $\ln(\sqrt{x})$ by simply recognizing that $\ln(\sqrt{x}) = \ln(x^{1/2}) = 1/2 \ln x$, and hence multiplying the Taylor series for $\ln x$ by $1/2$. This example was chosen to demonstrate other aspects of series, such as the fact that the interval of convergence changes.

Example 8.8.7   **Using Taylor series to evaluate definite integrals**

Use the Taylor series of e^{-x^2} to evaluate $\displaystyle\int_0^1 e^{-x^2}\, dx$.

SOLUTION   We learned, when studying Numerical Integration, that e^{-x^2} does not have an antiderivative expressible in terms of elementary functions. This means any definite integral of this function must have its value approximated, and not computed exactly.

We can quickly write out the Taylor series for e^{-x^2} using the Taylor series of e^x:

$$e^x = \sum_{n=0}^{\infty} \frac{x^n}{n!} = 1 + x + \frac{x^2}{2!} + \frac{x^3}{3!} + \cdots$$

and so

$$e^{-x^2} = \sum_{n=0}^{\infty} \frac{(-x^2)^n}{n!}$$

$$= \sum_{n=0}^{\infty} (-1)^n \frac{x^{2n}}{n!}$$

$$= 1 - x^2 + \frac{x^4}{2!} - \frac{x^6}{3!} + \cdots.$$

Notes:

We use Theorem 8.6.3 to integrate:

$$\int e^{-x^2}\, dx = C + x - \frac{x^3}{3} + \frac{x^5}{5\cdot 2!} - \frac{x^7}{7\cdot 3!} + \cdots + (-1)^n \frac{x^{2n+1}}{(2n+1)n!} + \cdots$$

This *is* the antiderivative of e^{-x^2}; while we can write it out as a series, we cannot write it out in terms of elementary functions. We can evaluate the definite integral $\int_0^1 e^{-x^2}\, dx$ using this antiderivative; substituting 1 and 0 for x and subtracting gives

$$\int_0^1 e^{-x^2}\, dx = 1 - \frac{1}{3} + \frac{1}{5\cdot 2!} - \frac{1}{7\cdot 3!} + \frac{1}{9\cdot 4!} \cdots.$$

Summing the 5 terms shown above give the approximation of 0.74749. Since this is an alternating series, we can use the Alternating Series Approximation Theorem, (Theorem 8.5.2), to determine how accurate this approximation is. The next term of the series is $1/(11\cdot 5!) \approx 0.00075758$. Thus we know our approximation is within 0.00075758 of the actual value of the integral. This is arguably much less work than using Simpson's Rule to approximate the value of the integral.

Example 8.8.8   Using Taylor series to solve differential equations
Solve the differential equation $y' = 2y$ in terms of a power series, and use the theory of Taylor series to recognize the solution in terms of an elementary function.

Solution   We found the first 5 terms of the power series solution to this differential equation in Example 8.6.5 in Section 8.6. These are:

$$a_0 = 1, \quad a_1 = 2, \quad a_2 = \frac{4}{2} = 2, \quad a_3 = \frac{8}{2\cdot 3} = \frac{4}{3}, \quad a_4 = \frac{16}{2\cdot 3\cdot 4} = \frac{2}{3}.$$

We include the "unsimplified" expressions for the coefficients found in Example 8.6.5 as we are looking for a pattern. It can be shown that $a_n = 2^n/n!$. Thus the solution, written as a power series, is

$$y = \sum_{n=0}^{\infty} \frac{2^n}{n!} x^n = \sum_{n=0}^{\infty} \frac{(2x)^n}{n!}.$$

Using Key Idea 8.8.1 and Theorem 8.8.2, we recognize $f(x) = e^{2x}$:

$$e^x = \sum_{n=0}^{\infty} \frac{x^n}{n!} \quad \Rightarrow \quad e^{2x} = \sum_{n=0}^{\infty} \frac{(2x)^n}{n!}.$$

Notes:

Finding a pattern in the coefficients that match the series expansion of a known function, such as those shown in Key Idea 8.8.1, can be difficult. What if the coefficients in the previous example were given in their reduced form; how could we still recover the function $y = e^{2x}$?

Suppose that all we know is that

$$a_0 = 1, \quad a_1 = 2, \quad a_2 = 2, \quad a_3 = \frac{4}{3}, \quad a_4 = \frac{2}{3}.$$

Definition 8.8.1 states that each term of the Taylor expansion of a function includes an $n!$. This allows us to say that

$$a_2 = 2 = \frac{b_2}{2!}, \quad a_3 = \frac{4}{3} = \frac{b_3}{3!}, \quad \text{and} \quad a_4 = \frac{2}{3} = \frac{b_4}{4!}$$

for some values b_2, b_3 and b_4. Solving for these values, we see that $b_2 = 4$, $b_3 = 8$ and $b_4 = 16$. That is, we are recovering the pattern we had previously seen, allowing us to write

$$f(x) = \sum_{n=0}^{\infty} a_n x^n = \sum_{n=0}^{\infty} \frac{b_n}{n!} x^n$$

$$= 1 + 2x + \frac{4}{2!}x^2 + \frac{8}{3!}x^3 + \frac{16}{4!}x^4 + \cdots$$

From here it is easier to recognize that the series is describing an exponential function.

There are simpler, more direct ways of solving the differential equation $y' = 2y$. We applied power series techniques to this equation to demonstrate its utility, and went on to show how *sometimes* we are able to recover the solution in terms of elementary functions using the theory of Taylor series. Most differential equations faced in real scientific and engineering situations are much more complicated than this one, but power series can offer a valuable tool in finding, or at least approximating, the solution.

This chapter introduced sequences, which are ordered lists of numbers, followed by series, wherein we add up the terms of a sequence. We quickly saw that such sums do not always add up to "infinity," but rather converge. We studied tests for convergence, then ended the chapter with a formal way of defining functions based on series. Such "series–defined functions" are a valuable tool in solving a number of different problems throughout science and engineering.

Coming in the next chapters are new ways of defining curves in the plane apart from using functions of the form $y = f(x)$. Curves created by these new methods can be beautiful, useful, and important.

Notes:

Exercises 8.8

Terms and Concepts

1. What is the difference between a Taylor polynomial and a Taylor series?

2. What theorem must we use to show that a function is equal to its Taylor series?

Problems

Key Idea 8.8.1 gives the nth term of the Taylor series of common functions. In Exercises 3 – 6, verify the formula given in the Key Idea by finding the first few terms of the Taylor series of the given function and identifying a pattern.

3. $f(x) = e^x$; $c = 0$

4. $f(x) = \sin x$; $c = 0$

5. $f(x) = 1/(1-x)$; $c = 0$

6. $f(x) = \tan^{-1} x$; $c = 0$

In Exercises 7 – 12, find a formula for the nth term of the Taylor series of $f(x)$, centered at c, by finding the coefficients of the first few powers of x and looking for a pattern. (The formulas for several of these are found in Key Idea 8.8.1; show work verifying these formula.)

7. $f(x) = \cos x$; $c = \pi/2$

8. $f(x) = 1/x$; $c = 1$

9. $f(x) = e^{-x}$; $c = 0$

10. $f(x) = \ln(1+x)$; $c = 0$

11. $f(x) = x/(x+1)$; $c = 1$

12. $f(x) = \sin x$; $c = \pi/4$

In Exercises 13 – 16, show that the Taylor series for $f(x)$, as given in Key Idea 8.8.1, is equal to $f(x)$ by applying Theorem 8.8.1; that is, show $\lim\limits_{n \to \infty} R_n(x) = 0$.

13. $f(x) = e^x$

14. $f(x) = \sin x$

15. $f(x) = \ln x$ (show equality only on $(1, 2)$)

16. $f(x) = 1/(1-x)$ (show equality only on $(-1, 0)$)

In Exercises 17 – 20, use the Taylor series given in Key Idea 8.8.1 to verify the given identity.

17. $\cos(-x) = \cos x$

18. $\sin(-x) = -\sin x$

19. $\frac{d}{dx}\left(\sin x \right) = \cos x$

20. $\frac{d}{dx}\left(\cos x \right) = -\sin x$

In Exercises 21 – 24, write out the first 5 terms of the Binomial series with the given k-value.

21. $k = 1/2$

22. $k = -1/2$

23. $k = 1/3$

24. $k = 4$

In Exercises 25 – 30, use the Taylor series given in Key Idea 8.8.1 to create the Taylor series of the given functions.

25. $f(x) = \cos\left(x^2\right)$

26. $f(x) = e^{-x}$

27. $f(x) = \sin\left(2x + 3\right)$

28. $f(x) = \tan^{-1}\left(x/2\right)$

29. $f(x) = e^x \sin x$ (only find the first 4 terms)

30. $f(x) = (1+x)^{1/2} \cos x$ (only find the first 4 terms)

In Exercises 31 – 32, approximate the value of the given definite integral by using the first 4 nonzero terms of the integrand's Taylor series.

31. $\displaystyle\int_0^{\sqrt{\pi}} \sin\left(x^2\right)\, dx$

32. $\displaystyle\int_0^{\pi^2/4} \cos\left(\sqrt{x}\right)\, dx$

A: Solutions To Selected Problems

Chapter 5

Section 5.1

1. Answers will vary.

3. Answers will vary.

5. Answers will vary.

7. velocity

9. $3/4x^4 + C$

11. $10/3x^3 - 2x + C$

13. $s + C$

15. $-3/(t) + C$

17. $\tan\theta + C$

19. $\sec x - \csc x + C$

21. $3^t/\ln 3 + C$

23. $4/3t^3 + 6t^2 + 9t + C$

25. $x^6/6 + C$

27. $ax + C$

29. $-\cos x + 3$

31. $x^4 - x^3 + 7$

33. $7^x/\ln 7 + 1 - 49/\ln 7$

35. $\frac{7x^3}{6} - \frac{9x}{2} + \frac{40}{3}$

37. $\theta - \sin(\theta) - \pi + 4$

39. $3x - 2$

41. $dy = (2xe^x\cos x + x^2 e^x\cos x - x^2 e^x \sin x)dx$

Section 5.2

1. Answers will vary.

3. 0

5. (a) 3
   (b) 4
   (c) 3
   (d) 0
   (e) -4
   (f) 9

7. (a) 4
   (b) 2
   (c) 4
   (d) 2
   (e) 1
   (f) 2

9. (a) π
   (b) π
   (c) 2π
   (d) 10π

11. (a) -59
    (b) -48

(c) -27
(d) -33

13. (a) 4
    (b) 4
    (c) -4
    (d) -2

15. (a) 2ft/s
    (b) 2ft
    (c) 1.5ft

17. (a) 64ft/s
    (b) 64ft
    (c) $t = 2$
    (d) $t = 2 + \sqrt{7} \approx 4.65$ seconds

19. 2

21. 16

23. 24

25. -7

27. $1/4x^4 - 2/3x^3 + 7/2x^2 - 9x + C$

29. $3/4t^{4/3} - 1/t + 2^t/\ln 2 + C$

Section 5.3

1. limits

3. Rectangles.

5. $2^2 + 3^2 + 4^2 = 29$

7. $0 - 1 + 0 + 1 + 0 = 0$

9. $1 + 1/2 + 1/3 + 1/4 + 1/5 = 137/60$

11. $1/2 + 1/6 + 1/12 + 1/20 = 4/5$

13. Answers may vary; $\sum_{i=1}^{5} 3i$

15. Answers may vary; $\sum_{i=1}^{4} \frac{i}{i+1}$

17. $5 \cdot 10 = 50$

19. 1045

21. -8525

23. 5050

25. 155

27. 24

29. 19

31. $\pi/3 + \pi/(2\sqrt{3}) \approx 1.954$

33. 0.388584

35. (a) Exact expressions will vary; $\frac{(1+n)^2}{4n^2}$.
    (b) 121/400, 10201/40000, 1002001/4000000
    (c) 1/4

37. (a) 8.
    (b) 8, 8, 8
    (c) 8

39. (a) Exact expressions will vary; $100 - 200/n$.
    (b) 80, 98, 499/5
    (c) 100

41. $F(x) = 5 \tan x + 4$

43. $G(t) = 4/6t^6 - 5/4t^4 + 8t + 9$

45. $G(t) = \sin t - \cos t - 78$

Section 5.4

1. Answers will vary.

3. T

5. 20

7. 0

9. 1

11. $(5 - 1/5)/\ln 5$

13. -4

15. $16/3$

17. $45/4$

19. $1/2$

21. $1/2$

23. $1/4$

25. 8

27. 0

29. Explanations will vary. A sketch will help.

31. $c = 2/\sqrt{3}$

33. $c = \ln(e - 1) \approx 0.54$

35. $2/\pi$

37. 2

39. 16

41. -300ft

43. 30ft

45. -1ft

47. -64ft/s

49. 2ft/s

51. $27/2$

53. $9/2$

55. $F'(x) = (3x^2 + 1)\frac{1}{x^3 + x}$

57. $F'(x) = 2x(x^2 + 2) - (x + 2)$

Section 5.5

1. F

3. They are superseded by the Trapezoidal Rule; it takes an equal amount of work and is generally more accurate.

5. (a) 3/4

    (b) 2/3

    (c) 2/3

7. (a) $\frac{1}{4}(1 + \sqrt{2})\pi \approx 1.896$

    (b) $\frac{1}{6}(1 + 2\sqrt{2})\pi \approx 2.005$

    (c) 2

9. (a) 38.5781

    (b) $147/4 \approx 36.75$

    (c) $147/4 \approx 36.75$

11. (a) 0

    (b) 0

    (c) 0

13. Trapezoidal Rule: 0.9006
Simpson's Rule: 0.90452

15. Trapezoidal Rule: 13.9604
Simpson's Rule: 13.9066

17. Trapezoidal Rule: 1.1703
Simpson's Rule: 1.1873

19. Trapezoidal Rule: 1.0803
Simpson's Rule: 1.077

21. (a) $n = 161$ (using max $(f''(x)) = 1$)

    (b) $n = 12$ (using max $(f^{(4)}(x)) = 1$)

23. (a) $n = 1004$ (using max $(f''(x)) = 39$)

    (b) $n = 62$ (using max $(f^{(4)}(x)) = 800$)

25. (a) Area is 30.8667 cm^2.

    (b) Area is 308, 667 yd^2.

Chapter 6

Section 6.1

1. Chain Rule.

3. $\frac{1}{8}(x^3 - 5)^8 + C$

5. $\frac{1}{18}(x^2 + 1)^9 + C$

7. $\frac{1}{2}\ln|2x + 7| + C$

9. $\frac{2}{3}(x + 3)^{3/2} - 6(x + 3)^{1/2} + C = \frac{2}{3}(x - 6)\sqrt{x + 3} + C$

11. $2e^{\sqrt{x}} + C$

13. $-\frac{1}{2x^2} - \frac{1}{x} + C$

15. $\frac{\sin^3(x)}{3} + C$

17. $-\frac{1}{6}\sin(3 - 6x) + C$

19. $\frac{1}{2}\ln|\sec(2x) + \tan(2x)| + C$

21. $\frac{\sin(x^2)}{2} + C$

23. The key is to rewrite $\cot x$ as $\cos x/\sin x$, and let $u = \sin x$.

25. $\frac{1}{3}e^{3x-1} + C$

27. $\frac{1}{2}e^{(x-1)^2} + C$

29. $\ln(e^x + 1) + C$

31. $\frac{27^x}{\ln 27} + C$

33. $\frac{1}{2}\ln^2(x) + C$

35. $\frac{3}{2}(\ln x)^2 + C$

37. $\frac{x^2}{2} + 3x + \ln|x| + C$

39. $\frac{x^3}{3} - \frac{x^2}{2} + x - 2\ln|x + 1| + C$

41. $\frac{3}{2}x^2 - 8x + 15\ln|x + 1| + C$

43. $\sqrt{7}\tan^{-1}\left(\frac{x}{\sqrt{7}}\right) + C$

45. $14\sin^{-1}\left(\frac{x}{\sqrt{5}}\right) + C$

47. $\frac{5}{4}\sec^{-1}(|x|/4) + C$

49. $\frac{\tan^{-1}\left(\frac{x-1}{\sqrt{7}}\right)}{\sqrt{7}} + C$

51. $3\sin^{-1}\left(\frac{x-4}{5}\right) + C$

53. $-\frac{1}{3(x^3+3)} + C$

55. $-\sqrt{1-x^2} + C$

57. $-\frac{2}{3}\cos^{\frac{3}{2}}(x) + C$

59. $\ln|x-5| + C$

61. $\frac{3x^2}{2} + \ln|x^2 + 3x + 5| - 5x + C$

63. $3\ln|3x^2 + 9x + 7| + C$

65. $\frac{1}{18}\tan^{-1}\left(\frac{x^2}{9}\right) + C$

67. $\sec^{-1}(|2x|) + C$

69. $\frac{3}{2}\ln|x^2 - 2x + 10| + \frac{1}{3}\tan^{-1}\left(\frac{x-1}{3}\right) + C$

71. $\frac{15}{2}\ln|x^2 - 10x + 32| + x + \frac{41\tan^{-1}\left(\frac{x-5}{\sqrt{7}}\right)}{\sqrt{7}} + C$

73. $\frac{x^2}{2} + 3\ln|x^2 + 4x + 9| - 4x + \frac{24\tan^{-1}\left(\frac{x+2}{\sqrt{5}}\right)}{\sqrt{5}} + C$

75. $\tan^{-1}(\sin(x)) + C$

77. $3\sqrt{x^2 - 2x - 6} + C$

79. $-\ln 2$

81. $2/3$

83. $(1-e)/2$

85. $\pi/2$

Section 6.2

1. T

3. Determining which functions in the integrand to set equal to "u" and which to set equal to "dv".

5. $\sin x - x\cos x + C$

7. $-x^2\cos x + 2x\sin x + 2\cos x + C$

9. $1/2e^{x^2} + C$

11. $-\frac{1}{2}xe^{-2x} - \frac{e^{-2x}}{4} + C$

13. $1/5e^{2x}(\sin x + 2\cos x) + C$

15. $1/10e^{5x}(\sin(5x) + \cos(5x)) + C$

17. $\sqrt{1-x^2} + x\sin^{-1}(x) + C$

19. $\frac{1}{2}x^2\tan^{-1}(x) - \frac{x}{2} + \frac{1}{2}\tan^{-1}(x) + C$

21. $\frac{1}{2}x^2\ln|x| - \frac{x^2}{4} + C$

23. $-\frac{x^2}{4} + \frac{1}{2}x^2\ln|x-1| - \frac{x}{2} - \frac{1}{2}\ln|x-1| + C$

25. $\frac{1}{3}x^3\ln|x| - \frac{x^3}{9} + C$

27. $2(x+1) + (x+1)(\ln(x+1))^2 - 2(x+1)\ln(x+1) + C$

29. $\ln|\sin(x)| - x\cot(x) + C$

31. $\frac{1}{3}(x^2 - 2)^{3/2} + C$

33. $x\sec x - \ln|\sec x + \tan x| + C$

35. $1/2x(\sin(\ln x) - \cos(\ln x)) + C$

37. $2\sin(\sqrt{x}) - 2\sqrt{x}\cos(\sqrt{x}) + C$

39. $2\sqrt{x}e^{\sqrt{x}} - 2e^{\sqrt{x}} + C$

41. π

43. 0

45. $1/2$

47. $\frac{3}{4e^2}$ $\frac{5}{4e^\pi}$

49. $1/5\left(e^\pi + e^{-\pi}\right)$

Section 6.3

1. F

3. F

5. $-\frac{1}{5}\cos^5(x) + C$

7. $\frac{1}{5}\cos^5 x - \frac{1}{3}\cos^3 x + C$

9. $\frac{1}{11}\sin^{11} x - \frac{2}{9}\sin^9 x + \frac{1}{7}\sin^7 x + C$

11. $\frac{x}{8} - \frac{1}{32}\sin(4x) + C$

13. $\frac{1}{2}\left(-\frac{1}{8}\cos(8x) - \frac{1}{2}\cos(2x)\right) + C$

15. $\frac{1}{2}\left(\frac{1}{4}\sin(4x) - \frac{1}{10}\sin(10x)\right) + C$

17. $\frac{1}{2}\left(\sin(x) + \frac{1}{3}\sin(3x)\right) + C$

19. $\frac{\tan^5(y)}{5} + C$

21. $\frac{\tan^6(x)}{6} + \frac{\tan^4(x)}{4} + C$

23. $\frac{\sec^5(x)}{5} - \frac{\sec^3(x)}{3} + C$

25. $\frac{1}{3}\tan^3 x - \tan x + x + C$

27. $\frac{1}{2}(\sec x \tan x - \ln|\sec x + \tan x|) + C$

29. $\frac{2}{5}$

31. $32/315$

33. $2/3$

35. $16/15$

Section 6.4

1. backwards

3. (a) $\tan^2\theta + 1 = \sec^2\theta$
   (b) $9\sec^2\theta$.

5. $\frac{1}{2}\left(x\sqrt{x^2 + 1} + \ln|\sqrt{x^2 + 1} + x|\right) + C$

7. $\frac{1}{2}\left(\sin^{-1} x + x\sqrt{1-x^2}\right) + C$

9. $\frac{1}{2}x\sqrt{x^2 - 1} - \frac{1}{2}\ln|x + \sqrt{x^2 - 1}| + C$

11. $x\sqrt{x^2 + 1/4} + \frac{1}{4}\ln|2\sqrt{x^2 + 1/4} + 2x| + C = \frac{1}{2}x\sqrt{4x^2 + 1} + \frac{1}{4}\ln|\sqrt{4x^2 + 1} + 2x| + C$

13. $4\left(\frac{1}{2}x\sqrt{x^2 - 1/16} - \frac{1}{32}\ln|4x + 4\sqrt{x^2 - 1/16}|\right) + C = \frac{1}{2}x\sqrt{16x^2 - 1} - \frac{1}{8}\ln|4x + \sqrt{16x^2 - 1}| + C$

15. $3\sin^{-1}\left(\frac{x}{\sqrt{7}}\right) + C$ (Trig. Subst. is not needed)

17. $\sqrt{x^2 - 11} - \sqrt{11}\sec^{-1}(x/\sqrt{11}) + C$

19. $\sqrt{x^2 - 3} + C$ (Trig. Subst. is not needed)

21. $-\frac{1}{\sqrt{x^2+9}} + C$ (Trig. Subst. is not needed)

23. $\frac{1}{18}\frac{x+2}{x^2+4x+13} + \frac{1}{54}\tan^{-1}\left(\frac{x+2}{2}\right) + C$

25. $\frac{1}{7}\left(-\frac{\sqrt{5-x^2}}{x} - \sin^{-1}(x/\sqrt{5})\right) + C$

27. $\pi/2$

29. $2\sqrt{2} + 2\ln(1 + \sqrt{2})$

31. $9\sin^{-1}(1/3) + \sqrt{8}$ Note: the new lower bound is $\theta = \sin^{-1}(-1/3)$ and the new upper bound is $\theta = \sin^{-1}(1/3)$. The final answer comes with recognizing that $\sin^{-1}(-1/3) = -\sin^{-1}(1/3)$ and that $\cos\left(\sin^{-1}(1/3)\right) = \cos\left(\sin^{-1}(-1/3)\right) = \sqrt{8}/3$.

Section 6.5

1. rational

3. $\frac{A}{x} + \frac{B}{x-3}$

5. $\frac{A}{x-\sqrt{7}} + \frac{B}{x+\sqrt{7}}$

7. $3 \ln |x-2| + 4 \ln |x+5| + C$

9. $\frac{1}{3}(\ln |x+2| - \ln |x-2|) + C$

11. $\ln |x+5| - \frac{2}{x+5} + C$

13. $\frac{5}{x+1} + 7 \ln |x| + 2 \ln |x+1| + C$

15. $-\frac{1}{5} \ln |5x-1| + \frac{2}{3} \ln |3x-1| + \frac{3}{7} \ln |7x+3| + C$

17. $\frac{x^2}{2} + x + \frac{125}{9} \ln |x-5| + \frac{64}{9} \ln |x+4| - \frac{35}{2} + C$

19. $\frac{1}{6}\left(-\ln |x^2+2x+3| + 2 \ln |x| - \sqrt{2} \tan^{-1}\left(\frac{x+1}{\sqrt{2}}\right)\right) + C$

21. $\ln |3x^2+5x-1| + 2 \ln |x+1| + C$

23. $\frac{9}{10} \ln |x^2+9| + \frac{1}{5} \ln |x+1| - \frac{4}{15} \tan^{-1}\left(\frac{x}{3}\right) + C$

25. $3\left(\ln |x^2-2x+11| + \ln |x-9|\right) + 3\sqrt{\frac{2}{5}} \tan^{-1}\left(\frac{x-1}{\sqrt{10}}\right) + C$

27. $\ln(2000/243) \approx 2.108$

29. $-\pi/4 + \tan^{-1} 3 - \ln(11/9) \approx 0.263$

Section 6.6

1. Because $\cosh x$ is always positive.

3.
$$\coth^2 x - \operatorname{csch}^2 x = \left(\frac{e^x+e^{-x}}{e^x-e^{-x}}\right)^2 - \left(\frac{2}{e^x-e^{-x}}\right)^2$$
$$= \frac{(e^{2x}+2+e^{-2x})-(4)}{e^{2x}-2+e^{-2x}}$$
$$= \frac{e^{2x}-2+e^{-2x}}{e^{2x}-2+e^{-2x}}$$
$$= 1$$

5.
$$\cosh^2 x = \left(\frac{e^x+e^{-x}}{2}\right)^2$$
$$= \frac{e^{2x}+2+e^{-2x}}{4}$$
$$= \frac{1}{2}\frac{(e^{2x}+e^{-2x})+2}{2}$$
$$= \frac{1}{2}\left(\frac{e^{2x}+e^{-2x}}{2}+1\right)$$
$$= \frac{\cosh 2x + 1}{2}.$$

7.
$$\frac{d}{dx}[\operatorname{sech} x] = \frac{d}{dx}\left[\frac{2}{e^x+e^{-x}}\right]$$
$$= \frac{-2(e^x-e^{-x})}{(e^x+e^{-x})^2}$$
$$= -\frac{2(e^x-e^{-x})}{(e^x+e^{-x})(e^x+e^{-x})}$$
$$= -\frac{2}{e^x+e^{-x}} \cdot \frac{e^x-e^{-x}}{e^x+e^{-x}}$$
$$= -\operatorname{sech} x \tanh x$$

9. $\int \tanh x \, dx = \int \frac{\sinh x}{\cosh x} \, dx$
Let $u = \cosh x$; $du = (\sinh x)dx$
$$= \int \frac{1}{u} \, du$$
$$= \ln |u| + C$$
$$= \ln(\cosh x) + C.$$

11. $2 \cosh 2x$

13. $2x \sec^2(x^2)$

15. $\sinh^2 x + \cosh^2 x$

17. $\frac{-2x}{(x^2)\sqrt{1-x^4}}$

19. $\frac{4x}{\sqrt{4x^4-1}}$

21. $-\csc x$

23. $y = x$

25. $y = \frac{9}{25}(x + \ln 3) - \frac{4}{5}$

27. $y = x$

29. $1/2 \ln(\cosh(2x)) + C$

31. $1/2 \sinh^2 x + C$ or $1/2 \cosh^2 x + C$

33. $x \cosh(x) - \sinh(x) + C$

35. $\cosh^{-1} x/3 + C = \ln\left(x + \sqrt{x^2-9}\right) + C$

37. $\cosh^{-1}(x^2/2) + C = \ln(x^2 + \sqrt{x^4-4}) + C$

39. $\frac{1}{16} \tan^{-1}(x/2) + \frac{1}{32} \ln |x-2| + \frac{1}{32} \ln |x+2| + C$

41. $\tan^{-1}(e^x) + C$

43. $x \tanh^{-1} x + 1/2 \ln |x^2-1| + C$

45. 0

47. 2

Section 6.7

1. $0/0, \infty/\infty, 0 \cdot \infty, \infty - \infty, 0^0, 1^\infty, \infty^0$

3. F

5. derivatives; limits

7. Answers will vary.

9. 3

11. -1

13. 5

15. 2/3

17. ∞

19. 0

21. 0

23. ∞

25. 0

27. -2

29. 0

31. 0

33. ∞

35. ∞

37. 0

39. 1

41. 1

43. 1

45. 1

47. 1

49. 2

51. $-\infty$

53. 0

Section 6.8

1. The interval of integration is finite, and the integrand is continuous on that interval.

3. converges; could also state < 10.

5. $p > 1$

7. $e^5/2$

9. $1/3$

11. $1/\ln 2$

13. diverges

15. 1

17. diverges

19. diverges

21. diverges

23. 1

25. 0

27. $-1/4$

29. diverges

31. 1

33. $1/2$

35. diverges; Limit Comparison Test with $1/x$.

37. diverges; Limit Comparison Test with $1/x$.

39. converges; Direct Comparison Test with e^{-x}.

41. converges; Direct Comparison Test with $1/(x^2 - 1)$.

43. converges; Direct Comparison Test with $1/e^x$.

Chapter 7

Section 7.1

1. T

3. Answers will vary.

5. $4\pi + \pi^2 \approx 22.436$

7. π

9. $1/2$

11. $1/\ln 4$

13. 4.5

15. $2 - \pi/2$

17. $1/6$

19. All enclosed regions have the same area, with regions being the reflection of adjacent regions. One region is formed on $[\pi/4, 5\pi/4]$, with area $2\sqrt{2}$.

21. 1

23. $9/2$

25. $1/12(9 - 2\sqrt{2}) \approx 0.514$

27. 1

29. 4

31. $219{,}000 \text{ ft}^2$

Section 7.2

1. T

3. Recall that "dx" does not just "sit there;" it is multiplied by $A(x)$ and represents the thickness of a small slice of the solid. Therefore dx has units of in, giving $A(x)\,dx$ the units of in^3.

5. $48\pi\sqrt{3}/5$ units^3

7. $\pi^2/4$ units^3

9. $9\pi/2$ units^3

11. $\pi^2 - 2\pi$ units^3

13.
  (a) $\pi/2$
  (b) $5\pi/6$
  (c) $4\pi/5$
  (d) $8\pi/15$

15.
  (a) $4\pi/3$
  (b) $2\pi/3$
  (c) $4\pi/3$
  (d) $\pi/3$

17.
  (a) $\pi^2/2$
  (b) $\pi^2/2 - 4\pi\sinh^{-1}(1)$
  (c) $\pi^2/2 + 4\pi\sinh^{-1}(1)$

19. Placing the tip of the cone at the origin such that the x-axis runs through the center of the circular base, we have $A(x) = \pi x^2/4$. Thus the volume is $250\pi/3$ units^3.

21. Orient the cone such that the tip is at the origin and the x-axis is perpendicular to the base. The cross-sections of this cone are right, isosceles triangles with side length $2x/5$; thus the cross-sectional areas are $A(x) - 2x^2/25$, giving a volume of $80/3$ units^3.

Section 7.3

1. T

3. F

5. $9\pi/2$ units^3

7. $\pi^2 - 2\pi$ units^3

9. $48\pi\sqrt{3}/5$ units^3

11. $\pi^2/4$ units^3

13.
  (a) $4\pi/5$
  (b) $8\pi/15$
  (c) $\pi/2$
  (d) $5\pi/6$

15.
  (a) $4\pi/3$
  (b) $\pi/3$
  (c) $4\pi/3$
  (d) $2\pi/3$

17.
  (a) $2\pi(\sqrt{2} - 1)$
  (b) $2\pi(1 - \sqrt{2} + \sinh^{-1}(1))$

Section 7.4

1. T

3. $\sqrt{2}$

5. $4/3$

7. $109/2$

9. $12/5$

11. $-\ln(2-\sqrt{3}) \approx 1.31696$

13. $\int_0^1 \sqrt{1+4x^2}\,dx$

15. $\int_0^1 \sqrt{1+\frac{1}{4x}}\,dx$

17. $\int_{-1}^1 \sqrt{1+\frac{x^2}{1-x^2}}\,dx$

19. $\int_1^2 \sqrt{1+\frac{1}{x^4}}\,dx$

21. 1.4790

23. Simpson's Rule fails, as it requires one to divide by 0. However, recognize the answer should be the same as for $y = x^2$; why?

25. Simpson's Rule fails.

27. 1.4058

29. $2\pi \int_0^1 2x\sqrt{5}\,dx = 2\pi\sqrt{5}$

31. $2\pi \int_0^1 x^3\sqrt{1+9x^4}\,dx = \pi/27(10\sqrt{10}-1)$

33. $2\pi \int_0^1 \sqrt{1-x^2}\sqrt{1+x/(1-x^2)}\,dx = 4\pi$

Section 7.5

1. In SI units, it is one joule, i.e., one Newton–meter, or kg·m/s^2·m. In Imperial Units, it is ft–lb.

3. Smaller.

5. (a) 500 ft–lb
   (b) $100 - 50\sqrt{2} \approx 29.29$ ft–lb

7. (a) $\frac{1}{2} \cdot d \cdot l^2$ ft–lb
   (b) 75 %
   (c) $\ell(1-\sqrt{2}/2) \approx 0.2929\ell$

9. (a) 756 ft–lb
   (b) 60,000 ft–lb
   (c) Yes, for the cable accounts for about 1% of the total work.

11. 575 ft–lb

13. 0.05 J

15. 5/3 ft–lb

17. $f \cdot d/2$ J

19. 5 ft–lb

21. (a) 52,929.6 ft–lb
    (b) 18,525.3 ft–lb
    (c) When 3.83 ft of water have been pumped from the tank, leaving about 2.17 ft in the tank.

23. 212,135 ft–lb

25. 187,214 ft–lb

27. 4,917,150 J

Section 7.6

1. Answers will vary.

3. 499.2 lb

5. 6739.2 lb

7. 3920.7 lb

9. 2496 lb

11. 602.59 lb

13. (a) 2340 lb

    (b) 5625 lb

15. (a) 1597.44 lb
    (b) 3840 lb

17. (a) 56.42 lb
    (b) 135.62 lb

19. 5.1 ft

Chapter 8

Section 8.1

1. Answers will vary.

3. Answers will vary.

5. $2, \frac{8}{3}, \frac{8}{3}, \frac{32}{15}, \frac{64}{45}$

7. $-\frac{1}{3}, -2, -\frac{81}{5}, -\frac{512}{3}, -\frac{15625}{7}$

9. $a_n = 3n+1$

11. $a_n = 10 \cdot 2^{n-1}$

13. $1/7$

15. 0

17. diverges

19. converges to 0

21. diverges

23. converges to e

25. converges to 0

27. converges to 2

29. bounded

31. bounded

33. neither bounded above or below

35. monotonically increasing

37. never monotonic

39. Let $\{a_n\}$ be given such that $\lim\limits_{n\to\infty} |a_n| = 0$. By the definition of the limit of a sequence, given any $\varepsilon > 0$, there is a m such that for all $n > m$, $\big| |a_n| - 0 \big| < \varepsilon$. Since $\big| |a_n| - 0 \big| = |a_n - 0|$, this directly implies that for all $n > m$, $|a_n - 0| < \varepsilon$, meaning that $\lim\limits_{n\to\infty} a_n = 0$.

41. A sketch of one proof method:

    Let any $\varepsilon > 0$ be given. Since $\{a_n\}$ and $\{b_n\}$ converge, there exists an $N > 0$ such that for all $n \geq N$, both a_n and b_n are within $\varepsilon/2$ of L; we can conclude that they are at most ε apart from each other. Since $a_n \leq c_n \leq b_n$, one can show that c_n is within ε of L, showing that $\{c_n\}$ also converges to L.

Section 8.2

1. Answers will vary.

3. One sequence is the sequence of terms $\{a_n\}$. The other is the sequence of nth partial sums, $\{S_n\} = \{\sum_{i=1}^n a_i\}$.

5. F

7. (a) $-1, -\frac{1}{2}, -\frac{5}{6}, -\frac{7}{12}, -\frac{47}{60}$
   (b) Plot omitted

9. (a) $-1, 0, -1, 0, -1$
   (b) Plot omitted

11. (a) $1, \frac{3}{2}, \frac{5}{3}, \frac{41}{24}, \frac{103}{60}$
    (b) Plot omitted

13. (a) $-0.9, -0.09, -0.819, -0.1629, -0.75339$

(b) Plot omitted

15. $\lim_{n\to\infty} a_n = 3$; by Theorem 8.2.4 the series diverges.

17. $\lim_{n\to\infty} a_n = \infty$; by Theorem 8.2.4 the series diverges.

19. $\lim_{n\to\infty} a_n = 1/2$; by Theorem 8.2.4 the series diverges.

21. Converges; p-series with $p = 5$.

23. Diverges; geometric series with $r = 6/5$.

25. Diverges; fails n^{th} term test

27. F

29. Diverges; by Theorem 8.2.3 this is half the Harmonic Series, which diverges by growing without bound. "Half of growing without bound" is still growing without bound.

31. (a) $S_n = \frac{1-(1/4)^n}{3/4}$

    (b) Converges to 4/3.

33. (a) $S_n = \left(\frac{n(n+1)}{2}\right)^2$

    (b) Diverges

35. (a) $S_n = 5\frac{1-1/2^n}{1/2}$

    (b) Converges to 10.

37. (a) $S_n = \frac{1-(-1/3)^n}{4/3}$

    (b) Converges to 3/4.

39. (a) With partial fractions, $a_n = \frac{3}{2}\left(\frac{1}{n} - \frac{1}{n+2}\right)$. Thus
    $S_n = \frac{3}{2}\left(\frac{3}{2} - \frac{1}{n+1} - \frac{1}{n+2}\right)$.

    (b) Converges to 9/4

41. (a) $S_n = \ln\left(1/(n+1)\right)$

    (b) Diverges (to $-\infty$).

43. (a) $a_n = \frac{1}{n(n+3)}$; using partial fractions, the resulting telescoping sum reduces to
    $S_n = \frac{1}{3}\left(1 + \frac{1}{2} + \frac{1}{3} - \frac{1}{n+1} - \frac{1}{n+2} - \frac{1}{n+3}\right)$

    (b) Converges to 11/18.

45. (a) With partial fractions, $a_n = \frac{1}{2}\left(\frac{1}{n-1} - \frac{1}{n+1}\right)$. Thus
    $S_n = \frac{1}{2}\left(3/2 - \frac{1}{n} - \frac{1}{n+1}\right)$.

    (b) Converges to 3/4.

47. (a) The nth partial sum of the odd series is $1 + \frac{1}{3} + \frac{1}{5} + \cdots + \frac{1}{2n-1}$. The nth partial sum of the even series is $\frac{1}{2} + \frac{1}{4} + \frac{1}{6} + \cdots + \frac{1}{2n}$. Each term of the even series is less than the corresponding term of the odd series, giving us our result.

    (b) The nth partial sum of the odd series is $1 + \frac{1}{3} + \frac{1}{5} + \cdots + \frac{1}{2n-1}$. The nth partial sum of 1 plus the even series is $1 + \frac{1}{2} + \frac{1}{4} + \cdots + \frac{1}{2(n-1)}$. Each term of the even series is now greater than or equal to the corresponding term of the odd series, with equality only on the first term. This gives us the result.

    (c) If the odd series converges, the work done in (a) shows the even series converges also. (The sequence of the nth partial sum of the even series is bounded and monotonically increasing.) Likewise, (b) shows that if the even series converges, the odd series will, too. Thus if either series converges, the other does.

    Similarly, (a) and (b) can be used to show that if either series diverges, the other does, too.

(d) If both the even and odd series converge, then their sum would be a convergent series. This would imply that the Harmonic Series, their sum, is convergent. It is not. Hence each series diverges.

Section 8.3

1. continuous, positive and decreasing

3. The Integral Test (we do not have a continuous definition of $n!$ yet) and the Limit Comparison Test (same as above, hence we cannot take its derivative).

5. Converges

7. Diverges

9. Converges

11. Converges

13. Converges; compare to $\sum_{n=1}^{\infty}\frac{1}{n^2}$, as $1/(n^2 + 3n - 5) \le 1/n^2$ for all $n > 1$.

15. Diverges; compare to $\sum_{n=1}^{\infty}\frac{1}{n}$, as $1/n \le \ln n/n$ for all $n \ge 3$.

17. Diverges; compare to $\sum_{n=1}^{\infty}\frac{1}{n}$. Since $n = \sqrt{n^2} > \sqrt{n^2 - 1}$,
    $1/n \le 1/\sqrt{n^2 - 1}$ for all $n \ge 2$.

19. Diverges; compare to $\sum_{n=1}^{\infty}\frac{1}{n}$:
    $$\frac{1}{n} = \frac{n^2}{n^3} < \frac{n^2 + n + 1}{n^3} < \frac{n^2 + n + 1}{n^3 - 5},$$
    for all $n \ge 1$.

21. Diverges; compare to $\sum_{n=1}^{\infty}\frac{1}{n}$. Note that
    $$\frac{n}{n^2 - 1} = \frac{n^2}{n^2 - 1}\cdot\frac{1}{n} > \frac{1}{n},$$
    as $\frac{n^2}{n^2-1} > 1$, for all $n \ge 2$.

23. Converges; compare to $\sum_{n=1}^{\infty}\frac{1}{n^2}$.

25. Diverges; compare to $\sum_{n=1}^{\infty}\frac{\ln n}{n}$.

27. Diverges; compare to $\sum_{n=1}^{\infty}\frac{1}{n}$.

29. Diverges; compare to $\sum_{n=1}^{\infty}\frac{1}{n}$. Just as $\lim_{n\to 0}\frac{\sin n}{n} = 1$,
    $\lim_{n\to\infty}\frac{\sin(1/n)}{1/n} = 1$.

31. Converges; compare to $\sum_{n=1}^{\infty}\frac{1}{n^{3/2}}$.

33. Converges; Integral Test

35. Diverges; the nth Term Test and Direct Comparison Test can be used.

37. Converges; the Direct Comparison Test can be used with sequence $1/3^n$.

39. Diverges; the nth Term Test can be used, along with the Integral Test.

41. (a) Converges; use Direct Comparison Test as $\frac{a_n}{n} < n$.

(b) Converges; since original series converges, we know $\lim_{n\to\infty} a_n = 0$. Thus for large n, $a_n a_{n+1} < a_n$.

(c) Converges; similar logic to part (b) so $(a_n)^2 < a_n$.

(d) May converge; certainly $na_n > a_n$ but that does not mean it does not converge.

(e) Does not converge, using logic from (b) and n^{th} Term Test.

Section 8.4

1. algebraic, or polynomial.

3. Integral Test, Limit Comparison Test, and Root Test

5. Converges

7. Converges

9. The Ratio Test is inconclusive; the p-Series Test states it diverges.

11. Converges

13. Converges; note the summation can be rewritten as $\displaystyle\sum_{n=1}^{\infty} \frac{2^n n!}{3^n n!}$, to which the Ratio Test or Geometric Series Test can be applied.

15. Converges

17. Converges

19. Diverges

21. Diverges. The Root Test is inconclusive, but the nth-Term Test shows divergence. (The terms of the sequence approach e^2, not 0, as $n \to \infty$.)

23. Converges

25. Diverges; Limit Comparison Test with $1/n$.

27. Converges; Ratio Test or Limit Comparison Test with $1/3^n$.

29. Diverges; n^{th}-Term Test or Limit Comparison Test with 1.

31. Diverges; Direct Comparison Test with $1/n$

33. Converges; Root Test

Section 8.5

1. The signs of the terms do not alternate; in the given series, some terms are negative and the others positive, but they do not necessarily alternate.

3. Many examples exist; one common example is $a_n = (-1)^n/n$.

5. (a) converges
   (b) converges (p-Series)
   (c) absolute

7. (a) diverges (limit of terms is not 0)
   (b) diverges
   (c) n/a; diverges

9. (a) converges
   (b) diverges (Limit Comparison Test with $1/n$)
   (c) conditional

11. (a) diverges (limit of terms is not 0)
    (b) diverges
    (c) n/a; diverges

13. (a) diverges (terms oscillate between ± 1)
    (b) diverges
    (c) n/a; diverges

15. (a) converges

(b) converges (Geometric Series with $r = 2/3$)

(c) absolute

17. (a) converges
    (b) converges (Ratio Test)
    (c) absolute

19. (a) converges
    (b) diverges (p-Series Test with $p = 1/2$)
    (c) conditional

21. $S_5 = -1.1906$; $S_6 = -0.6767$;
$$-1.1906 \leq \sum_{n=1}^{\infty} \frac{(-1)^n}{\ln(n+1)} \leq -0.6767$$

23. $S_6 = 0.3681$; $S_7 = 0.3679$;
$$0.3681 \leq \sum_{n=0}^{\infty} \frac{(-1)^n}{n!} \leq 0.3679$$

25. $n = 5$

27. Using the theorem, we find $n = 499$ guarantees the sum is within 0.001 of $\pi/4$. (Convergence is actually faster, as the sum is within ε of $\pi/24$ when $n \geq 249$.)

Section 8.6

1. 1

3. 5

5. $1 + 2x + 4x^2 + 8x^3 + 16x^4$

7. $1 + x + \frac{x^2}{2} + \frac{x^3}{6} + \frac{x^4}{24}$

9. (a) $R = \infty$
   (b) $(-\infty, \infty)$

11. (a) $R = 1$
    (b) $(2, 4]$

13. (a) $R = 2$
    (b) $(-2, 2)$

15. (a) $R = 1/5$
    (b) $(4/5, 6/5)$

17. (a) $R = 1$
    (b) $(-1, 1)$

19. (a) $R = \infty$
    (b) $(-\infty, \infty)$

21. (a) $R = 1$
    (b) $[-1, 1]$

23. (a) $R = 0$
    (b) $x = 0$

25. (a) $f'(x) = \displaystyle\sum_{n=1}^{\infty} n^2 x^{n-1}$; $(-1, 1)$

(b) $\displaystyle\int f(x)\, dx = C + \sum_{n=0}^{\infty} \frac{n}{n+1} x^{n+1}$; $(-1, 1)$

27. (a) $f'(x) = \displaystyle\sum_{n=1}^{\infty} \frac{n}{2^n} x^{n-1}$; $(-2, 2)$

(b) $\displaystyle\int f(x)\, dx = C + \sum_{n=0}^{\infty} \frac{1}{(n+1)2^n} x^{n+1}$; $[-2, 2)$

29. (a) $f'(x) = \displaystyle\sum_{n=1}^{\infty} \frac{(-1)^n x^{2n-1}}{(2n-1)!} = \sum_{n=0}^{\infty} \frac{(-1)^{n+1} x^{2n+1}}{(2n+1)!}$;
$(-\infty, \infty)$

(b) $\int f(x)\,dx = C + \sum_{n=0}^{\infty} \dfrac{(-1)^n x^{2n+1}}{(2n+1)!}$; $(-\infty, \infty)$

31. $1 + 3x + \frac{9}{2}x^2 + \frac{9}{2}x^3 + \frac{27}{8}x^4$

33. $1 + x + x^2 + x^3 + x^4$

35. $0 + x + 0x^2 - \frac{1}{6}x^3 + 0x^4$

Section 8.7

1. The Maclaurin polynomial is a special case of Taylor polynomials. Taylor polynomials are centered at a specific x-value; when that x-value is 0, it is a Maclauring polynomial.

3. $p_2(x) = 6 + 3x - 4x^2$.

5. $p_3(x) = 1 - x + \frac{1}{2}x^2 - \frac{1}{6}x^3$

7. $p_5(x) = x + x^2 + \frac{1}{2}x^3 + \frac{1}{6}x^4 + \frac{1}{24}x^5$

9. $p_4(x) = \frac{2x^4}{3} + \frac{4x^3}{3} + 2x^2 + 2x + 1$

11. $p_4(x) = x^4 - x^3 + x^2 - x + 1$

13. $p_4(x) = 1 + \frac{1}{2}(-1+x) - \frac{1}{8}(-1+x)^2 + \frac{1}{16}(-1+x)^3 - \frac{5}{128}(-1+x)^4$

15. $p_6(x) = \frac{1}{\sqrt{2}} - \frac{-\frac{\pi}{4}+x}{\sqrt{2}} - \frac{\left(-\frac{\pi}{4}+x\right)^2}{2\sqrt{2}} + \frac{\left(-\frac{\pi}{4}+x\right)^3}{6\sqrt{2}} + \frac{\left(-\frac{\pi}{4}+x\right)^4}{24\sqrt{2}} - \frac{\left(-\frac{\pi}{4}+x\right)^5}{120\sqrt{2}} - \frac{\left(-\frac{\pi}{4}+x\right)^6}{720\sqrt{2}}$

17. $p_5(x) = \frac{1}{2} - \frac{x-2}{4} + \frac{1}{8}(x-2)^2 - \frac{1}{16}(x-2)^3 + \frac{1}{32}(x-2)^4 - \frac{1}{64}(x-2)^5$

19. $p_3(x) = \frac{1}{2} + \frac{1+x}{2} + \frac{1}{4}(1+x)^2$

21. $p_3(x) = x - \frac{x^3}{6}$; $p_3(0.1) = 0.09983$. Error is bounded by $\pm \frac{1}{4!} \cdot 0.1^4 \approx \pm 0.000004167$.

23. $p_2(x) = 3 + \frac{1}{6}(-9+x) - \frac{1}{216}(-9+x)^2$; $p_2(10) = 3.16204$.
The third derivative of $f(x) = \sqrt{x}$ is bounded on $(8, 11)$ by 0.003.
Error is bounded by $\pm \frac{0.003}{3!} \cdot 1^3 = \pm 0.0005$.

25. The nth derivative of $f(x) = e^x$ is bounded by 3 on intervals containing 0 and 1. Thus $|R_n(1)| \le \frac{3}{(n+1)!}1^{(n+1)}$. When $n = 7$, this is less than 0.0001.

27. The nth derivative of $f(x) = \cos x$ is bounded by 1 on intervals containing 0 and $\pi/3$. Thus $|R_n(\pi/3)| \le \frac{1}{(n+1)!}(\pi/3)^{(n+1)}$.
When $n = 7$, this is less than 0.0001. Since the Maclaurin polynomial of $\cos x$ only uses even powers, we can actually just use $n = 6$.

29. The nth term is $\frac{1}{n!}x^n$.

31. The nth term Is: when n even, 0; when n is odd, $\frac{(-1)^{(n-1)/2}}{n!}x^n$.

33. The nth term is $(-1)^n x^n$.

35. $1 + x + \frac{1}{2}x^2 + \frac{1}{6}x^3 + \frac{1}{24}x^4$

37. $1 + 2x - 2x^2 + 4x^3 - 10x^4$

Section 8.8

1. A Taylor polynomial is a **polynomial**, containing a finite number of terms. A Taylor series is a **series**, the summation of an infinite number of terms.

3. All derivatives of e^x are e^x which evaluate to 1 at $x = 0$.
The Taylor series starts $1 + x + \frac{1}{2}x^2 + \frac{1}{3!}x^3 + \frac{1}{4!}x^4 + \cdots$;
the Taylor series is $\sum_{n=0}^{\infty} \frac{x^n}{n!}$

5. The nth derivative of $1/(1-x)$ is $f^{(n)}(x) = (n)!/(1-x)^{n+1}$, which evaluates to $n!$ at $x = 0$.
The Taylor series starts $1 + x + x^2 + x^3 + \cdots$;
the Taylor series is $\sum_{n=0}^{\infty} x^n$

7. The Taylor series starts
$0 - (x - \pi/2) + 0x^2 + \frac{1}{6}(x - \pi/2)^3 + 0x^4 - \frac{1}{120}(x - \pi/2)^5$;
the Taylor series is $\sum_{n=0}^{\infty}(-1)^{n+1}\dfrac{(x - \pi/2)^{2n+1}}{(2n+1)!}$

9. $f^{(n)}(x) = (-1)^n e^{-x}$; at $x = 0$, $f^{(n)}(0) = -1$ when n is odd and $f^{(n)}(0) = 1$ when n is even.
The Taylor series starts $1 - x + \frac{1}{2}x^2 - \frac{1}{3!}x^3 + \cdots$;
the Taylor series is $\sum_{n=0}^{\infty}(-1)^n \dfrac{x^n}{n!}$.

11. $f^{(n)}(x) = (-1)^{n+1}\frac{n!}{(x+1)^{n+1}}$; at $x = 1$, $f^{(n)}(1) = (-1)^{n+1}\frac{n!}{2^{n+1}}$
The Taylor series starts
$\frac{1}{2} + \frac{1}{4}(x - 1) - \frac{1}{8}(x - 1)^2 + \frac{1}{16}(x - 1)^3 \cdots$;
the Taylor series is $\sum_{n=0}^{\infty}(-1)^{n+1}\dfrac{(x - 1)^n}{2^{n+1}}$.

13. Given a value x, the magnitude of the error term $R_n(x)$ is bounded by
$$|R_n(x)| \le \frac{\max\left|f^{(n+1)}(z)\right|}{(n+1)!}\left|x^{(n+1)}\right|,$$
where z is between 0 and x.
If $x > 0$, then $z < x$ and $f^{(n+1)}(z) = e^z < e^x$. If $x < 0$, then $x < z < 0$ and $f^{(n+1)}(z) = e^z < 1$. So given a fixed x value, let $M = \max\{e^x, 1\}$; $f^{(n)}(z) < M$. This allows us to state
$$|R_n(x)| \le \frac{M}{(n+1)!}\left|x^{(n+1)}\right|.$$
For any x, $\displaystyle\lim_{n\to\infty}\frac{M}{(n+1)!}\left|x^{(n+1)}\right| = 0$. Thus by the Squeeze Theorem, we conclude that $\displaystyle\lim_{n\to\infty}R_n(x) = 0$ for all x, and hence
$$e^x = \sum_{n=0}^{\infty}\frac{x^n}{n!} \quad \text{for all } x.$$

15. Given a value x, the magnitude of the error term $R_n(x)$ is bounded by
$$|R_n(x)| \le \frac{\max\left|f^{(n+1)}(z)\right|}{(n+1)!}\left|(x - 1)^{(n+1)}\right|,$$
where z is between 1 and x.
Note that $\left|f^{(n+1)}(x)\right| = \frac{n!}{x^{n+1}}$.
Per the statement of the problem, we only consider the case $1 < x < 2$.
If $1 < x < 2$, then $1 < z < x$ and $f^{(n+1)}(z) = \frac{n!}{z^{n+1}} < n!$. Thus
$$|R_n(x)| \le \frac{n!}{(n+1)!}\left|(x - 1)^{(n+1)}\right| = \frac{(x-1)^{n+1}}{n+1} < \frac{1}{n+1}.$$
Thus
$$\lim_{n\to\infty}|R_n(x)| < \lim_{n\to\infty}\frac{1}{n+1} = 0,$$
hence
$$\ln x = \sum_{n=1}^{\infty}(-1)^{n+1}\frac{(x-1)^n}{n} \quad \text{on } (1, 2).$$

17. Given $\cos x = \sum_{n=0}^{\infty}(-1)^n \dfrac{x^{2n}}{(2n)!}$,
$$\cos(-x) = \sum_{n=0}^{\infty}(-1)^n \frac{(-x)^{2n}}{(2n)!} = \sum_{n=0}^{\infty}(-1)^n \frac{x^{2n}}{(2n)!} = \cos x, \text{ as all}$$
powers in the series are even.

19. Given $\sin x = \displaystyle\sum_{n=0}^{\infty}(-1)^n\frac{x^{2n+1}}{(2n+1)!}$,

$\dfrac{d}{dx}\left(\sin x\right) = \dfrac{d}{dx}\left(\displaystyle\sum_{n=0}^{\infty}(-1)^n\frac{x^{2n+1}}{(2n+1)!}\right) =$

$\displaystyle\sum_{n=0}^{\infty}(-1)^n\frac{(2n+1)x^{2n}}{(2n+1)!} = \sum_{n=0}^{\infty}(-1)^n\frac{x^{2n}}{(2n)!} = \cos x.$ (The summation still starts at $n=0$ as there was no constant term in the expansion of $\sin x$).

21. $1 + \dfrac{x}{2} - \dfrac{x^2}{8} + \dfrac{x^3}{16} - \dfrac{5x^4}{128}$

23. $1 + \dfrac{x}{3} - \dfrac{x^2}{9} + \dfrac{5x^3}{81} - \dfrac{10x^4}{243}$

25. $\displaystyle\sum_{n=0}^{\infty}(-1)^n\frac{(x^2)^{2n}}{(2n)!} = \sum_{n=0}^{\infty}(-1)^n\frac{x^{4n}}{(2n)!}.$

27. $\displaystyle\sum_{n=0}^{\infty}(-1)^n\frac{(2x+3)^{2n+1}}{(2n+1)!}.$

29. $x + x^2 + \dfrac{x^3}{3} - \dfrac{x^5}{30}$

31. $\displaystyle\int_0^{\sqrt{\pi}}\sin\left(x^2\right)dx \approx \int_0^{\sqrt{\pi}}\left(x^2 - \dfrac{x^6}{6} + \dfrac{x^{10}}{120} - \dfrac{x^{14}}{5040}\right)dx = 0.8877$

Index

Differentiation Rules

1. $\dfrac{d}{dx}(cx) = c$

2. $\dfrac{d}{dx}(u \pm v) = u' \pm v'$

3. $\dfrac{d}{dx}(u \cdot v) = uv' + u'v$

4. $\dfrac{d}{dx}\left(\dfrac{u}{v}\right) = \dfrac{vu' - uv'}{v^2}$

5. $\dfrac{d}{dx}(u(v)) = u'(v)v'$

6. $\dfrac{d}{dx}(c) = 0$

7. $\dfrac{d}{dx}(x) = 1$

8. $\dfrac{d}{dx}(x^n) = nx^{n-1}$

9. $\dfrac{d}{dx}(e^x) = e^x$

10. $\dfrac{d}{dx}(a^x) = \ln a \cdot a^x$

11. $\dfrac{d}{dx}(\ln x) = \dfrac{1}{x}$

12. $\dfrac{d}{dx}(\log_a x) = \dfrac{1}{\ln a} \cdot \dfrac{1}{x}$

13. $\dfrac{d}{dx}(\sin x) = \cos x$

14. $\dfrac{d}{dx}(\cos x) = -\sin x$

15. $\dfrac{d}{dx}(\csc x) = -\csc x \cot x$

16. $\dfrac{d}{dx}(\sec x) = \sec x \tan x$

17. $\dfrac{d}{dx}(\tan x) = \sec^2 x$

18. $\dfrac{d}{dx}(\cot x) = -\csc^2 x$

19. $\dfrac{d}{dx}(\sin^{-1} x) = \dfrac{1}{\sqrt{1-x^2}}$

20. $\dfrac{d}{dx}(\cos^{-1} x) = \dfrac{-1}{\sqrt{1-x^2}}$

21. $\dfrac{d}{dx}(\csc^{-1} x) = \dfrac{-1}{|x|\sqrt{x^2-1}}$

22. $\dfrac{d}{dx}(\sec^{-1} x) = \dfrac{1}{|x|\sqrt{x^2-1}}$

23. $\dfrac{d}{dx}(\tan^{-1} x) = \dfrac{1}{1+x^2}$

24. $\dfrac{d}{dx}(\cot^{-1} x) = \dfrac{-1}{1+x^2}$

25. $\dfrac{d}{dx}(\cosh x) = \sinh x$

26. $\dfrac{d}{dx}(\sinh x) = \cosh x$

27. $\dfrac{d}{dx}(\tanh x) = \sech^2 x$

28. $\dfrac{d}{dx}(\sech x) = -\sech x \tanh x$

29. $\dfrac{d}{dx}(\csch x) = -\csch x \coth x$

30. $\dfrac{d}{dx}(\coth x) = -\csch^2 x$

31. $\dfrac{d}{dx}(\cosh^{-1} x) = \dfrac{1}{\sqrt{x^2-1}}$

32. $\dfrac{d}{dx}(\sinh^{-1} x) = \dfrac{1}{\sqrt{x^2+1}}$

33. $\dfrac{d}{dx}(\sech^{-1} x) = \dfrac{-1}{x\sqrt{1-x^2}}$

34. $\dfrac{d}{dx}(\csch^{-1} x) = \dfrac{-1}{|x|\sqrt{1+x^2}}$

35. $\dfrac{d}{dx}(\tanh^{-1} x) = \dfrac{1}{1-x^2}$

36. $\dfrac{d}{dx}(\coth^{-1} x) = \dfrac{1}{1-x^2}$

Integration Rules

1. $\displaystyle\int c \cdot f(x)\, dx = c \int f(x)\, dx$

2. $\displaystyle\int f(x) \pm g(x)\, dx =$
   $\displaystyle\int f(x)\, dx \pm \int g(x)\, dx$

3. $\displaystyle\int 0\, dx = C$

4. $\displaystyle\int 1\, dx = x + C$

5. $\displaystyle\int x^n\, dx = \dfrac{1}{n+1}x^{n+1} + C,\ n \neq -1$

6. $\displaystyle\int e^x\, dx - e^x + C$

7. $\displaystyle\int \ln x\, dx = x \ln x - x + C$

8. $\displaystyle\int a^x\, dx = \dfrac{1}{\ln a} \cdot a^x + C$

9. $\displaystyle\int \dfrac{1}{x}\, dx = \ln|x| + C$

10. $\displaystyle\int \cos x\, dx = \sin x + C$

11. $\displaystyle\int \sin x\, dx = -\cos x + C$

12. $\displaystyle\int \tan x\, dx = -\ln|\cos x| + C$

13. $\displaystyle\int \sec x\, dx = \ln|\sec x + \tan x| + C$

14. $\displaystyle\int \csc x\, dx = -\ln|\csc x + \cot x| + C$

15. $\displaystyle\int \cot x\, dx = \ln|\sin x| + C$

16. $\displaystyle\int \sec^2 x\, dx = \tan x + C$

17. $\displaystyle\int \csc^2 x\, dx = -\cot x + C$

18. $\displaystyle\int \sec x \tan x\, dx = \sec x + C$

19. $\displaystyle\int \csc x \cot x\, dx = -\csc x + C$

20. $\displaystyle\int \cos^2 x\, dx = \dfrac{1}{2}x + \dfrac{1}{4}\sin(2x) + C$

21. $\displaystyle\int \sin^2 x\, dx = \dfrac{1}{2}x - \dfrac{1}{4}\sin(2x) + C$

22. $\displaystyle\int \dfrac{1}{x^2+a^2}\, dx = \dfrac{1}{a}\tan^{-1}\left(\dfrac{x}{a}\right) + C$

23. $\displaystyle\int \dfrac{1}{\sqrt{a^2-x^2}}\, dx = \sin^{-1}\left(\dfrac{x}{a}\right) + C$

24. $\displaystyle\int \dfrac{1}{x\sqrt{x^2-a^2}}\, dx = \dfrac{1}{a}\sec^{-1}\left(\dfrac{|x|}{a}\right) + C$

25. $\displaystyle\int \cosh x\, dx = \sinh x + C$

26. $\displaystyle\int \sinh x\, dx = \cosh x + C$

27. $\displaystyle\int \tanh x\, dx = \ln(\cosh x) + C$

28. $\displaystyle\int \coth x\, dx = \ln|\sinh x| + C$

29. $\displaystyle\int \dfrac{1}{\sqrt{x^2-a^2}}\, dx = \ln\left|x + \sqrt{x^2-a^2}\right| + C$

30. $\displaystyle\int \dfrac{1}{\sqrt{x^2+a^2}}\, dx = \ln\left|x + \sqrt{x^2+a^2}\right| + C$

31. $\displaystyle\int \dfrac{1}{a^2-x^2}\, dx = \dfrac{1}{2a}\ln\left|\dfrac{a+x}{a-x}\right| + C$

32. $\displaystyle\int \dfrac{1}{x\sqrt{a^2-x^2}}\, dx = \dfrac{1}{a}\ln\left(\dfrac{x}{a+\sqrt{a^2-x^2}}\right) + C$

33. $\displaystyle\int \dfrac{1}{x\sqrt{x^2+a^2}}\, dx = \dfrac{1}{a}\ln\left|\dfrac{x}{a+\sqrt{x^2+a^2}}\right| + C$

The Unit Circle

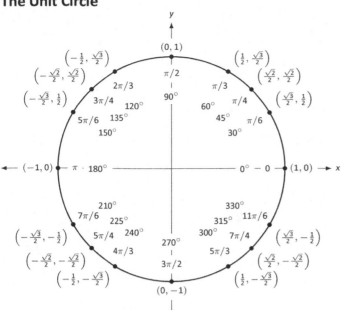

Definitions of the Trigonometric Functions

Unit Circle Definition

$$\sin \theta = y \qquad \cos \theta = x$$

$$\csc \theta = \frac{1}{y} \qquad \sec \theta = \frac{1}{x}$$

$$\tan \theta = \frac{y}{x} \qquad \cot \theta = \frac{x}{y}$$

Right Triangle Definition

$$\sin \theta = \frac{O}{H} \qquad \csc \theta = \frac{H}{O}$$

$$\cos \theta = \frac{A}{H} \qquad \sec \theta = \frac{H}{A}$$

$$\tan \theta = \frac{O}{A} \qquad \cot \theta = \frac{A}{O}$$

Common Trigonometric Identities

Pythagorean Identities

$$\sin^2 x + \cos^2 x = 1$$
$$\tan^2 x + 1 = \sec^2 x$$
$$1 + \cot^2 x = \csc^2 x$$

Cofunction Identities

$$\sin\left(\frac{\pi}{2} - x\right) = \cos x \qquad \csc\left(\frac{\pi}{2} - x\right) = \sec x$$

$$\cos\left(\frac{\pi}{2} - x\right) = \sin x \qquad \sec\left(\frac{\pi}{2} - x\right) = \csc x$$

$$\tan\left(\frac{\pi}{2} - x\right) = \cot x \qquad \cot\left(\frac{\pi}{2} - x\right) = \tan x$$

Double Angle Formulas

$$\sin 2x = 2 \sin x \cos x$$
$$\cos 2x = \cos^2 x - \sin^2 x$$
$$= 2\cos^2 x - 1$$
$$= 1 - 2\sin^2 x$$
$$\tan 2x = \frac{2 \tan x}{1 - \tan^2 x}$$

Sum to Product Formulas

$$\sin x + \sin y = 2 \sin\left(\frac{x+y}{2}\right) \cos\left(\frac{x-y}{2}\right)$$

$$\sin x - \sin y = 2 \sin\left(\frac{x-y}{2}\right) \cos\left(\frac{x+y}{2}\right)$$

$$\cos x + \cos y = 2 \cos\left(\frac{x+y}{2}\right) \cos\left(\frac{x-y}{2}\right)$$

$$\cos x - \cos y = -2 \sin\left(\frac{x+y}{2}\right) \sin\left(\frac{x-y}{2}\right)$$

Power–Reducing Formulas

$$\sin^2 x = \frac{1 - \cos 2x}{2}$$

$$\cos^2 x = \frac{1 + \cos 2x}{2}$$

$$\tan^2 x = \frac{1 - \cos 2x}{1 + \cos 2x}$$

Even/Odd Identities

$$\sin(-x) = -\sin x$$
$$\cos(-x) = \cos x$$
$$\tan(-x) = -\tan x$$
$$\csc(-x) = -\csc x$$
$$\sec(-x) = \sec x$$
$$\cot(-x) = -\cot x$$

Product to Sum Formulas

$$\sin x \sin y = \frac{1}{2}\left(\cos(x-y) - \cos(x+y)\right)$$

$$\cos x \cos y = \frac{1}{2}\left(\cos(x-y) + \cos(x+y)\right)$$

$$\sin x \cos y = \frac{1}{2}\left(\sin(x+y) + \sin(x-y)\right)$$

Angle Sum/Difference Formulas

$$\sin(x \pm y) = \sin x \cos y \pm \cos x \sin y$$

$$\cos(x \pm y) = \cos x \cos y \mp \sin x \sin y$$

$$\tan(x \pm y) = \frac{\tan x \pm \tan y}{1 \mp \tan x \tan y}$$

Areas and Volumes

Triangles

$h = a \sin \theta$

Area $= \frac{1}{2}bh$

Law of Cosines:
$c^2 = a^2 + b^2 - 2ab \cos \theta$

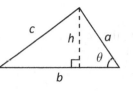

Parallelograms

Area $= bh$

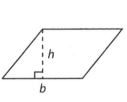

Trapezoids

Area $= \frac{1}{2}(a + b)h$

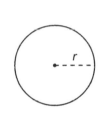

Circles

Area $= \pi r^2$

Circumference $= 2\pi r$

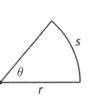

Sectors of Circles

θ in radians

Area $= \frac{1}{2}\theta r^2$

$s = r\theta$

Right Circular Cone

Volume $= \frac{1}{3}\pi r^2 h$

Surface Area $=$
$\pi r\sqrt{r^2 + h^2} + \pi r^2$

Right Circular Cylinder

Volume $= \pi r^2 h$

Surface Area $=$
$2\pi rh + 2\pi r^2$

Sphere

Volume $= \frac{4}{3}\pi r^3$

Surface Area $= 4\pi r^2$

General Cone

Area of Base $= A$

Volume $= \frac{1}{3}Ah$

General Right Cylinder

Area of Base $= A$

Volume $= Ah$

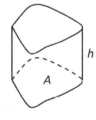

Algebra

Factors and Zeros of Polynomials

Let $p(x) = a_n x^n + a_{n-1} x^{n-1} + \cdots + a_1 x + a_0$ be a polynomial. If $p(a) = 0$, then a is a *zero* of the polynomial and a solution of the equation $p(x) = 0$. Furthermore, $(x - a)$ is a *factor* of the polynomial.

Fundamental Theorem of Algebra

An *n*th degree polynomial has *n* (not necessarily distinct) zeros. Although all of these zeros may be imaginary, a real polynomial of odd degree must have at least one real zero.

Quadratic Formula

If $p(x) = ax^2 + bx + c$, and $0 \le b^2 - 4ac$, then the real zeros of p are $x = (-b \pm \sqrt{b^2 - 4ac})/2a$

Special Factors

$$x^2 - a^2 = (x - a)(x + a) \qquad\qquad x^3 - a^3 = (x - a)(x^2 + ax + a^2)$$
$$x^3 + a^3 = (x + a)(x^2 - ax + a^2) \qquad x^4 - a^4 = (x^2 - a^2)(x^2 + a^2)$$
$$(x + y)^n = x^n + nx^{n-1}y + \frac{n(n-1)}{2!}x^{n-2}y^2 + \cdots + nxy^{n-1} + y^n$$
$$(x - y)^n = x^n - nx^{n-1}y + \frac{n(n-1)}{2!}x^{n-2}y^2 - \cdots \pm nxy^{n-1} \mp y^n$$

Binomial Theorem

$$(x + y)^2 = x^2 + 2xy + y^2 \qquad\qquad (x - y)^2 = x^2 - 2xy + y^2$$
$$(x + y)^3 = x^3 + 3x^2y + 3xy^2 + y^3 \qquad (x - y)^3 = x^3 - 3x^2y + 3xy^2 - y^3$$
$$(x + y)^4 = x^4 + 4x^3y + 6x^2y^2 + 4xy^3 + y^4 \qquad (x - y)^4 = x^4 - 4x^3y + 6x^2y^2 - 4xy^3 + y^4$$

Rational Zero Theorem

If $p(x) = a_n x^n + a_{n-1} x^{n-1} + \cdots + a_1 x + a_0$ has integer coefficients, then every *rational zero* of p is of the form $x = r/s$, where r is a factor of a_0 and s is a factor of a_n.

Factoring by Grouping

$$acx^3 + adx^2 + bcx + bd = ax^2(cs + d) + b(cx + d) = (ax^2 + b)(cx + d)$$

Arithmetic Operations

$$ab + ac = a(b + c) \qquad \frac{a}{b} + \frac{c}{d} = \frac{ad + bc}{bd} \qquad \frac{a + b}{c} = \frac{a}{c} + \frac{b}{c}$$

$$\frac{\left(\dfrac{a}{b}\right)}{\left(\dfrac{c}{d}\right)} = \left(\frac{a}{b}\right)\left(\frac{d}{c}\right) = \frac{ad}{bc} \qquad \frac{\left(\dfrac{a}{b}\right)}{c} = \frac{a}{bc} \qquad \frac{a}{\left(\dfrac{b}{c}\right)} = \frac{ac}{b}$$

$$a\left(\frac{b}{c}\right) = \frac{ab}{c} \qquad \frac{a - b}{c - d} = \frac{b - a}{d - c} \qquad \frac{ab + ac}{a} = b + c$$

Exponents and Radicals

$$a^0 = 1, \ a \ne 0 \qquad (ab)^x = a^x b^x \qquad a^x a^y = a^{x+y} \qquad \sqrt{a} = a^{1/2} \qquad \frac{a^x}{a^y} = a^{x-y} \qquad \sqrt[n]{a} = a^{1/n}$$

$$\left(\frac{a}{b}\right)^x = \frac{a^x}{b^x} \qquad \sqrt[n]{a^m} = a^{m/n} \qquad a^{-x} = \frac{1}{a^x} \qquad \sqrt[n]{ab} = \sqrt[n]{a}\sqrt[n]{b} \qquad (a^x)^y = a^{xy} \qquad \sqrt[n]{\frac{a}{b}} = \frac{\sqrt[n]{a}}{\sqrt[n]{b}}$$

Additional Formulas

Summation Formulas:

$$\sum_{i=1}^{n} c = cn \qquad\qquad \sum_{i=1}^{n} i = \frac{n(n+1)}{2}$$

$$\sum_{i=1}^{n} i^2 = \frac{n(n+1)(2n+1)}{6} \qquad \sum_{i=1}^{n} i^3 = \left(\frac{n(n+1)}{2}\right)^2$$

Trapezoidal Rule:

$$\int_a^b f(x)\,dx \approx \frac{\Delta x}{2}\left[f(x_1) + 2f(x_2) + 2f(x_3) + \ldots + 2f(x_n) + f(x_{n+1})\right]$$

with Error $\leq \dfrac{(b-a)^3}{12n^2}\left[\max\left|f''(x)\right|\right]$

Simpson's Rule:

$$\int_a^b f(x)\,dx \approx \frac{\Delta x}{3}\left[f(x_1) + 4f(x_2) + 2f(x_3) + 4f(x_4) + \ldots + 2f(x_{n-1}) + 4f(x_n) + f(x_{n+1})\right]$$

with Error $\leq \dfrac{(b-a)^5}{180n^4}\left[\max\left|f^{(4)}(x)\right|\right]$

Arc Length:

$$L = \int_a^b \sqrt{1 + f'(x)^2}\ dx$$

Surface of Revolution:

$$S = 2\pi \int_a^b f(x)\sqrt{1 + f'(x)^2}\ dx$$

(where $f(x) \geq 0$)

$$S = 2\pi \int_a^b x\sqrt{1 + f'(x)^2}\ dx$$

(where $a, b \geq 0$)

Work Done by a Variable Force:

$$W = \int_a^b F(x)\ dx$$

Force Exerted by a Fluid:

$$F = \int_a^b w\,d(y)\,\ell(y)\ dy$$

Taylor Series Expansion for $f(x)$:

$$p_n(x) = f(c) + f'(c)(x-c) + \frac{f''(c)}{2!}(x-c)^2 + \frac{f'''(c)}{3!}(x-c)^3 + \ldots + \frac{f^{(n)}(c)}{n!}(x-c)^n$$

Maclaurin Series Expansion for $f(x)$, where $c = 0$:

$$p_n(x) = f(0) + f'(0)x + \frac{f''(0)}{2!}x^2 + \frac{f'''(0)}{3!}x^3 + \ldots + \frac{f^{(n)}(0)}{n!}x^n$$

Summary of Tests for Series:

Test	Series	Condition(s) of Convergence	Condition(s) of Divergence	Comment
nth-Term	$\displaystyle\sum_{n=1}^{\infty} a_n$		$\displaystyle\lim_{n\to\infty} a_n \neq 0$	This test cannot be used to show convergence.
Geometric Series	$\displaystyle\sum_{n=0}^{\infty} r^n$	$\lvert r \rvert < 1$	$\lvert r \rvert \geq 1$	$\text{Sum} = \dfrac{1}{1-r}$
Telescoping Series	$\displaystyle\sum_{n=1}^{\infty} (b_n - b_{n+a})$	$\displaystyle\lim_{n\to\infty} b_n = L$		$\text{Sum} = \left(\displaystyle\sum_{n=1}^{a} b_n\right) - L$
p-Series	$\displaystyle\sum_{n=1}^{\infty} \dfrac{1}{(an+b)^p}$	$p > 1$	$p \leq 1$	
Integral Test	$\displaystyle\sum_{n=0}^{\infty} a_n$	$\displaystyle\int_{1}^{\infty} a(n)\, dn$ is convergent	$\displaystyle\int_{1}^{\infty} a(n)\, dn$ is divergent	$a_n = a(n)$ must be continuous
Direct Comparison	$\displaystyle\sum_{n=0}^{\infty} a_n$	$\displaystyle\sum_{n=0}^{\infty} b_n$ converges and $0 \leq a_n \leq b_n$	$\displaystyle\sum_{n=0}^{\infty} b_n$ diverges and $0 \leq b_n \leq a_n$	
Limit Comparison	$\displaystyle\sum_{n=0}^{\infty} a_n$	$\displaystyle\sum_{n=0}^{\infty} b_n$ converges and $\displaystyle\lim_{n\to\infty} a_n/b_n \geq 0$	$\displaystyle\sum_{n=0}^{\infty} b_n$ diverges and $\displaystyle\lim_{n\to\infty} a_n/b_n > 0$	Also diverges if $\displaystyle\lim_{n\to\infty} a_n/b_n = \infty$
Ratio Test	$\displaystyle\sum_{n=0}^{\infty} a_n$	$\displaystyle\lim_{n\to\infty} \dfrac{a_{n+1}}{a_n} < 1$	$\displaystyle\lim_{n\to\infty} \dfrac{a_{n+1}}{a_n} > 1$	$\{a_n\}$ must be positive. Also diverges if $\displaystyle\lim_{n\to\infty} a_{n+1}/a_n = \infty$
Root Test	$\displaystyle\sum_{n=0}^{\infty} a_n$	$\displaystyle\lim_{n\to\infty} (a_n)^{1/n} < 1$	$\displaystyle\lim_{n\to\infty} (a_n)^{1/n} > 1$	$\{a_n\}$ must be positive. Also diverges if $\displaystyle\lim_{n\to\infty} (a_n)^{1/n} = \infty$